SERIES ON SEMICONDUCTOR SCIENCE AND TECHNOLOGY

Series Editors

R. J. Nicholas University of Oxford
H. Kamimura University of Tokyo

SERIES ON SEMICONDUCTOR SCIENCE AND TECHNOLOGY

---

1. M. Jaros: *Physics and applications of semiconductor microstructures*
2. V.N. Dobrovolsky and V.G. Litovchenko: *Surface electronic transport phenomena in semiconductors*
3. M.J. Kelly: *Low-dimensional semiconductors*
4. P.K. Basu: *Theory of optical processes in semiconductors*
5. N. Balkan: *Hot electrons in semiconductors*
6. B. Gil: *Group III nitride semiconductor compounds: physics and applications*
7. M. Sugawara: *Plasma etching*
8. M. Balkanski, R.F. Wallis: *Semiconductor physics and applications*
9. B. Gil: *Low-dimensional nitride semiconductors*
10. L. Challis: *Electron-phonon interactions in low-dimensional structures*
11. V. Ustinov, A. Zhukov, A. Egorov, N. Maleev: *Quantum dot lasers*
12. H. Spieler: *Semiconductor detector systems*
13. S. Maekawa: *Concepts in spin electronics*
14. S. D. Ganichev, W. Prettl: *Intense terahertz excitation of semiconductors*
15. N. Miura: *Physics of semiconductors in high magnetic fields*
16. A.V. Kavokin, J. J. Baumberg, G. Malpuech, F. P. Laussy: *Microcavities*
17. S. Maekawa, S. O. Valenzuela, E. Saitoh, T. Kimura: *Spin current*
18. B. Gil: *III-nitride semiconductors and their modern devices*
19. A. Toropov, T. Shubina: *Plasmonic Effects in Metal-Semiconductor Nanostructures*
20. B.K. Ridley: *Hybrid Phonons in Nanostructures*
21. A.V. Kavokin, J. J. Baumberg, G. Malpuech, F. P. Laussy: *Microcavities, Second edition*
22. S. Maekawa, S. O. Valenzuela, E. Saitoh, T. Kimura: *Spin current, Second Edition*
23. M. M. Glazov: *Electron and nuclear spin dynamics in semiconductor nanostructures*

Electron and Nuclear Spin Dynamics in Semiconductor Nanostructures

M. M. Glazov

Ioffe Institute, 194021, St. Petersburg, Russia

OXFORD
UNIVERSITY PRESS

OXFORD
UNIVERSITY PRESS

Great Clarendon Street, Oxford, OX2 6DP,
United Kingdom

Oxford University Press is a department of the University of Oxford.
It furthers the University's objective of excellence in research, scholarship,
and education by publishing worldwide. Oxford is a registered trade mark of
Oxford University Press in the UK and in certain other countries

First Edition published in 2018

Published in the United States of America by Oxford University Press
198 Madison Avenue, New York, NY 10016, United States of America

British Library Cataloguing in Publication Data
Data available

Library of Congress Control Number: 2018945550

ISBN 978–0–19–880730–8

DOI: 10.1093/oso/9780198807308.001.0001

In memory of my father

Preface

In recent decades, the physics community has experienced a revival of interest in spin effects in solid state systems. This interest is two-fold: On one hand, the solid state systems, particularly semiconductors and semiconductor nanosystems, allow one to perform benchtop studies of the quantum and relativistic phenomena. The spin–orbit interaction effects on electron spin states and the hyperfine interaction of electron and nuclear spins are among the most prominent examples of such a class of effects. On the other hand, interest is supported by the prospects of realizing spin-based electronics where electron or nuclear spins may play the role of quantum or classical information carriers. This has resulted in the formation of a novel research field; spin electronics or, in short, spintronics.

The range of spin effects is vast. It extends from the effects of spin orientation and detection to the processes of spin transfer and manipulation. Moreover, a variety of spin systems in semiconductors have been studied including the spins of charge carriers, that is of electrons and holes, spins of nuclei, spins of impurities, and so on. Needless to mention, each semiconductor system brings about certain specific aspects of spin phenomena resulting in a multitude of spectacular and unique facets of spin physics in semiconductors. Many of these effects are observed experimentally and have firm theoretical grounds. As a result, spin physics in semiconductors has turned into a mature topic of research. To date, various aspects of spin effects in semiconductors have been concisely reviewed in collective books *Optical Orientation*, edited by F. Meier and B. P. Zakharchenya (North-Holland, 1984) and *Spin Physics in Semiconductors*, edited by M. I. Dyakonov (Springer, 2008), sometimes called “Old Testament” and “New Testament” of semiconductor spin physics, respectively. Each of the topics addressed in these references could become the subject of a separate monograph reflecting the state of the art in the corresponding area of spin physics.

This book aims at a rather detailed presentation of intriguing physics in the interacting systems of electron and nuclear spins in semiconductors, with particular emphasis on low-dimensional structures. The two spin systems naturally appear in practically all widespread semiconductor compounds. For instance, in most prominent GaAs, AlAs, or InAs materials that serve as building blocks for many semiconductor nanostructures, all host lattice nuclei have nonzero spins. The hyperfine interaction of charge carrier and nuclear spins gives rise to the spin exchange between these two systems. As a result, the understanding of the intertwined spin systems of electrons and nuclei is crucial for in-depth study and control of the spin phenomena in semiconductors. This is one of the reasons why the physics of electron and nuclear spins takes a special place in semiconductor spintronics. From the point of view of possible applications, the extremely long lived (for up to days), nuclear spin polarization attracts a lot of interest in the community in issues related to the deliberate polarization of nuclei and detection of spin coherence. Interest in the physics of electron and nuclear spins is also strongly motivated by the beautiful and complex physics occurring in this many-body and nonlinear system.

Hopefully, the present book will be useful for postgraduates and researchers willing to enter the field of electron and nuclear spin dynamics in semiconductors, and comprehend the general physical picture of the interacting electrons and nuclei. Here, I have attempted to present the phenomenology and qualitative analysis together with the more rigorous models. The basics of experiments are outlined, however, the details on particular experimental methods and setups are not included because this goes far beyond my expertise. A certain amount of reference information and details on theoretical approaches included in this book might make it useful for specialists in the field as well.

It is a great pleasure to thank those who have helped me with writing the book. I am grateful to the students of Academic University in St. Petersburg (Russia) who have attended my lectures on spin effects in semiconductors. Their interest and stimulating questions have helped me to deeper understand things, which I thought were trivial. I am very grateful to my colleagues both at Ioffe Institute and all over the world for collaboration and support. I am indebted to my colleagues T. Amand, I. D. Avdeev, E. Chekhovich, M. V. Durnev, A. Greilich, I. V. Ignatiev, E. L. Ivchenko, K. V. Kavokin, V. L. Korenev, X. Marie, M. Yu. Petrov, A. V. Poshakinskiy, D. S. Smirnov, A. Tartakovskii, B. Urbaszek, D. R. Yakovlev, I. A. Yugova, and V. S. Zapasskii for valuable discussions and critical reading of the manuscript. I owe my wife, Marina Semina, and my mother, Nina Glazova, a debt of gratitude for their continuing support and help.

St. Petersburg, Russia
January, 2018

Contents

1 Introduction

Creation, detection, and manipulation of the spin degrees of freedom of electrons and nuclei; phenomena of spin relaxation, decoherence, and dephasing; and the processes of spin transfer between different subsystems are among the most important problems studied in semiconductor spintronics. These effects taking place in semiconductor nanostructures with localized charge carriers, and the underlying interactions in spin systems are the focus of this book. The aim of this short introductory chapter is to provide an outline of the material contained in the book and explain the logic of its presentation.

Chapter 2 on "Spin Systems in Semiconductor Nanostructures" contains an introduction to a rich variety of phenomena taking place in the interacting system of electrons and nuclei in semiconductors. It includes the basics of the electronic properties of nanostructures and of spin physics, an overview of the fundamental interactions in the electron and nuclear spin systems, selection rules at optical transitions in semiconductors, and spin resonance phenomenon as well as optical orientation and dynamical nuclear polarization effects. The last section 2.6 contains an analysis of the particular features of spin dynamics arising in structures with localized electrons, such as quantum dots, which are studied later in the book. The aim of this chapter is to provide the basic minimum of information needed to read the remaining chapters. In the following chapters, we provide references to the basic notions and concepts introduced in Chapter 2 that might be helpful for readers interested in the contents of just a few particular chapters.

Chapter 3 on "Spin Resonance" is devoted to one of key phenomena in the field of spin physics; namely, the resonant absorption of an electromagnetic wave under conditions where the splitting between spin levels in a magnetic field is equal to the photon energy. This method is particularly relevant for identification of nuclear spin effects, because the resonance spectra provide "fingerprints" of different involved spin species and make it possible to distinguish different nuclear isotopes. As discussed in this chapter, nuclear magnetic resonance also provides access to local magnetic fields acting on nuclear spins and is caused by the magnetic interactions between the nuclei or by the quadrupole splittings of nuclear spin states caused by the environment. We also outline specific manifestations of spin resonance in the optical response of semiconductors; that is, optically detected

Electron and Nuclear Spin Dynamics in Semiconductor Nanostructures. M. M. Glazov, Oxford University Press (2018). © M. M. Glazov. DOI: 10.1093/oso/9780198807308.001.0001

magnetic resonance, Section 3.3. This technique has became an indispensable tool due to its high sensitivity and relative simplicity of realization in semiconductors.

Chapter 4 on "Hyperfine Interaction of Electron and Nuclear Spins" discusses the interaction that underlies most of phenomena in the field of electron and nuclear spin dynamics. This interaction originates from the magnetic interaction between nuclear and electron spins. In many cases, it is reduced to the Fermi contact interaction whose strength is proportional to the probability of finding an electron at the nucleus. A more involved situation is realized for the valence band holes in III-V or II-VI materials where the hole Bloch functions vanish at the nuclei, see Section 4.2. The modification of the hyperfine coupling Hamiltonian in nanosystems is also analyzed. The chapter contains also an overview of experimental data aimed at determination of the hyperfine interaction parameters in semiconductors and semiconductor nanostructures.

One of the main manifestations of the hyperfine interaction is the transfer of spin between the electron and nuclear systems. This effect is studied in detail in Chapter 5 entitled "Dynamical Nuclear Polarization." While thermal orientation of nuclei in a magnetic field is negligible due to their small magnetic moments, hyperfine coupling can effectively induce the spin flux from electrons to nuclei, or vice versa, provided that the electron spin system is driven out of equilibrium; for example, by means of optical orientation. In this chapter, we present the microscopic approach to the dynamical nuclear polarization effect based on the kinetic equation method followed by a more phenomenological but very powerful description of dynamical nuclear polarization in terms of the nuclear spin temperature concept. In this way, one can account for interactions between neighbouring nuclei without solving a complex many-body problem. The hyperfine interaction also induces the feedback of polarized nuclei on the electron spin system giving rise to a number of nonlinear effects such as bistability of nuclear spin polarization and the anomalous Hanle effect. These and other nonlinear effects in the interacting electron and nuclear system are presented in Sections 5.3–5.5 both from theoretical and experimental viewpoints.

Chapter 6 focuses on details of the optical manipulation of electron spin states. It also addresses manifestations of the electron and nuclear spin dynamics in the optical response of semiconductor nanostructures via spin-Faraday and spin-Kerr effects. Coupling of spins with light provides the most efficient way for nonmagnetic spin manipulation. The main aim of this chapter is to provide theoretical grounds for optical spin injection, ultrafast spin control, and readout of spin states by means of circularly and linearly polarized light pulses. The Faraday and Kerr effects induced by electron and nuclear spin polarization are analyzed both by means of a macroscopic, semi-phenomenological approach, and by using the microscopic quantum mechanical model. Theoretical analysis is supported by experimental data.

After establishing the main interactions in the electron and nuclear spin systems and describing the protocols for electron spin initialization, control, and detection, we turn to the discussion of the electron spin decoherence and relaxation phenomena. These processes limit the conservation time of spin states as well as the response time of the spin system to external perturbations. In particular, Chapter 7 on "Electron Spin Decoherence By Nuclei" focuses on the hyperfine interaction induced decoherence of localized electron spins. After formulation of the central spin model where the spin of charge carrier interacts with the bath of nuclear spins, we present different methods to calculate the spin dynamics within this model. A simple but physically transparent semiclassical treatment where the nuclear spins are considered as largely static classical magnetic moments is followed by a more advanced quantum mechanical approach where the feedback of electron spin dynamics on the nuclei is taken into account. The chapter is concluded by Section 7.4 containing an overview of experimental data and its comparison with the models.

Although the hyperfine interaction serves as a major source of electron spin dephasing, it can be strongly suppressed by an external magnetic field. This is because a substantial difference between the electron and nuclei Zeeman splittings gives rise to the vanishing probabilities of electron-nuclei flip-flop processes making hyperfine coupling-induced spin relaxation inefficient. Hence, other interactions provide electron spin relaxation in strong fields. Electron spin relaxation in sufficiently strong magnetic fields is addressed in Chapter 8 on "Electron Spin Relaxation Beyond the Hyperfine Interaction." Spin relaxation mechanisms related to the joint action of electron-phonon interaction and spin-orbit coupling are discussed in detail. Experimental manifestations of such processes and comparison between the experiment and theory are presented in Section 8.4. We also discuss in Section 8.5 a very specific situation where the spin-flip is caused by the magneto-dipole radiative transitions between the Zeeman-split levels.

Chapter 9 on "Electron Spin Precession Mode-Locking and Nuclei-Induced Frequency Focusing" is devoted to a rich variety of effects in spin dynamics arising from the conditions of pump-probe experiments. Here, we consider the situation where electron spin is injected by a periodic train of circularly polarized pump pulses and precesses between the pulses in an external magnetic field. Nontrivial effects such as resonant spin amplification and spin coherence mode-locking take place due to the commensurability of the repetition period of pump pulses and the charge carrier spin precession period. Theoretical approaches to describing the electron and nuclear spin coherence in such a case and experimental manifestations of these unusual regimes of spin dynamics are discussed in detail.

While the standard approach to studying electron and nuclear spin dynamics experimentally can be briefly summarized as excitation of spin system and monitoring its evolution (or detecting the time-integrated response), a lot can be learned from the equilibrium behavior of spins. In thermal equilibrium, both electron and nuclear spin systems are unpolarized on average, but characterized by nonzero fluctuations. The intensity and dynamics of these inevitable stochastic fluctuations of spins contain information on spin relaxation and decoherence times, spin precession period, and interactions in spin systems. The theory of spin fluctuations in semiconductor nanosystems as well as experimental advances in the field of spin noise spectroscopy are reviewed in Chapter 10. Specific situations where the spin noise spectroscopy can be particularly useful for spin dynamics studies are discussed, the analysis of recent progress in the field of nonequilibrium spin fluctuations is presented as well.

The book is concluded by Chapter 11 on "Strong Coupling of Electron and Nuclear Spins: Outlook and Prospects" where some prospects in the field of electron and nuclear spin dynamics are outlined. Particular emphasis is put on a situation where the hyperfine interaction is so strong that it leads to a qualitative rearrangement of the energy spectrum resulting in the coherent excitation transfer between the electron and nucleus. The strong coupling between the spin of the charge carrier and of the nucleus is realized, for example, in the case of deep centers in semiconductors or in isotopically purified systems. We also discuss the effect of nuclear spin polaron (i.e., ordered state) formation at low enough temperatures of nuclear spins, where the orientation of the carrier spin results in alignment of the spins of nuclei interacting with the electron or hole.

2

Spin Systems in Semiconductor Nanostructures

2.1 Brief overview of spin systems in semiconductors

Spin is the *intrinsic magnetic moment* or *intrinsic angular momentum* of the particle. In contrast to the orbital angular momentum and the magnetic moment related to the orbital motion of charged particles, the presence of spin is not caused by any classical motion. The electrons, protons, and neutrons are fermions and possess a spin of $1/2$. Strictly speaking, the intrinsic angular momentum associated with the spin for these particles equals $\hbar/2$, but the reduced Planck constant $\hbar$ is often omitted, and hereafter we follow this convenient practice.

The protons and neutrons combine and form the nuclei that arrange in the crystalline lattices of semiconductors. In the prominent semiconductor material of GaAs, the relevant (stable) nuclei isotopes are ^{69}Ga, ^{71}Ga, and ^{75}As, all of which have nonzero spins of $3/2$. We recall that the superscript at the upper left of the chemical symbol denotes the mass number; that is, the total number of protons and neutrons constituting a nucleus. The most relevant isotopes of In are ^{113}In (stable, about 4% of natural abundance) and ^{115}In (weakly radiative with a lifetime of >400 trillion years, 95% abundance; i.e., stable from the view point of practical applications) have spins of $9/2$. Hence, in GaAs and InAs semiconductors and nanostructures all host lattice nuclei that have nonzero spin. This might not be the case for other important semiconductors, for example, the stable isotopes of Si: ^{28}Si, ^{29}Si, and ^{30}Si with abundances of about 92, 5, and 3%, respectively, have spins of 0 (^{28}Si, ^{30}Si) and $1/2$ (^{29}Si). Therefore, the majority of nuclei in silicon have zero spin. Some of the isotopes relevant to the semiconductors under study, together with their abundance and spins, are listed in Table 2.1.

Hence, in semiconductors we naturally have the spin system of charge carriers, electrons and holes, and of the host lattice nuclei. One of the specific features of semiconductors is the strong interaction of the charge carriers with light. The polarization state of the electromagnetic wave can be associated with the spin (*spin angular momentum*) of the photon. Unlike electrons, protons and neutrons, the photons are bosons with spin equal to 1. The right- and left-circularly

Electron and Nuclear Spin Dynamics in Semiconductor Nanostructures. M. M. Glazov, Oxford University Press (2018). © M. M. Glazov. DOI: 10.1093/oso/9780198807308.001.0001

Table 2.1 *Nuclear isotopes with nonzero spin in several group-IV, III–VI, and II–VI semiconductors. The asterisk denotes unstable but abundant isotopes. Magnetic moments, gyromagnetic ratios, and quadrupole moments (for $I > 1/2$ in barns $= 10^{-28}$ m^2) are given. From (Fuller, 1976) and Evaluated Nuclear Structure Data File http://www-nds.iaea.org/.*

Isotope	Abundance %	Spin I	Magnetic moment μ_I/μ_N	Gyromagnetic ratio γ_I (MHz/T)	Quadrupole moment Q (barn)
^{27}Al	100	5/2	+3.6415069	+11.103	+0.147
^{75}As	100	3/2	+1.439475	+7.315	+0.314
^{13}C	1.1	1/2	+0.7024118	+10.708	
^{111}Cd	12.8	1/2	−0.5948861	−9.069	
^{113}Cd(∗)	12.2	1/2	−0.6223009	−9.487	
^{19}F	100	1/2	+2.6288	+40.077	
^{69}Ga	60.1	3/2	+2.01659	+10.248	+0.171
^{71}Ga	39.9	3/2	+2.562266	+13.021	+0.107
^{199}Hg	16.9	1/2	+0.5058855	+7.712	
^{201}Hg	13.2	3/2	−0.5602257	−2.846	+0.387
^{113}In	4.3	9/2	+5.5229	+9.355	+0.799
^{115}In(∗)	95.7	9/2	+5.5408	+9.386	+0.86
^{95}Mo	15.9	5/2	−0.9142	−2.787	−0.022
^{97}Mo	9.6	5/2	−0.9335	−2.846	+0.255
^{14}N	99.6	1	+0.403761	+3.078	+0.0193
^{15}N	0.4	1/2	−0.28319	−4.317	
^{31}P	100	1/2	+1.130903	+17.24	
^{33}S	0.76	3/2	+0.6438	+3.272	−0.084
^{121}Sb	57.21	5/2	+3.3634	+10.255	−0.36
^{123}Sb	42.79	7/2	+2.5498	+5.553	−0.49
^{77}Se	7.6	1/2	+0.535	+8.156	
^{29}Si	4.7	1/2	−0.5553	−8.465	
^{123}Te	0.9	1/2	−0.7369478	−11.235	
^{125}Te	7	1/2	−0.8885051	−13.545	
^{67}Zn	4	5/2	+0.875479	+2.669	+0.150
^{183}W	14.3	1/2	+0.11778	+1.796	

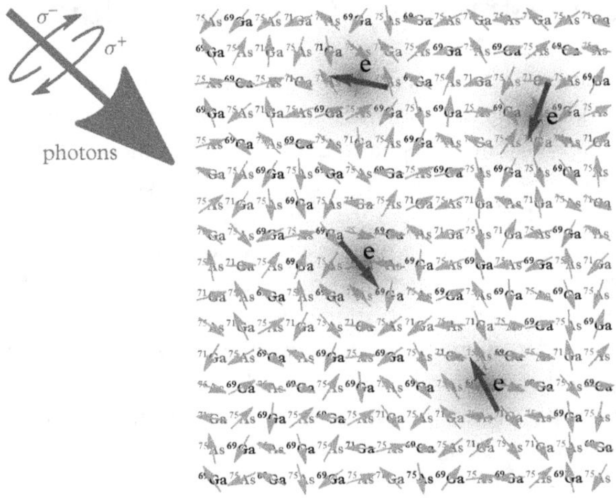

Figure 2.1 *Schematics of the basic spin systems in a semiconductor, which are the focus of this book, illustrated with the example of GaAs. The large arrow denotes* σ^+ *and* σ^- *polarized photons carrying angular moment. GaAs crystalline structure is sketched as a lattice of* ^{69}Ga, ^{71}Ga, *and* ^{75}As *isotopes, each carries a nonzero spin shown by the small gray arrows. The localized electron wavefunctions (e.g., bound to neutral donors) are illustrated by shaded circles, dark thick arrows show electron spins. The dimensions of the bound state wavefunctions and the lattice constant are not shown to scale.*

polarized light corresponds to the photon spin components on the light propagation direction being $+1$ and -1, respectively. The spin component of 0 is forbidden for photons because the electromagnetic field is transversal.

Figure 2.1 sketches these three spin systems, which are the most relevant to the following. The interactions in the spin systems outlined in Section 2.3 allow for spin transfer between photons, electrons, and nuclei. In many cases, the charge carriers act as intermediaries between the spins of photons and lattice nuclei: The electron spins can be efficiently polarized in the process of light absorption due to the *spin-orbit interaction* (this is the process known as *optical orientation*) and electrons can transfer spin to the lattice nuclei via the *hyperfine interaction*. In this chapter, we start with a concise introduction to the electronic properties of semiconductors and semiconductor nanostructures (Section 2.2), then discuss interactions in spin systems in Section 2.3, outline the main effects in spin systems in semiconductors in Sections 2.4 and 2.5, and conclude this chapter with the motivation for the study of spin dynamics in the interacting system of the lattice nuclei and localized electron spins (Section 2.6).

2.2 Basics of semiconductor nanostructure physics

The motion of electrons in bulk crystals resembles electron dynamics in free space. There are allowed energy bands with a continuous energy spectrum separated by the forbidden intervals known as band gaps. The formation of forbidden bands is because the electron potential energy is periodic in space. Due to the *translational symmetry* of bulk materials the electron states are characterized by the following quantum numbers: The continuous variable $\boldsymbol{k}$ being the quasi-wavevector or, in short, the wavevector of the electron, and discrete parameters n being the band label and $s = \pm 1/2$ being the electron spin index. Usually, the spin-orbit coupling mixes orbital and spin degrees of freedom, therefore, it is convenient to use a single subscript n that accounts both for the band and spin states. Accordingly, the wavefunction $\Psi_{\boldsymbol{k},n}(\boldsymbol{r})$ of the electron in semiconductor can be presented in the form of the Bloch function

$$\Psi_{\boldsymbol{k},n}(\boldsymbol{r}) = \frac{1}{\sqrt{v}} \exp(\mathrm{i}\boldsymbol{k}\cdot\boldsymbol{r}) U_{\boldsymbol{k},n}(\boldsymbol{r}). \tag{2.1}$$

Here, v is the normalization volume, $\boldsymbol{r} = (x, y, z)$ is the electron position vector, and $U_{\boldsymbol{k},n}(\boldsymbol{r})$ is the periodic Bloch amplitude: At all translations that match the crystalline lattice with itself, the function $U_{\boldsymbol{k},n}(\boldsymbol{r})$ remains invariant. Equation (2.1) is the Bloch theorem that follows from the translational symmetry arguments. The values of the $\boldsymbol{k}$ are usually selected within the first *Brillouin zone* of the $\boldsymbol{k}$-space, since the addition to $\boldsymbol{k}$ or subtraction from $\boldsymbol{k}$ of a reciprocal lattice vector can be accounted for by the change of the periodic amplitude in Eq. (2.1). The functions (2.1) satisfy the Schrödinger equation

$$\mathcal{H}_0 \Psi_{\boldsymbol{k},n} = E_n(\boldsymbol{k}) \Psi_{\boldsymbol{k},n}, \quad \mathcal{H}_0 = \frac{\hat{\boldsymbol{p}}^2}{2m_0} + V(\boldsymbol{r}) + \mathcal{H}_{so}. \tag{2.2}$$

Here $\hat{\boldsymbol{p}} = -\mathrm{i}\hbar\nabla_{\boldsymbol{r}} = -\mathrm{i}\hbar(\partial/\partial x, \partial/\partial y, \partial/\partial z)$ is the electron momentum operator, m_0 is the free electron mass, $V(\boldsymbol{r})$ is the *periodic* single-particle potential and $\mathcal{H}_{so}$ is the spin-orbit Hamiltonian, whose form is presented later in Eq. (2.11).

Equation (2.1) demonstrates that the electron state in a crystal is, like in free space, a plane wave, but, unlike in free space, modulated with a function $U_{\boldsymbol{k},n}(\boldsymbol{r})$ oscillating on the scale of the lattice constant. Just like in free space, the electron energy $E_n(\boldsymbol{k})$ is a continuous function of the wavevector $\boldsymbol{k}$ in the allowed bands, however, the $E_n(\boldsymbol{k})$ dependence can be both qualitatively and quantitatively different from that of the electron in vacuum. Figure 2.2 illustrates some of the relevant energy bands in the prototypical system of GaAs: the lowest empty band, that is, the *conduction band*, nearest excited conduction bands as well as the topmost occupied bands, the *valence bands*. The conduction and valence bands are separated by the forbidden band gap, E_g, for instance, in GaAs $E_g \approx 1.5$ eV. Of course, the band filling by the charge carriers depends on the temperature and

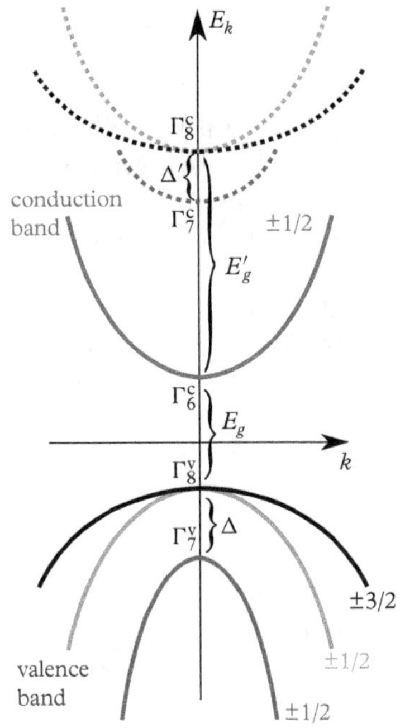

Figure 2.2 *Sketch of the conduction and valence bands in a GaAs-like semiconductor. The remote conduction bands are shown by dotted lines. E_g and E_g' are the band gaps, Δ and Δ' are the spin-orbit splittings of bands, the bands are labelled by the irreducible representations Γ_6, Γ_7, and Γ_8 relevant at $\boldsymbol{k} = 0$, with the superscripts c and v denoting the conduction and valence bands respectively. Additionally, the bands, are labelled by the spin components, see Section 2.3 for details. The electron representation for labelling the valence band states is used.*

presence of impurities; that is, doping. Hence, the classification of bands in terms of conduction and valence is performed for pure semiconductors in the limit of zero temperature.

Calculations of electron dispersion is a complex task where various atomistic methods such as such as density functional theory, pseudopotential method, tight-binding approximation are used. However, as a rule, one is interested in the electron states in the vicinity of some point $\boldsymbol{k} = \boldsymbol{K}_0$ in the Brillouin zone where in the extrema of the conduction band, $E_c(\boldsymbol{k})$, or the valence band, $E_v(\boldsymbol{k})$, energy dispersions are realized. GaAs is the direct band gap semiconductor, where the nearest in energy extrema of conduction and valence bands are formed at $\boldsymbol{K}_0 = 0$. This point is also denoted as the Γ-point, see Figure 2.2, and we focus on this situation hereafter. The Bloch amplitude $U_{\boldsymbol{k},n}$ is decomposed in the series of the periodic amplitudes at the Γ-point $U_{\boldsymbol{k},n} = \sum_{n'} C_{n'}^{(n)}(\boldsymbol{k}) U_{\boldsymbol{0},n'}$, where the subscript $\boldsymbol{0}$ denotes $\boldsymbol{k} = 0$. Making use of the $\boldsymbol{k}\cdot\boldsymbol{p}$*-perturbation theory* (see Appendix B) one can show that the column vector $\hat{C}^{(n)}(\boldsymbol{k})$ formed of the decomposition coefficients $C_{n'}(\boldsymbol{k})$ satisfies the matrix equation

$$\mathcal{H}(\boldsymbol{k})\hat{C}^{(n)}(\boldsymbol{k}) = E_n(\boldsymbol{k})\hat{C}^{(n)}(\boldsymbol{k}). \tag{2.3}$$

Here, $\mathcal{H}(\boldsymbol{k})$ is the *effective Hamiltonian*. Within the second-order $\boldsymbol{k}\cdot\boldsymbol{p}$-perturbation theory, the matrix elements $\mathcal{H}_{n;n'}(\boldsymbol{k})$ of the effective Hamiltonian $\mathcal{H}(\boldsymbol{k})$ are linear and quadratic in the components of the wavevector $\boldsymbol{k}$. In the simplest possible case where only one band is taken into consideration and the spin degree of freedom is disregarded, the effective Hamiltonian takes the form $\mathcal{H}(\boldsymbol{k}) = \hbar^2 k^2/2m$, where m is the *effective mass*.

Within the $\boldsymbol{k}\cdot\boldsymbol{p}$ approach one usually includes only the nearest bands in the Hamiltonian $\mathcal{H}(\boldsymbol{k})$, while contributions from the remote bands are accounted for by the quadratic in the wavevector, $\sim k^2$, and sometimes, higher order terms in the $\mathcal{H}(\boldsymbol{k})$. The form of the effective Hamiltonian in the bulk material can be established by means of symmetry analysis: The invariants method allows one to find a general form of $\mathcal{H}_{n;n'}$ while the parameters of the Hamiltonian are determined by comparison with atomistic calculations and experiments. For further details on the $\boldsymbol{k}\cdot\boldsymbol{p}$-method we refer the reader to Appendix B and the books (Bir and Pikus, 1974; Ivchenko and Pikus, 1997; Voon and Willatzen, 2009) where this method is justified and presented in every detail.

In the presence of external electric and magnetic fields, impurities, lattice deformations, as well as in nanostructures, such as *quantum wells* (QWs), *quantum wires* (QWRs), and *quantum dots* (QDs), the electron potential energy is no longer periodic in space and the Bloch theorem does not apply anymore. For example, in quantum wells the electron motion is free in the well plane and restricted along the well normal; in quantum wires the electrons are free to move along one direction, the wire axis; and in quantum dots the motion of charge carriers is restricted in all three spatial directions. Generally, the atomistic methods should be used to calculate the wavefunctions and energies of size-quantized states. These methods have high accuracy but, as a rule, relatively high complexity, which does not allow one to trace the physics of the effects easily. Therefore, in many cases it is instructive to use approximate methods, namely the *effective mass method* or effective Hamiltonian method, which is valid if the charge carrier localization in nanostructures or in the impurity potential takes place on the length scale, which exceeds by far the lattice constant.

In the effective Hamiltonian method, the electron wavefunction is sought in each homogeneous part of the system as a linear combination of bulk solutions, see Eq. (2.1),

$$\Psi(\boldsymbol{r}) = \sum_n \varphi_n(\boldsymbol{r}) U_{0,n}(\boldsymbol{r}). \tag{2.4}$$

The exponential function $\exp(\mathrm{i}\boldsymbol{k}\cdot\boldsymbol{r})$ describing a plane wave in Eq. (2.1) is replaced by an *envelope function* $\varphi_n(\boldsymbol{r})$ describing the size quantization effect, similar to the case of a free particle where the plane wave is replaced by the appropriate wavefunction in the given potential. The envelope functions are smooth on the scale of the lattice constant; they obey the Schrödinger equation in bulk layers with the effective matrix Hamiltonian, Eq. (2.3), where $\boldsymbol{k} = -\mathrm{i}\nabla$. The functions $\varphi_n(\boldsymbol{r})$ are matched at the heterointerfaces by the appropriate boundary conditions.

The amount included in the consideration bands depends on the phenomenon studied, required accuracy, and so on. If just one band is accounted for, as an example, the conduction band, and the spin is disregarded for a moment, the effective Hamiltonian takes the particularly simple form,

$$\mathcal{H}(\boldsymbol{k}) = \frac{\hbar^2 k^2}{2m} + V_{\text{eff}}, \tag{2.5}$$

where $V_{\text{eff}} = V_{\text{eff}}(\boldsymbol{r})$ is the effective potential energy. This model is referred to as the effective mass approximation. For the conduction band in GaAs $m \approx 0.067 m_0$, where m_0 is the free electron mass. In heterostructures formed of homogeneous bulk semiconductors, the potential energy $V_{\text{eff}}(\boldsymbol{r})$ takes a constant value within each homogeneous region, the discontinuities of $V_{\text{eff}}(\boldsymbol{r})$ at the heterointerfaces equal to the band offsets between the materials. The effective potential can also include contributions due to external fields. The effective mass differs in different homogeneous regions as well. In nanostructures formed by a smooth variation of composition, both V_{eff} and m are smooth functions of coordinates, and the kinetic energy is usually represented as

$$-\frac{\hbar^2}{2}\nabla \frac{1}{m(\boldsymbol{r})} \nabla.$$

Similarly, for an electron interacting with the ionized donor in a bulk semiconductor, the potential $V_{\text{eff}}(\boldsymbol{r})$ is usually approximated by the attractive Coulomb potential, $V_{\text{eff}}(\boldsymbol{r}) = -e^2/(\varkappa r)$, where $r = |\boldsymbol{r}|$ and $\varkappa$ is the static dielectric constant. Equation (2.5) is similar to the Hamiltonian of a free electron interacting with a potential field V_{eff}. Just as for an electron bound to an attractive potential, in semiconductors the attractive potential due to the electric field of the ionized donor or due to the band offsets results in the formation of localized states with a discrete energy spectrum. The engineering of the potential $V_{\text{eff}}(\boldsymbol{r})$ by means of advanced growth techniques opens up possibilities to design the wavefunctions of the size-quantized states in nanostructures and tune their energies on demand. The details on the size quantization effect in nanostructures can be found, for example, in (Bastard, 1988; Davies, 1998; Ivchenko and Pikus, 1997; Winkler, 2003; Ivchenko, 2005) where various semiconductor systems are considered.

2.3 Interactions in spin systems

Free electron spin states are described by the spinor wavefunctions denoted as $|\uparrow\rangle, |\downarrow\rangle$. These are eigenfunctions of the spin-z component operator, $\hat{s}_z$, with the eigenvalues being $\pm 1/2$, respectively. In this basis, the electron spin operator is conveniently represented by the Pauli matrices:

$$\hat{\boldsymbol{s}} = \frac{1}{2}\boldsymbol{\sigma}, \quad \boldsymbol{\sigma} = (\sigma_x, \sigma_y, \sigma_z), \tag{2.6}$$

where

$$\sigma_x = \begin{pmatrix} 0 & 1 \\ 1 & 0 \end{pmatrix}, \quad \sigma_y = \begin{pmatrix} 0 & -\mathrm{i} \\ \mathrm{i} & 0 \end{pmatrix}, \quad \sigma_z = \begin{pmatrix} 1 & 0 \\ 0 & -1 \end{pmatrix}. \tag{2.7}$$

The intrinsic magnetic moment $\hat{\boldsymbol{\mu}}$ of a free electron is related to its spin as $\hat{\boldsymbol{\mu}} = -2\mu_B\hat{\boldsymbol{s}}$, where $\mu_B = |e|\hbar/(2m_0c)$ is the Bohr magneton with c being the speed of light. As a result, for a free electron the Hamiltonian describing its spin interaction with the external magnetic field $\boldsymbol{B}$, that is, the *Zeeman effect*, has a particularly simple form

$$\mathcal{H}_B = -(\hat{\boldsymbol{\mu}} \cdot \boldsymbol{B}) = \frac{g_0}{2}\mu_B(\boldsymbol{\sigma} \cdot \boldsymbol{B}), \tag{2.8}$$

where $g_0 = 2$ is the free electron g-factor (also known as the *Landé factor*).

In crystals and nanostructures, the electron Zeeman effect is described by a similar Hamiltonian

$$\mathcal{H}_B = \frac{1}{2}\sum_{\alpha,\beta} g_{\alpha\beta}\sigma_\alpha B_\beta, \tag{2.9}$$

where we introduced the tensor of g-factors $g_{\alpha\beta}$. Here $\alpha, \beta = x, y, z$ are the Cartesian coordinates. Equation (2.9) can be recast in the form

$$\mathcal{H}_B = \frac{\hbar}{2}(\boldsymbol{\sigma} \cdot \boldsymbol{\Omega_B}), \tag{2.10}$$

where the pseudovector $\boldsymbol{\Omega_B}$ with the components

$$\Omega_{\boldsymbol{B},\alpha} = \frac{\mu_B}{\hbar}\sum_\beta g_{\alpha\beta}B_\beta,$$

has a transparent physical meaning: It is the frequency of the spin precession in the external field $\boldsymbol{B}$. The $\boldsymbol{\Omega}_B$ is the *Larmor frequency* of electron spin precession. The number of independent components of $g_{\alpha\beta}$ is determined by the symmetry of the system. For instance, in cubic semiconductors such as GaAs or InAs, the tensor $g_{\alpha\beta} = g\delta_{\alpha\beta}$ reduces to a scalar, g, and $\delta_{\alpha\beta}$ is the Kronecker δ-symbol. The difference between the free electron g-factor and the g-factor in many semiconductors is quite significant; for example, in GaAs for conduction band electrons $g \approx -0.44$, while in InSb $g \approx -50$. The difference results from the *spin-orbit interaction*. The size quantization of electron states in nanosystems results both in renormalization of the g-factor with respect to the bulk values, and to the appearance of new independent components of the g-factor tensor (Ivchenko and Kiselev, 1992; Kalevich and Korenev, 1993; Ivchenko, 2005; Yugova *et al.*, 2007*a*).

The spin-orbit interaction or *spin-orbit coupling* provides the coupling between the spin states of the electron and its orbital motion. The spin-orbit interaction is a relativistic effect but, for the semiconductors under study, it is sufficient to account for the relativistic effects in the lowest order. The spin-orbit coupling is thus described by the Hamiltonian (Bir and Pikus, 1974; Landau and Lifshitz, 1977)

$$\mathcal{H}_{so} = -\frac{i\hbar^2}{4m_0^2c^2}\left(\boldsymbol{\sigma} \cdot [\nabla V(\boldsymbol{r}) \times \nabla]\right), \tag{2.11}$$

Here, $V(\boldsymbol{r})$ is the periodic potential in the crystal, Eq. (2.2), and the operator $\mathcal{H}_{so}$ acts in the basis of Bloch functions, $\Psi_{\boldsymbol{k},n}(\boldsymbol{r})$, arranged in the form of spinors,

$$\hat{\Psi}_{\boldsymbol{k},n}(\boldsymbol{r}) = \Psi_{\boldsymbol{k},n,+1/2}(\boldsymbol{r})|\uparrow\rangle + \Psi_{\boldsymbol{k},n,-1/2}(\boldsymbol{r})|\downarrow\rangle = \begin{pmatrix} \Psi_{\boldsymbol{k},n,+1/2}(\boldsymbol{r}) \\ \Psi_{\boldsymbol{k},n,-1/2}(\boldsymbol{r}) \end{pmatrix}.$$

The interaction in the form of Eq. (2.11) has a profound effect on the electron states in semiconductors and acts in many cases as a driving force for electron spin dynamics. Here we recall the most important manifestations of the spin-orbit coupling.

In semiconductors like GaAs, the conduction band in the vicinity of the Γ point is formed mainly of s-like atomic states (with the orbital momentum being 0), while the valence band is mainly formed of the p-like atomic states (with the orbital momentum being 1). Formally, the orbital part of the Bloch amplitude of the conduction band at $\boldsymbol{k} = 0$, $\mathcal{S}(\boldsymbol{r})$, is invariant under all operations in the point symmetry group T_d relevant to the zinc-blende lattice crystals. It corresponds to the irreducible representation A_1; see Appendix A for details. Accounting for the spin the pair of Bloch amplitudes $U_{\boldsymbol{0},n,+1/2} = \mathcal{S}(\boldsymbol{r})|\uparrow\rangle$, $U_{\boldsymbol{0},n,-1/2} = \mathcal{S}(\boldsymbol{r})|\downarrow\rangle$ transforms according to the irreducible spinor representation Γ_6 of the T_d point group. There are three orbital Bloch amplitudes, $\mathcal{X}(\boldsymbol{r})$, $\mathcal{Y}(\boldsymbol{r})$, $\mathcal{Z}(\boldsymbol{r})$, of the valence band at the Γ-point, which form the basis of the irreducible representation F_2 of the T_d point symmetry group. These three functions transform at all symmetry operations just like the components of the polar vector, such as the position vector $\boldsymbol{r}$, or the electric field $\boldsymbol{E}$. Accounting for the spin, we have six functions, for example, $\mathcal{X}(\boldsymbol{r})|\uparrow\rangle$, $\mathcal{Y}(\boldsymbol{r})|\downarrow\rangle$, which form a basis of the representation $F_2 \otimes \Gamma_6$. This representation is reducible, it can be decomposed in the sum of irreducible spinor representations $\Gamma_8 \oplus \Gamma_7$. The dimensions of these representations are 4 and 2, respectively. Hence, accounting for the spin-orbit interaction, one expects the splitting of the valence band top into the quadruplet (Γ_8 representation) and the doublet (Γ_7 representation), see Figure 2.2. The splitting between the Γ_8 and Γ_7

states is called the spin-orbit splitting of the valence band and denoted as Δ. It amounts to about 340 meV in GaAs, 390 meV in InAs, 750 meV in GaSb, and reaches almost 1 eV in InSb (Vurgaftman *et al.*, 2001; Adachi, 2005, 2009).

In fact, this fine structure of the valence band states is similar to the fine structure of the $2p$-level of the hydrogen atom. Indeed, the periodic potential $V(\boldsymbol{r})$ in Eq. (2.11), is invariant under all transformations of the crystal point group, the derivatives $\partial V/\partial x, \partial V/\partial y, \partial V/\partial z$ transform as components of a vector $\boldsymbol{r} = (x, y, z)$. Hence, the selection rules for the matrix elements of the operator $\mathcal{H}_{so}$ are exactly the same as in the atomic case and, correspondingly, the matrix elements of the $\mathcal{H}_{so}$ operator in the basis of the valence band Bloch amplitudes $\mathcal{X}(\boldsymbol{r})|\uparrow\rangle, \mathcal{X}(\boldsymbol{r})|\downarrow\rangle, \ldots$ differ from those in the atom by a common factor only. By analogy with the hydrogen atom, the valence band basic functions at the Γ point are usually labelled by the total angular momentum $\mathcal{J}$ and its component $\mathcal{J}_z$. The angular momentum $\mathcal{J}$ is the sum of the valence band electron spin (1/2) and orbital (1) momenta. It runs through the values from $-3/2$ to $+3/2$ for the Γ_8 states:

$$|\Gamma_8, +3/2\rangle = -\frac{\mathcal{X} + \mathrm{i}\mathcal{Y}}{\sqrt{2}}|\uparrow\rangle, \tag{2.12a}$$

$$|\Gamma_8, +1/2\rangle = \sqrt{\frac{2}{3}}\mathcal{Z}|\uparrow\rangle - \frac{\mathcal{X} + \mathrm{i}\mathcal{Y}}{\sqrt{6}}|\downarrow\rangle, \tag{2.12b}$$

$$|\Gamma_8, -1/2\rangle = \sqrt{\frac{2}{3}}\mathcal{Z}|\downarrow\rangle + \frac{\mathcal{X} - \mathrm{i}\mathcal{Y}}{\sqrt{6}}|\uparrow\rangle, \tag{2.12c}$$

$$|\Gamma_8, -3/2\rangle = \frac{\mathcal{X} - \mathrm{i}\mathcal{Y}}{\sqrt{2}}|\downarrow\rangle, \tag{2.12d}$$

and takes the values $\pm 1/2$ for the Γ_7 states:

$$|\Gamma_7, +1/2\rangle = -\frac{1}{\sqrt{3}}\left[\mathcal{Z}|\uparrow\rangle + (\mathcal{X} + \mathrm{i}\mathcal{Y})|\downarrow\rangle\right], \tag{2.13a}$$

$$|\Gamma_7, -1/2\rangle = \frac{1}{\sqrt{3}}\left[\mathcal{Z}|\downarrow\rangle - (\mathcal{X} + \mathrm{i}\mathcal{Y})|\uparrow\rangle\right]. \tag{2.13b}$$

Here, we omitted the argument of the orbital functions, $\boldsymbol{r}$, and used the canonical basis for the angular momentum basic functions, see; for example, (Edmonds, 1974; Varshalovich *et al.*, 1988).

It is noteworthy that at $\boldsymbol{k} \neq 0$ the Γ_8 quadruplet splits into two branches, usually denoted as the heavy-hole band (with a larger absolute value of effective mass) and the light-hole band (with a smaller magnitude of effective mass), see Figure 2.2. As illustrated in Figure 2.2, for $\boldsymbol{k}$ directed along one of the cubic axes $\langle 100\rangle$, these bands correspond to the states with the projections of the total angular momentum onto the $\boldsymbol{k}$ being $\pm 3/2$ and $\pm 1/2$, respectively. Note that the situation is somewhat more complex for the arbitrary orientation of $\boldsymbol{k}$, moreover, in some semiconductors the signs of the heavy- and light-hole effective masses can be different. The details of the valence band fine structure can be found, for example, in (Bir and Pikus, 1974; Ivchenko and Pikus, 1997; Yu and Cardona, 2010).

Furthermore, the spin-orbit interaction affects the electron and hole dispersion in semiconductor and low-dimensional structures where the free motion is possible at least along one spatial direction. Indeed, in noncentrosymmetric systems, such as in zinc-blende lattice semiconductors and in many nanostructures, the energies of two spin states with a given wavevector $\boldsymbol{k}$, $E_{n\pm1/2}(\boldsymbol{k})$, are not equal. The spin-dependent contribution in the conduction band effective Hamiltonian of a semiconductor with the T_d point symmetry is cubic in the wavevector (Dresselhaus, 1955)

$$\mathcal{H}_D = \gamma_c\left[\sigma_x k_x\left(k_y^2 - k_z^2\right) + \sigma_y k_y\left(k_z^2 - k_x^2\right) + \sigma_z k_z\left(k_x^2 - k_y^2\right)\right], \tag{2.14}$$

with γ_c being the coefficient, $\gamma_c \approx 20$ eVÅ^3 for GaAs. The contribution of Eq. (2.14) to the effective Hamiltonian is termed the *Dresselhaus term*. Here, we use the set of cubic axes, $x \parallel [100]$, $y \parallel [010]$, $z \parallel [001]$. Such a form of spin splitting of the conduction band is readily understood from the symmetry arguments: The set of the Pauli matrices forms the basis of the F_1 representation of the T_d point symmetry group, the combinations $k_x(k_y^2 - k_z^2)$, $k_y(k_z^2 - k_x^2)$, $k_z(k_x^2 - k_y^2)$ also transform according to the same irreducible representation. For the valence band states, both $\boldsymbol{k}$-linear and $\boldsymbol{k}$-cubic terms are allowed (Pikus *et al.*, 1988; Ivchenko and Pikus, 1997). In wurtzite semiconductors and in nanostructures the symmetry is, as a rule, lower and allows for the coupling between the spin and the linear combinations of the wavevector components (Rashba and Sheka, 1959; Ohkawa and Uemura, 1974; Vas'ko, 1980; Bychkov and Rashba, 1984; Dyakonov and Kachorovskii, 1986). In any case, the spin-orbit contributions to the effective electron Hamiltonian can be written in a form similar to Eq. (2.10) (Winkler, 2003; Ivchenko, 2005; Dyakonov, 2017):

$$\mathcal{H}_{so} = \frac{\hbar}{2}\left(\boldsymbol{\sigma}\cdot\boldsymbol{\Omega}_{\boldsymbol{k}}\right). \tag{2.15}$$

Here, the pseudovector, $\boldsymbol{\Omega}_{\boldsymbol{k}}$, is an odd function of the electron wavevector, $\boldsymbol{\Omega}_{\boldsymbol{k}} = -\boldsymbol{\Omega}_{-\boldsymbol{k}}$. This is the result of the time reversal, $t \to -t$, symmetry: in nonmagnetic systems at the $t \to -t$ transformation both spin and wavevector change their signs ensuring the invariant form of the interaction. The spin-dependent combinations involving even powers of the wavevector components are thus forbidden. Equation (2.15) demonstrates that the electron moving in a noncentrosymmetric structure experiences an effective magnetic field, whose strength and direction depend on the electron wavevector. Hence, the electron orbital motion becomes coupled with the electron spin dynamics.

So far, we have discussed the single particle states and the single particle energy dispersion. Importantly, the interaction between electrons is spin-dependent as well. This is because of the Pauli principle, which imposes the antisymmetry of the total wavefunction of fermions with respect to the particle permutations. Even if the spin-orbit interaction is absent and, correspondingly, the spin part of the two-electron wavefunction can be separated from the orbital part, the symmetry of the orbital part of the wavefunction is enforced by the symmetry of its spin part. According to the momentum summation rules, the total spin S of the electron pair can be either 0 or 1. The spin part of the function with $S=0$, that is, *spin singlet*, reads

$$\frac{1}{\sqrt{2}}(|\uparrow\rangle_1|\downarrow\rangle_2 - |\uparrow\rangle_1|\downarrow\rangle_2), \tag{2.16}$$

and it is antisymmetric with respect to the particle permutations (the subscripts 1 and 2 enumerate the electrons). Hence, the orbital function of the spin singlet is symmetric with respect to the electron permutations. By contrast, the spin parts of the $S=1$ *triplet* have the form

$$|\uparrow\rangle_1|\uparrow\rangle_2, \quad \frac{1}{\sqrt{2}}(|\uparrow\rangle_1|\downarrow\rangle_2 + |\uparrow\rangle_1|\downarrow\rangle_2), \quad |\downarrow\rangle_1|\downarrow\rangle_2, \tag{2.17}$$

being symmetric with respect to the permutations, making the corresponding orbital wavefunction antisymmetric. Difference of the orbital wavefunctions results in the difference of the Coulomb interaction energies. Hence, the states with $S=0$ and $S=1$ are split due to the Coulomb interaction. In many cases it is convenient to avoid the antisymmetrization procedure and introduce the effective *exchange interaction* in the form of interaction between the spins as

$$\mathcal{H}_{exch} = -\mathcal{J}\hat{\boldsymbol{s}}_1 \cdot \hat{\boldsymbol{s}}_2, \tag{2.18}$$

where $\hat{\boldsymbol{s}}_i$ is the spin operator of ith electron, $i=1,2$, and $\mathcal{J}$ is the exchange interaction constant. One can readily check that the eigenstates of the $\mathcal{H}_{exch}$ operator are indeed the singlet and the triplet states, see Eqs (2.16) and (2.17), with the energies $3\mathcal{J}/4$ and $-\mathcal{J}/4$, respectively. In the presence of the spin-orbit interaction, the scalar product of the spin operators in Eq. (2.18) is replaced by the combination (Kavokin, 2004; Bădescu *et al.*, 2005; Gangadharaiah *et al.*, 2008; Glazov and Kulakovskii, 2009)

$$-\sum_{\alpha\beta}\mathcal{J}_{\alpha\beta}\hat{s}_{1,\alpha}\hat{s}_{2,\beta},$$

where the components of the exchange constant tensor $\mathcal{J}_{\alpha\beta}$ are determined by the symmetry of the system, orbital wavefunctions of the electrons and the spin-orbit coupling. It is worth stressing that the exchange interaction is just a consequence of the permutation symmetry requirements and the Coulomb interaction. The exchange interaction is of major importance in diluted magnetic semiconductors, for example GaAs:Mn, where the paramagnetic ions with unpaired electron spins, such as manganese, are incorporated. It is also very important for *excitons*, Coulomb-correlated electron-hole pairs: The exchange interaction between an electron and a hole determines to a large extent the exciton fine structure and serves as a driving force for the spin dynamics of excitons (Bir and Pikus, 1974; Ivchenko, 2005). It is noteworthy that, for electronic excitations in semiconductors, it is by far more important compared with the magnetic interaction between the charge carriers.

Magnetic interactions play a role where the spins of the lattice nuclei are involved. For a brief introduction to the physics of nuclei we refer the reader to the concise lecture course by Landau and Smorodinsky (2014). The basic interactions involving nuclear spins in solids are presented in (Meier and Zakharchenya, 1984; Abragam, 2002; Dyakonov, 2017). Consider a nucleus with the spin I. The nucleus can be treated as a point-like magnetic moment $\boldsymbol{\mu}$ since the size of the nucleus is in the range of $(1\ldots10)\times10^{-5}$ Å, so it is far smaller than the lattice constant in semiconductors, as well as, naturally, the de Broglie wavelength of the electron. The magnetic moment operator $\hat{\boldsymbol{\mu}}$ is related to the nuclear spin operator $\hat{\boldsymbol{I}}$ as

$$\hat{\boldsymbol{\mu}} = \hbar\gamma_I \hat{\boldsymbol{I}} = \frac{\mu_I}{I}\hat{\boldsymbol{I}}, \tag{2.19}$$

where γ_I is the gyromagnetic ratio of the nucleus and μ_I is the nuclear magnetic moment. The interaction of nuclear spin with the magnetic field $\boldsymbol{B}$ is described by the Hamiltonian

$$\mathcal{H}_{B,N} = -(\hat{\boldsymbol{\mu}}\cdot\boldsymbol{B}). \tag{2.20}$$

Analogous to Eq. (2.9), one also introduces the nuclear g-factor, g_n, as $g_n = \hbar\gamma_I/\mu_N$, where $\mu_N = |e|\hbar/(2m_pc)$ is the nuclear magneton and m_p is the proton mass, recasting Eq. (2.20) in the form:

$$\mathcal{H}_{B,N} = -g_n\mu_N(\hat{\boldsymbol{I}}\cdot\boldsymbol{B}), \quad g_n = \frac{\mu_I}{I\mu_N}. \tag{2.21}$$

Note the difference in signs in Eqs (2.9) and (2.21): Formally, it comes from the fact the the nucleus is positively charged, such a sign convention is used in (Abragam and Bleaney, 1970; Kalevich *et al.*, 2017) and is quite convenient. Typically, the nuclear g-factors are in the order of unity, hence, in the same magnetic field the Zeeman splittings of nuclear spin sublevels are about three orders of magnitude smaller than those of the electron; see Table 2.1 where the magnetic moments and gyromagnetic ratios for some relevant isotopes are listed.

The nuclear magnetic moment $\boldsymbol{\mu}$ produces the magnetic field that acts on the electron spins and on the spins of the neighboring nuclei. The *hyperfine interaction* Hamiltonian of the conduction band electron and nuclear spin in III–V and II–VI semiconductors can be written in the form

$$\mathcal{H}_{hf} = Av_0(\hat{\boldsymbol{I}}\cdot\hat{\boldsymbol{s}})|\varphi_c(\boldsymbol{R})|^2, \tag{2.22}$$

where v_0 is the *primitive cell* volume and $\varphi_c(\boldsymbol{R})$ is the conduction band envelope function, Eq. (2.4), at the nucleus. Note that primitive cell of GaAs-type semiconductor contains two atoms. Sometimes the term unit cell is used in the literature. Here and in what follows we use primitive cell term to avoid confusion. The constant A depends on the material and nucleus type; it is, as a rule, in the order of 100 μeV. For the conduction band electrons in GaAs-type semiconductors, the hyperfine interaction results from the contact Fermi interaction of the electron and nuclear spins because the orbital Bloch amplitude $\mathcal{S}(\boldsymbol{r})$ is nonzero at the host lattice nuclei. For the valence band holes, the orbital Bloch amplitudes $\mathcal{X}(\boldsymbol{r})$, $\mathcal{Y}(\boldsymbol{r})$ and $\mathcal{Z}(\boldsymbol{r})$ vanish at the nucleus and the hyperfine interaction of hole, and nuclear spins is provided by the dipole–dipole interaction. It is about an order of magnitude weaker compared with the electrons. The hyperfine interaction results in the transfer of spin between the electron and nuclear spin systems in semiconductors. Equation (2.22) demonstrates that the polarized nuclei exert the effective magnetic field, known as the *Overhauser field*, onto the electron spin. The polarized electrons also create an effective magnetic field acting on the nuclear spins known as the *Knight field*. The theory of the hyperfine interaction of the charge carrier spins with the host lattice nuclei is presented in Chapter 4.

The interaction between the lattice nuclei spins is simply the *dipole–dipole interaction* of the localized magnetic moments. The Hamiltonian of interaction between the nuclear spins $\boldsymbol{I}$ and $\boldsymbol{I}'$ has the form

$$\mathcal{H}_{dd} = \frac{\mu_I \mu_{I'}}{II'} \left(\frac{\hat{\boldsymbol{I}} \cdot \hat{\boldsymbol{I}}'}{r^3} - 3 \frac{(\hat{\boldsymbol{I}} \cdot \boldsymbol{r}) \cdot (\hat{\boldsymbol{I}}' \cdot \boldsymbol{r})}{r^5} \right), \tag{2.23}$$

where $\boldsymbol{r}$ is the vector connecting the nuclei. A typical magnetic field (also known as the *local magnetic field*, B_L) created at a given nucleus by the other host lattice nuclei is in the order of $B_L \sim 0.1$ mT. In contrast to electrons in semiconductors, an overlap of nuclei is negligible and the exchange interaction between the nuclei is irrelevant. Physically, the dipole–dipole interaction between the nuclear spins results in the redistribution of the spin polarization among the ensemble of nuclei, that is, in the nuclear spin diffusion and in the thermalization of the nuclear spin system. It follows from Eq. (2.23) that the dipole–dipole interaction does not conserve the total spin of the nuclei and, as a result, may serve as a source of nuclear spin relaxation.

Moreover, a nucleus with spin I higher than $1/2$ has a nonzero electric *quadrupole moment*. Qualitatively, its presence is caused by the non-spherical charge distribution in the nucleus, hence, the quadrupole moment serves as a measure of nuclei being prolate or oblate. As a result, the nuclear spin interacts with the electric field gradients:

$$\mathcal{H}_Q = \sum_{\alpha\beta} \frac{\partial^2 \mathcal{V}}{\partial r_\alpha \partial r_\beta} \frac{|e|Q}{6I(2I-1)} \left[\frac{3}{2}(\hat{I}_\alpha \hat{I}_\beta + \hat{I}_\beta \hat{I}_\alpha) - I(I+1) \right]. \tag{2.24}$$

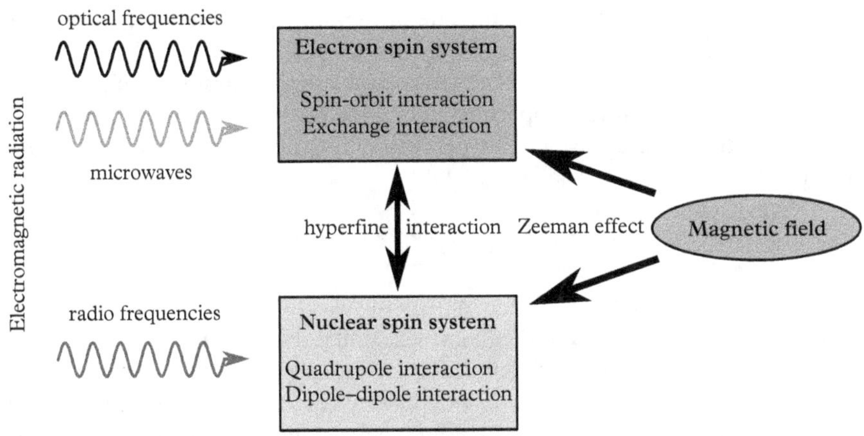

Figure 2.3 *Chart of the important interactions in spin systems.*

Here, $\mathcal{V}$ is the potential of the electric field, and Q is the constant, namely the quadrupole moment, which depends on the isotope, see Table 2.1. The form of the *quadrupole interaction* Hamiltonian, Eq. (2.24), can be easily understood from the symmetry arguments: The products of the pseudovector components $\hat{I}_\alpha \hat{I}_\beta + \hat{I}_\beta \hat{I}_\alpha$ transform exactly in the same way as the components of the rank-two tensor $\partial^2 \mathcal{V}/\partial r_\alpha \partial r_\beta$. It is noteworthy that for a spin 1/2, the symmetrized products $\hat{I}_\alpha \hat{I}_\beta + \hat{I}_\beta \hat{I}_\alpha$ either vanish or reduce to the unit 2×2 matrix, and the interaction in Eq. (2.24) vanishes. The quadrupole interaction defines the eigenstates of the nuclear spin system in an absence or in weak magnetic fields. It is especially important in strained systems, such as quantum dots or other heterostructures with lattice constant mismatch, where the elastic deformations induce the field gradients (Sundfors, 1969*a*; Bulutay, 2012). In solid solutions such as $Ga_xAl_{1-x}As$ some of the Ga nuclei are substituted with the Al ones, as a result, the arsenic nuclei experience the electric field gradients due to the asymmetric environment (Meier and Zakharchenya, 1984). Figure 2.3 illustrates the main interactions in the spin systems in semiconductors and impact of electromagnetic radiation with different frequencies on the electron and nuclear spins.

2.4 Optical orientation of electron spins and the Hanle effect

Nonmagnetic injection, detection, and manipulation of electron and nuclear spins are among the most essential tasks of semiconductor spintronics. The possibilities to control the electron spins by static or alternating electric field, including the electric field of the light wave, are enabled by the spin-orbit interaction. The *optical orientation* of the electron spins is the one of the most prominent manifestations of the spin-orbit interaction in semiconductors.

We consider the semiconductor or semiconductor nanosystem excited by the plane electromagnetic wave propagating along the positive direction of the z-axis. The electric field in the wave can be presented in the complex form as (Landau and Lifshitz, 1975; Berestetskii *et al.*, 1999)

$$\boldsymbol{E}(\boldsymbol{r},t) = \boldsymbol{e}E_0 \mathrm{e}^{\mathrm{i}\boldsymbol{qr}-\mathrm{i}\omega t} + \mathrm{c.c.} \tag{2.25}$$

Here, ω is the frequency of radiation, $\boldsymbol{q}$ is its wavevector, $\boldsymbol{q} = (0,0,q_z)$ with $q_z > 0$ in the geometry in question, $\boldsymbol{e}$ is the (complex) unit polarization vector of the wave, E_0 is the field amplitude and c.c. denotes the complex conjugate. We recall that the polarization state of light can be described in terms of three *Stokes parameters*: P_c is the circular polarization degree, and P_l and P_l' are the linear polarization degrees in two frames of axis rotated by 45° with respect to each other. For a monochromatic wave propagating in the positive direction of the z-axis, the Stokes parameters can be expressed via the components of the complex polarization vector, $\boldsymbol{e}$, as

$$P_c = \mathrm{i}[\boldsymbol{e} \times \boldsymbol{e}^*]_z = \mathrm{i}(e_x e_y^* - e_x^* e_y), \tag{2.26a}$$

$$P_l = |e_x|^2 - |e_y|^2, \quad P_l' = e_x e_y^* + e_x^* e_y. \tag{2.26b}$$

For fully polarized light $P_c + P_l + P_l' = 1$.

Figure 2.4 illustrates the selection rules for the interband optical transitions at the Γ-point of the Brillouin zone in a GaAs-type semiconductor. The basic states of the conduction, Γ_6, and valence, Γ_8, Γ_7, bands are labelled by the z-component of the spin, s_z, or total angular momentum, J_z, respectively. The component of the total angular momentum of the system "photon+electron" is conserved in the course of the optical transition. The σ^+ photon propagating along the z-axis ($P_c = +1$) possesses the z-component of the spin angular momentum being +1

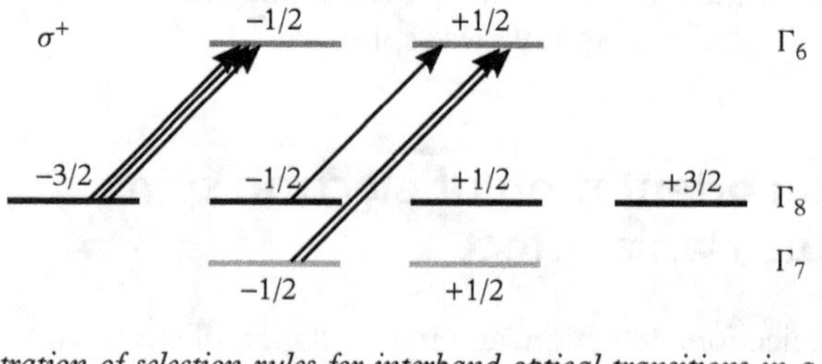

Figure 2.4 *Illustration of selection rules for interband optical transitions in a bulk GaAs-type semiconductor at the Γ-point. The number of arrows shows relative intensity of the transitions induced by the σ⁺ light propagating along the z ∥ [001] axis. The corresponding band structure is illustrated in Figure 2.2. The electron representation for labelling the valence band states is used.*

and the σ^- photon ($P_c = -1$) possesses the spin z-component of -1 (Berestetskii *et al.*, 1999). Therefore, for σ^+ polarized light the transitions from $|\Gamma_8, -3/2\rangle$ to $|\Gamma_6, -1/2\rangle$ as well as from $|\Gamma_8, -1/2\rangle$, $|\Gamma_7, -1/2\rangle$ to $|\Gamma_6, +1/2\rangle$ are allowed, see Figure 2.4. By contrast, for the σ^- polarized light the transitions from $|\Gamma_8, +3/2\rangle$ to $|\Gamma_6, +1/2\rangle$ as well as from $|\Gamma_8, +1/2\rangle$, $|\Gamma_7, +1/2\rangle$ to $|\Gamma_6, -1/2\rangle$ take place. In order to evaluate the rates of electro-dipole transitions, we present the perturbation induced by the plane wave in the form (Berestetskii *et al.*, 1999; Ivchenko, 2005)

$$-\frac{e}{m_0 c}\hat{\boldsymbol{p}} \cdot \boldsymbol{A},$$

where e is the electron charge, $\hat{\boldsymbol{p}}$ is the electron momentum operator, and $\boldsymbol{A}$ is the vector potential of the incident radiation. For optical transitions involving the valence and conduction band states at the Γ-point of the Brillouin zone, one can make use of the expressions (2.12) and (2.13) for the valence band Bloch functions and take into account the following relation between the momentum matrix elements calculated for the orbital states:

$$\langle \mathcal{S}|\hat{p}_x|\mathcal{X}\rangle = \langle \mathcal{S}|\hat{p}_y|\mathcal{Y}\rangle = \langle \mathcal{S}|\hat{p}_z|\mathcal{Z}\rangle = p_{cv}. \tag{2.27}$$

Here, p_{cv}, the *interband momentum matrix element*, is the only independent parameter describing the matrix elements of optical transitions in cubic semiconductors. This relation (2.27) immediately follows from the fact that the orbital Bloch amplitude $\mathcal{S}$ is invariant, while the three valence band Bloch amplitudes $\mathcal{X}$, $\mathcal{Y}$, $\mathcal{Z}$ transform in the same way as the Cartesian coordinates. As a result, one has for the ratios of the transition rates W^{σ}_{fi} from the state i to the state f in the circular polarization σ:

$$\frac{W^{\sigma^+}_{\Gamma_6,-1/2;\Gamma_8,-3/2}}{W^{\sigma^+}_{\Gamma_6,1/2;\Gamma_8,-1/2}} = 3, \qquad \frac{W^{\sigma^+}_{\Gamma_6,1/2;\Gamma_7,1/2}}{W^{\sigma^+}_{\Gamma_6,1/2;\Gamma_8,-1/2}} = 2. \tag{2.28}$$

For the opposite circular polarization, σ^-, the spin components should be inverted in Eq. (2.28). It follows from Eq. (2.28) that the transitions from the topmost (Γ_8) valence band to the conduction band induced by σ^+ circularly polarized light result in the predominant occupation of the electron state with $s_z = -1/2$, see Figure 2.4. In contrast, the excitation by the σ^- polarized light results in the predominant generation of the $s_z = +1/2$ electrons. In both cases the polarization degree of photoexcited electrons is 50%. It is noteworthy that the optical orientation of electron spins is possible due to the spin-orbit interaction: One can readily check that if the spin-orbit splitting of the valence band is zero then the generation rates of the spin-up and spin-down electrons are equal due to the additional contribution of the Γ_7 band (Dyakonov and Perel, 1971).

In quantum well and quantum dot structures the symmetry is lower, and due to the size-quantization and the elastic strain effects the four-fold degeneracy of the valence band is, as a rule, lifted. In many cases, the heavy-hole states with the total angular momentum projection onto the structure growth axis z, $J_z = \pm 3/2$, form the valence band top states (Ivchenko and Pikus, 1997; Ivchenko, 2005). In this case, the selection rules are strict: Under σ^+ polarized light intraband absorption only spin-down electrons are generated, while under σ^- polarized excitation only spin-up electrons are generated resulting in 100% spin polarization of photoelectrons.

Before we turn to the quantitative analysis of optical orientation and related phenomena, it is instructive to introduce the theoretical tools commonly used to describe spin dynamics. In bulk semiconductors and nanostructures like quantum wells and quantum dot ensembles, we are dealing with a macroscopic number of electrons; for example, typical electron densities in bulk GaAs vary from 10^{13} to 10^{19} cm^{-3}. Even in the case of a single quantum dot with only one resident electron, the experiments involve averaging over many repetitions of the same process or over a sufficiently long time interval. Therefore, both quantum mechanical and statistical averaging should be performed. This can be naturally done within the *spin density matrix* formalism. Let the index n enumerate all orbital states of the electron (e.g., n includes the band index and the wavevector $\boldsymbol{k}$ for free electrons in bulk semiconductor). For a given n, the electron spin density matrix is the 2×2 matrix conveniently presented in the form

$$\rho_n = f_n \hat{I} + \boldsymbol{s}_n \cdot \hat{\boldsymbol{\sigma}}, \tag{2.29}$$

where f_n is the occupancy of the state n averaged over the spin and $\boldsymbol{s}_n$ is the average spin in a given state. The scalar and pseudovector functions, f_n and $\boldsymbol{s}_n$, are the particle and spin distribution functions, respectively. In accordance with the definition (2.29), the spin and particle distribution functions can be expressed as

$$f_n = \frac{1}{2}\mathrm{Sp}\{\rho_n\}, \quad \boldsymbol{s}_n = \mathrm{Sp}\{\rho_n \hat{\boldsymbol{s}}\} = \frac{1}{2}\mathrm{Sp}\{\rho_n \boldsymbol{\sigma}\}. \tag{2.30}$$

The density matrix satisfies the kinetic equation

$$\frac{\partial \rho_n}{\partial t} + \frac{\mathrm{i}}{\hbar}[\mathcal{H}, \rho_n] + \mathcal{Q}_n\{\rho\} = 0, \tag{2.31}$$

where $\mathcal{H}$ is the system Hamiltonian and the collision integral (or relaxation operator) $\mathcal{Q}_n\{\rho\}$ is the operator that describes various incoherent processes, such as scattering, generation, and recombination. The kinetic Eq. (2.31) can be presented as a set of coupled equations for the particle and spin distribution functions. Particularly, making use of Eq. (2.30) we obtain

$$\frac{\partial \boldsymbol{s}_n}{\partial t} + \boldsymbol{s}_n \times \boldsymbol{\Omega}_n + \boldsymbol{Q}_n\{\boldsymbol{s}, f\} = 0. \tag{2.32}$$

Here, $\boldsymbol{\Omega}_n$ is the effective spin precession frequency in the state n, which includes contributions from the external magnetic field, $\boldsymbol{\Omega}_{\boldsymbol{B}}$, Eq. (2.10), the hyperfine interaction with nuclear spins, $\boldsymbol{\Omega}_N$, Eq. (2.22), and the contributions caused by the spin-orbit interaction for free electrons in noncentrosymmetric systems, $\boldsymbol{\Omega}_{\boldsymbol{k}}$, Eq. (2.15); $\boldsymbol{Q}_n$ is the collision integral acting on the spin distribution function. Equation (2.32) allows for a transparent quasiclassical interpretation: The electron spin, $\boldsymbol{s}_n$, in a given state n precesses around the magnetic field (either real or effective) with the frequency $\boldsymbol{\Omega}_n$, while the collision integral $\boldsymbol{Q}_n$ describes the spin generation; for example, in the process of optical orientation, the spin relaxation, that is, the decay of nonequilibrium spin polarization, and the spin redistribution between orbital states.

Under excitation by circularly polarized light the total spin $\boldsymbol{S} = \sum_n \boldsymbol{s}_n$ of the electron ensemble is controlled by the interplay of the generation, recombination, and relaxation processes. Generally, it can be found by solving the kinetic Eq. (2.32) with the appropriate form of $\boldsymbol{Q}_n$; see (Dyakonov and Perel, 1971; Meier and Zakharchenya, 1984) for details. To illustrate the basic physics, we consider the phenomenological kinetic equation for the total spin $\boldsymbol{S}$, which takes the simple form:

$$\frac{d\boldsymbol{S}}{dt} + \boldsymbol{S} \times \boldsymbol{\Omega} + \frac{\boldsymbol{S}}{\tau_s} + \frac{\boldsymbol{S}}{\tau_r} = \boldsymbol{G}. \tag{2.33}$$

Here, $\boldsymbol{\Omega} = \boldsymbol{\Omega}_{\boldsymbol{B}} + \boldsymbol{\Omega}_N$, $\boldsymbol{G}$ is the spin generation rate, and τ_s and τ_r are the phenomenological spin relaxation and recombination times, respectively. In Eq. (2.33) the simplest possible model of recombination is assumed; physically, this regime can be realized in p-doped structures where electrons are the minority charge carriers. The pseudovector $\boldsymbol{G}$ is related to the intensity and polarization of the incident electromagnetic field. For single photon absorption, which is usually the case, the spin generation rate is proportional to the light intensity $I = (cn_\omega/2\pi)E_0^2$ with n_ω being the refraction index at the radiation frequency. The polarization dependence of the spin generation rate $\boldsymbol{G}$ can be readily understood by means of the symmetry arguments. Correspondingly, we present $\boldsymbol{G}$ in the form:

$$G_\alpha = I \sum_{\beta\gamma} R_{\alpha\beta\gamma} e_\beta e_\gamma^* = I\left(R_c^{(\alpha)} P_c + R_l^{(\alpha)} P_l + R_l^{(\alpha)\prime} P_l'\right). \tag{2.34}$$

Here, the parameters $R_{\alpha\beta\gamma}$ are determined by the selection rules. The second equality in Eq. (2.34) is written for the specific geometry of light propagating along the z-axis. It can be readily obtained by recasting the bilinear combinations $e_\beta e_\gamma^*$ in the symmetric and antisymmetric forms with respect to the permutation of α and β and making use of Eq. (2.26). Corresponding combinations of $R_{\alpha\beta\gamma}$ are thus denoted as $R_c^{(\alpha)}$, $R_l^{(\alpha)}$, and $R_l^{(\alpha)\prime}$. In bulk semiconductors described by the O_h and T_d point symmetry groups the parameters R_l and R_l' are zero and, in accordance with qualitative arguments presented previously and Figure 2.4, the orientation of spins is possible by the circularly polarized light only. Symmetry-wise, this relation is obvious because both $\boldsymbol{G}$ and the combination $\mathrm{i}[\boldsymbol{e}\times\boldsymbol{e}^*]$ are pseudovectors. In low-dimensional structures, for example, in zinc-blende lattice quantum wells grown along [110] or [113] directions, because of the low symmetry the electron spins can be polarized by the linearly polarized or unpolarized light in low dimensional structures (Tarasenko, 2005; Crankshaw *et al.*, 2009).

Let us discuss the spin polarization of charge carriers in semiconductors and semiconductor nanosystems in more quantitative detail. In accordance with the selection rules depicted in Figure 2.4 and summarized in Eq. (2.28) for the optical transitions at the absorption edge in bulk GaAs-like semiconductors, the photo-electron created by the σ^+ (σ^-) polarized photon has the spin-z component s_z of $-1/2$ ($+1/2$) with a probability of 3/4 and a $s_z = +1/2$ ($-1/2$) with a probability of 1/4. Hence, the polarization degree of photoelectrons,

$$P_s^0 = 2S_z/N_e,$$

where N_e is the number of photoelectrons generated in the course of light absorption and equals

$$P_s^0 = -\frac{P_c}{2}. \tag{2.35}$$

In low-dimensional structures where the valence band top is formed of the $|\Gamma_8, \pm 3/2\rangle$ states instead of Eq. (2.35), one has

$$P_s^0 = -P_c,$$

that is, the photoelectrons are fully polarized at the circularly polarized excitation. The steady-state electron spin polarization P_s under continuous wave (*cw*) excitation can be found from the kinetic equation (2.32). Particularly, in absence of a magnetic field, $\boldsymbol{\Omega} = 0$, and under the assumption that the electrons recombine with unpolarized holes, one has

$$P_s = \frac{P_s^0}{1 + \tau_r/\tau_s}. \tag{2.36}$$

Hence, if the spin relaxation is rapid enough, $\tau_s \ll \tau_r$, the electron spins disappear before the electron recombines and the resulting spin polarization approaches zero $P_s \approx P_s^0 \tau_s/\tau_r \ll P_s^0$. By contrast, if the spin relaxation is slow compared with recombination, $\tau_s \gg \tau_r$, the electrons retain their initial polarization degree, $P_s \approx P_s^0$. If the optical orientation is performed by a short pulse then, in accordance with Eq. (2.33), the electron spin decays exponentially with time as

$$S_z(t) = S_z^0 \mathrm{e}^{-t/T_s}, \tag{2.37}$$

where S_z^0 is the initial spin at $t = 0$ and we introduced the spin lifetime T_s as

$$\frac{1}{T_s} = \frac{1}{\tau_s} + \frac{1}{\tau_r}. \tag{2.38}$$

The number of electrons decays with the time constant equal to τ_r.

It is noteworthy that the same selection rules apply both for the electron generation and radiative recombination processes. As a result, for spin polarized electrons the luminescence of a semiconductor is also polarized. Its polarization degree is controlled by the selection rules as well as the electron and hole spin polarization. Typically, the hole spin relaxation is faster than for electrons (Meier and Zakharchenya, 1984). In a bulk zinc-blende lattice semiconductor the circular polarization of the luminescence, $P_c^{\text{lum}} = -P_s/2$, see Eq. (2.35), so the polarization degree of emission is limited by 1/4. In low-dimensional systems, where the selection rules are strict, the circular polarization degree of emission can reach unity.

In presence of a relatively weak magnetic field applied perpendicular to the light propagation axis (*Voigt geometry*), $\boldsymbol{\Omega_B} \perp z$, so that neither selection rules nor electron spin relaxation are affected by the field, the electron spin initially aligned along the z-axis starts to precess around $\boldsymbol{\Omega}_B$. In the time domain instead of Eq. (2.37) we have

$$S_z(t) = S_z^0 e^{-t/T_s} \cos(\Omega_B t), \tag{2.39}$$

while under steady-state excitation the electron spin polarization gets suppressed in accordance with

$$P_s(B) = \frac{P_s(0)}{1 + \Omega_B^2 T_s^2}, \tag{2.40}$$

where $P_s(0)$ is given by Eq. (2.36). The electron depolarization caused by the transverse magnetic field is known as the *Hanle effect*.

As a result, by monitoring the polarized luminescence in time resolved or *cw* experiments, one can access the main parameters controlling electron spin dynamics. For analysis of many important particular cases of optical orientation and optical spin pumping, we refer the reader to the monographs (Meier and Zakharchenya, 1984; Dyakonov, 2017). Spin orientation by short pulses of light in nanosystems is discussed in Chapter 6.

Besides electron spin dynamics studies by detection of luminescence polarization, one can use the two-pulse "pump-probe" technique, where the first circularly polarized pulse results in the electron spin orientation and the second linearly polarized pulse detects the electron spin coherence via the *spin-Faraday* (transmission geometry) or *spin-Kerr* (reflection geometry) effect. In the presence of magnetization, the polarization plane of the probe light propagating through the medium rotates and its rotation angle, ϑ, is directly proportional to the spin in the system

$$\vartheta \propto S_z. \tag{2.41}$$

In the pump-probe technique, one can study temporal dynamics of the electron and hole spins as well as to manipulate the spin dynamics by additional, control, light pulses (Dyakonov, 2017). Further details on spin detection and manipulation by optical pulses are presented in Chapter 6. It turns out that the studies of spin fluctuations by means of monitoring the noise of the spin-Faraday or spin-Kerr angles gives access to further details of electron and nuclear spin dynamics hardly detectable otherwise, as discussed in Chapter 10.

The electrons lose their spin orientation mainly by transferring it to the crystalline lattice. Both the spin-orbit and hyperfine interactions provide such a transfer. In bulk semiconductors, quantum well, and quantum wire structures where the electrons are free to move along at least one axis, spin-orbit coupling provides the dominant channel of spin relaxation (Dyakonov, 2017). Moreover, in noncentrosymmetric systems where the electron effective Hamiltonian contains the spin-dependent terms proportional to the odd powers of $\boldsymbol{k}$, Eq. (2.15), the effective magnetic field associated with conduction band spin splitting provides the major contribution to spin relaxation rate τ_s^{-1}. The corresponding spin relaxation mechanism is termed the *Dyakonov-Perel'* or *precession spin relaxation mechanism* (Dyakonov and Perel, 1972). Indeed, the electron spin precesses around the effective field $\boldsymbol{\Omega_k}$ and the scattering processes that change the magnitude and direction of the electron momentum randomize the spin precession. Introducing the effective scattering time τ and assuming that for typical electron wavevectors $\Omega_{\boldsymbol{k}}\tau \ll 1$, that is, the collision dominated regime is realized, one has an estimate for the spin relaxation rate

$$\frac{1}{\tau_s} \sim \left\langle \Omega_{\boldsymbol{k}}^2 \tau \right\rangle. \tag{2.42}$$

Here, the averaging denoted by the angular brackets $\langle \ldots \rangle$ is carried out over the electron distribution. The estimate (2.42) immediately follows from the fact that the root mean square of the electron spin rotation angle between the subsequent collisions is $\delta\Phi \sim \Omega_{\boldsymbol{k}}\tau$, and the number of collisions needed for spin to rotate by a significant angle, $\tau_s/\tau \sim \delta\Phi^{-2}$. In the Dyakonov–Perel' mechanism the spin relaxation takes place between collisions and is slowed down by the electron scattering.

It is noteworthy, that the expression in the form of Eq. (2.42) describes spin relaxation in many situations where the relaxation process can be considered a spin precession in a fluctuating real or effective magnetic field. In this situation, $\Omega_{\boldsymbol{k}}^2$ should be replaced by the mean square of the precession frequency in the fluctuating field and τ by the *correlation time* of the random field; that is, the time during which the field changes considerably.

In some cases, for example, in centrosymmetric semiconductors like Si or Ge, in some narrow-gap semiconductors, or for the valence band holes, the spin relaxation may be caused by the *Elliott–Yafet* mechanism (Elliott, 1954; Yafet, 1963), where the spin is lost due to spin-orbit electron scattering by an impurity or phonon. Here, in contrast to the Dyakonov–Perel' process, the stronger the scattering, the faster the spin relaxation and

$$\frac{1}{\tau_s} = \frac{C_{EY}}{\tau}. \tag{2.43}$$

The dimensionless proportionality constant, C_{EY}, in Eq. (2.43) is determined by the particular form of the spin-orbit interaction in the system and scattering mechanism and, as a rule, much smaller than unity. In optical orientation experiments where both electrons and holes are generated, the exchange interaction between the electrons and holes may provide an efficient channel of spin relaxation known as the *Bir–Aronov–Pikus* mechanism (Bir *et al.*, 1975). This mechanism is also important for excitons (Maialle *et al.*, 1993). Various aspects of free electron spin relaxation are reviewed in (Meier and Zakharchenya, 1984; Averkiev *et al.*, 2002; Zutic *et al.*, 2004; Wu *et al.*, 2010; Dyakonov, 2017) and are beyond the scope of the present book.

2.5 Spin effects in interacting electron-nuclear systems

Owing to small magnetic moments and a relatively weak interaction with light, the magnetic field-induced and optical orientation of nuclear spin systems in semiconductors is very difficult. However, due to hyperfine interaction, the electron spin polarization can be transferred to the host lattice nuclei. This process is known as the *dynamical nuclear polarization*, its details are presented in Chapter 5. Here, we just outline the essentials of the effect following (Meier and Zakharchenya 1984).

We consider a semiconductor in the presence of the magnetic field $\boldsymbol{B}$ applied in the *Faraday geometry*, that is, along the light propagation axis: $\boldsymbol{B} \parallel \boldsymbol{q} \parallel z$. We characterize the spin state of the nucleus by its spin-z component $I_z = -I, -I+1, \dots, I$ and spin state of the electron by its spin-z component $s_z = \pm 1/2$. The hyperfine Hamiltonian (2.22) allows for the simultaneous spin flips of the electron and nucleus, where I_z and s_z change by unity such that $I_z + s_z$ remains constant; for example, $I_z \to I_z' = I_z + 1$ and $s_z \to s_z' = s_z - 1$. This is the so-called *flip-flop* process. Disregarding the quadrupole effects and the dipole–dipole interaction between the nuclei we have in the steady-state the following detailed balance of the nuclear spin flip processes:

$$W_{I_z-1,I_z} N_{I_z} f_{-1/2} = W_{I_z,I_z-1} N_{I_z-1} f_{+1/2}, \tag{2.44}$$

where N_{I_z} is the occupation of the nuclear state with the spin component I_z, and f_{s_z} is the occupation of the spin component, s_z (both quantities are averaged over the ensemble of electrons and nuclei), $W_{I'_z,I_z}$ are the rates of the nuclear spin flip, $I_z \to I'_z$, processes. Equation (2.44) demonstrates that the nuclear spin component is changed only due to the interaction with the electron, the total spin in the system of nuclei and electrons is conserved by the hyperfine interaction. Under the assumption that the electron spin system interacts with the lattice far more efficiently than with nuclei (i.e., the electron spin relaxation is *not* controlled by the hyperfine interaction), for the spin flip rates one has

$$\frac{W_{I_z-1,I_z}}{W_{I_z,I_z-1}} = \exp\left(-\frac{g\mu_B B}{k_B T}\right). \tag{2.45}$$

Here, k_B is the Boltzmann constant and T is the lattice temperature. Equation (2.45) can be justified by taking into account that electron spin relaxation is by far more efficient than that of nuclei. In derivation of Eq. (2.45) we have neglected the Zeeman splitting of the nuclei compared with the electron one. Equation (2.45) demonstrates that the electron spin flip requires the transfer of the energy $g\mu_B B$. Taking into account that $f_{+1/2}/f_{-1/2} = (1+2S_z)/(1-2S_z)$, where $S_z = (f_{+1/2} - f_{-1/2})/2$ is the average spin-z component of the electron ensemble, we obtain (Meier and Zakharchenya, 1984)

$$\frac{N_{I_z}}{N_{I_z-1}} = \frac{1+2S_z}{1-2S_z}\frac{1-2S_T}{1+2S_T}. \tag{2.46}$$

Here, S_T is the thermal spin orientation of electrons,

$$S_T = -\frac{1}{2}\tanh\left(\frac{g\mu_B B}{2k_B T}\right). \tag{2.47}$$

Hence, for $S_z \neq S_T$, the nuclei become polarized as a result of the hyperfine interaction with electron spins. The average nuclear spin-z component reads

$$\langle I_z\rangle = \sum_{I_z} I_z N_{I_z} = \frac{4}{3}I(I+1)S_z \tag{2.48}$$

In derivation of the last equality we assumed that $|S_T| \ll |S_z| \ll 1$. The dynamic nuclear polarization degree, as a rule, exceeds by far the thermal (equilibrium) orientation of nuclear spins in the external magnetic field ($\sim g_n\mu_N B/k_B T$).

It follows from Eq. (2.46) that the ratio of the occupancies of nuclear states with the neighboring spin components I_z and $I_z - 1$ does not depend on I_z, just as it should be for the Boltzmann distribution. Formally, one can introduce the *nuclear spin temperature* Θ in accordance with

$$\frac{N_{I_z}}{N_{I_z-1}} = \exp\left(\frac{g_n\mu_N B}{k_B\Theta}\right). \tag{2.49}$$

In terms of the nuclear spin temperature, Eq. (2.48) can be recast as

$$\langle I_z \rangle = \frac{I(I+1)}{3}\frac{g_n\mu_N B}{k_B\Theta}, \tag{2.50}$$

where, again, we made use of the fact that $|S_z| \ll 1$ and, accordingly, $|g_n\mu_N B/(k_B\Theta)| \ll 1$. Under optical orientation conditions the nuclear spin temperature Θ can be positive or negative depending on the nuclear g-factor sign and relative direction of the electron spin polarization and external magnetic field. The absolute value of the nuclear spin temperature can be considerably smaller than the lattice temperature, $|\Theta| \ll T$. It is noteworthy that the concept of the nuclear spin temperature is not only a convenient mathematical tool to describe the polarized nuclei but also has a deep physical meaning (Abragam, 2002). Qualitatively, this is because the nuclear spin system is quite well isolated from the electrons and phonons in semiconductors. This concept will be further developed in Chapter 5.

The dynamic nuclear polarization results in the build-up of the effective Overhauser field acting on the electron spin. For homogeneous in space nuclear spin polarization, one can average Eq. (2.22) over the nuclei positions $\boldsymbol{R}$ and replace $\hat{\boldsymbol{I}}$ by the average quantity $\langle \boldsymbol{I} \rangle$. The nuclear field reads

$$\boldsymbol{B}_N = \frac{\xi A}{g\mu_B}\langle \boldsymbol{I} \rangle, \tag{2.51}$$

where ξ is the number of nuclei of a given isotope in the primitive cell. For completely polarized nuclei, the effective splitting of the electron spin sublevels is $\xi AI \sim 100$ μeV (Chapter 4). In other words, the Overhauser field created by the fully polarized nuclei can reach several tesla.

The build-up of nuclear polarization is provided by the hyperfine interaction of the electron and nuclear spins. The hyperfine interaction is also responsible for the relaxation of the nuclear spin polarization. The nuclear relaxation time, T_1, can be easily estimated taking into account random fluctuations of the Knight field experienced by the nucleus from the electron (cf. Eq. [2.42]):

$$\frac{1}{T_1} \sim \left(\frac{A}{\hbar N_n}\right)^2 F\tau_c. \tag{2.52}$$

Here, $N_n \sim V_c/v_0$ is the number of nuclei effectively interacting with the electron within the coherence volume V_c, F is the probability of finding the electron there, and τ_c is the electron spin correlation time at a given nucleus (Meier and Zakharchenya, 1984). For free electrons, $V_c \sim \lambda^3$ is the volume corresponding to the cube of the de Broglie wavelength, λ, $F = n\lambda^3$, and n is the electron density, τ_c is the electron flight time through the volume V_c. The time τ_c is usually called *electron correlation time*. For the localized carriers V_c plays the role of electron localization volume and τ_c is determined by the dynamics of the localized electron. One can arrive at Eq. (2.52) taking into account that the mean square of nuclear spin precession frequency in the Knight field is $\sim (A/\hbar N_n)^2 F$ and the correlation time of the Knight field is τ_c. Usually, the nuclear spin relaxation time T_1 is macroscopic ranging from fractions of seconds to hours, depending on the particular semiconductor, electron density, localization, presence, or absence of optical excitation, and so on. Note that the rate of the dynamic nuclear polarization is determined by the same expression, Eq. (2.52), but the value of the electron correlation time τ_c can be affected by the photoexcitation processes.

Previously, we outlined the basic effects behind the manipulation of the nuclear spin system by light: First, absorption of the polarized radiation results in the polarization of the electron spin system and second, the polarized electrons transfer a part of their spin polarization to the nuclear system. Obviously, the direct interaction of nuclear spins with optical radiation is not possible because there are no resonances in the nuclear system that correspond to the photon energies in the order of eV. In contrast, nuclei can effectively respond to the photons of much lower energies, with the frequencies typically lying in the radio frequency range. Indeed, as illustrated in Figure 2.5(a), in the presence of magnetic field $\boldsymbol{B}$ the nuclear spin states are split into equidistant sublevels that correspond to definite projections of the nuclear spin onto the field axis. In agreement with Eq. (2.21) the splitting between the neighboring states equals $\Delta E = g_n \mu_N B$. To be specific, let us assume that the field $\boldsymbol{B} \parallel z$, hence, the eigenstates of the system are characterized by the nuclear spin z-component, I_z. Typically, in the field of 1 T the splitting $\Delta E \sim 0.1$ μeV, see Table 2.1. The transitions between the spin sublevels can be induced by the alternating magnetic field of the electromagnetic wave incident on the sample. In accordance with the Hamiltonian (2.21) the perturbation in the form

$$\hat{I}_x \tilde{B}_x + \hat{I}_y \tilde{B}_y,$$

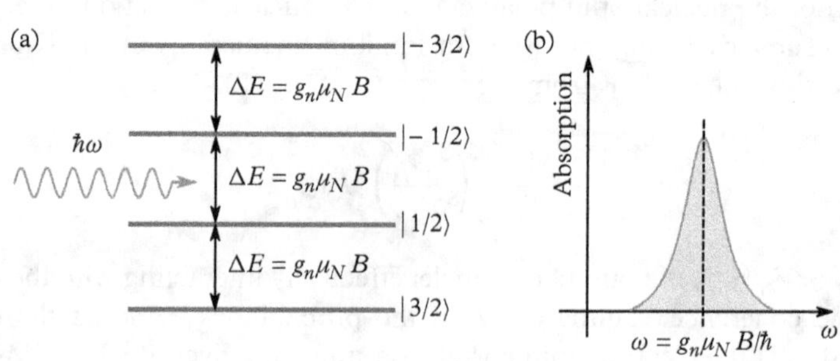

Figure 2.5 *Schematic illustration of the nuclear magnetic resonance. (a) States of the $I = 3/2$ nucleus in the magnetic field $\boldsymbol{B} \parallel z$. (b) Sketch of the absorption spectrum of the electromagnetic wave by the nuclei ($g_n > 0$).*

where $\tilde{B}_\alpha$ ($\alpha = x,y$) are the in-plane components of the oscillating magnetic field, mixes the sublevels with the spin components I_z and $I_z \pm 1$. The absorption of the electromagnetic wave is most efficient provided that the incident quantum energy equals to the distance between the sublevels; that is, $\hbar\omega = \Delta E$. Such a resonant absorption is known as the *nuclear magnetic resonance* (NMR). This phenomenon is the basic tool for studying nuclear spin physics. Due to the fact that each isotope has its own gyromagnetic ratio, see Table 2.1, the NMR spectrum contains "fingerprints" of the nuclei that allow one to unambiguously identify the involved nuclei. Furthermore, the efficiency of the transitions between the sublevels is proportional to the difference of their occupancies, $N_{I_z} - N_{I_z-1}$. Within the concept of the nuclear spin temperature, Eq. (2.49), the smaller the $|\Theta|$, the stronger the NMR signal is. This provides a route toward evaluation of the nuclear spin temperature. Moreover, the NMR allows one to manipulate nuclear spin states efficiently. We do not go into detail here (see Chapter 3 for further information) and consider one of simplest examples: The resonant transitions between the spin sublevels tend to equalize the occupancies of sublevels, erasing the nuclear spin polarization and, consequently, increasing the nuclear spin temperature.

NMR also provides very important information about nuclear spin dynamics. Here, the basic quantity is the linewidth of the NMR spectrum. In contrast with the nuclear spin flip processes caused mainly by the hyperfine interaction, Eq. (2.52), the broadening of the NMR line does not generally require an energy transfer: Any process that causes the loss of coherence between the sublevels contributes to the linewidth of the resonant absorption (Abragam, 2002). Obvious candidates are the dipole–dipole interaction between the nuclei, Eq. (2.23), and the quadrupole splittings, Eq. (2.24). For example, in bulk GaAs the fluctuating local field due to the dipole–dipole interaction produces the decoherence of nuclear spins with the rate $T_2 \sim 10^{-4}$ s. Hence, $\hbar/T_2$ gives a lower bound to the width of the NMR line. In nanostructures quadrupole interactions become very important and provide major contributions to the broadening of resonance (Kalevich *et al.*, 2017), see Chapter 3 for details.

Typically, in semiconductors, NMR is detected optically via the variation of the Overhauser field acting on the electron spin, which affects, for example, the Zeeman splittings in optical spectra, the polarization of emitted light, and so on. The details on the manifestations of nuclear spin resonance in the optical response are given in Chapter 3.

2.6 Specifics of spin dynamics of localized charge carriers

There are various ways to quench the orbital motion of electrons and holes in semiconductor and provide their efficient localization. The classical example of a system with localized electrons is a bulk semiconductor doped by donor impurities; for example, GaAs doped with Si. Provided that the donor density N_d is sufficiently low, such that

$$N_d a_B^3 \ll 1, \tag{2.53}$$

where a_B is the Bohr radius for the donor-bound electron, the donors are effectively isolated. Under this condition each donor-bound electron can be, at least in the first approximation, considered independent. Hence, each donor provides the Coulomb potential. The envelope function of the donor-bound electron $\varphi(\boldsymbol{r})$, Eq. (2.4), which satisfies the Schrödinger equation with the effective Hamiltonian in the form of Eq. (2.5) and $V_{\text{eff}}(r) = -e^2/(\varkappa r)$, can be presented in the hydrogen-like form

$$\varphi(\boldsymbol{r}) = \frac{\exp(-r/a_B)}{\sqrt{\pi a_B^3}}, \quad a_B = \frac{\varkappa \hbar^2}{me^2}. \tag{2.54}$$

Here, we recall that $\varkappa$ is the static dielectric constant of the host material, m is the electron effective mass, and $\boldsymbol{r}$ is the electron radius vector reckoned from the donor position. The binding energy of the donor in such an approximation is

$$E_b = \frac{me^4}{2\varkappa^2\hbar^2}. \tag{2.55}$$

In GaAs, the binding energy of a donor in this approximation is approximately several meV and the Bohr radius $a_B \sim 100$ Å. The description of the donor-bound electron in the framework of the hydrogen-like function is valid for shallow donors where $E_b \ll E_g$, E_g is the band gap. One can also see that the bound state wavefunction extends over many unit cells of the crystal supporting the applicability of the envelope function method to shallow donors. However, the impurity may strongly affect the crystalline lattice in its immediate vicinity. Within the envelope function method, this effect can be included in a short-range part of the potential induced by the donor impurity. The strength of the short-range interaction can be determined by comparison with experimental data or atomistic calculations. In the specific case of Si impurity in GaAs this effect is almost negligible. The situation becomes more involved for donors in multivalley semiconductors like silicon or germanium where the electron dispersion in each valley of energy spectrum is anisotropic, there are shallow acceptor impurities in zinc-blende and diamond lattice semiconductors because one has to account for the complex valence band structure, and for deep impurities where the binding energy is not much smaller than the band gap. The advanced models that describe donor and acceptor bound electrons in these listed situations are presented in a number of references, including, for example, (Bir and Pikus, 1974; Shklovskii and Efros, 1984; Abakumov *et al.*, 1991), and are beyond the scope of the present description.

Another prominent example of a system with localized charge carriers is a semiconductor quantum dot, also referred to as a semiconductor *nanocrystal*. A great variety of quantum dot structures are currently studied both experimentally and theoretically (Bimberg *et al.*, 1999; Efros and Rosen, 2000; Masumoto and Takagahara, 2002; Ivchenko, 2005; Ling, 2005; Gywat *et al.*, 2010; Tartakovskii Tartakovskii, 2012). Within the effective mass approach the electron wavefunction in a quantum dot is found from the Schrödinger equation with the effective Hamiltonian (2.5) and with the appropriate potential and boundary conditions. The form of the wavefunctions and the energies of size-quantized states depends on the quantum dot composition, size, and shape. The energy gaps between the levels in quantum dots are, as a rule, larger than for donor-bound electrons, typically about several tens of meV; moreover, the barriers are sufficiently high ($\gtrsim 100$ meV), making quantum dots well isolated one from another, although tunneling effects in quantum dot structures are also actively studied. It is worth stressing that the symmetry of quantum dots is typically lower than that of host semiconductors due to shape, interface, and elastic strain effects. This affects the states of the charge carriers, optical selection rules, and the form of Zeeman and hyperfine interactions. The corresponding manifestations of symmetry reduction are addressed in later chapters of the book.

Localization of electrons and holes qualitatively changes the spin dynamics of charge carriers compared with the case of free electrons. This is because the role of the spin-orbit interaction is suppressed for localized charge carriers. Consequently, the dynamics of nuclei coupled with localized electrons is also quite different from those interacting with free charge carriers. Let us discuss the major specific features of spin dynamics of localized electrons and holes interacting with lattice nuclei.

We start with the analysis of the spin-orbit interaction role. As we discussed in Section 2.4 the spin-orbit interaction provides the main contribution to electron spin relaxation via Dyakonov–Perel' and Elliott–Yafet mechanisms. It is important that these mechanisms require electron motion. In the Elliott–Yafet mechanism scattering is needed to change the electron orbital state and, hence, flip its spin. In Dyakonov–Perel' relaxation the spin is lost during electron free propagation as a result of spin precession in the effective field $\boldsymbol{\Omega}_{\boldsymbol{k}}$, see Eq. (2.15). Obviously, such a precession is quenched for a localized electron. This can be illustrated in the framework of a toy model, where we assume that the localized state wavefunction is continuous and smooth and that the parameters describing odd-in-$\boldsymbol{k}$ terms in Eq. (2.15) are independent of coordinates. In such a case, quantum mechanical averages of odd powers of the wavevector components vanish, as one can readily check integrating several times by parts, the expression

$$\langle k_\alpha^{2n+1}\rangle \propto \int d\boldsymbol{r}\,\varphi(\boldsymbol{r})\frac{\partial^{2n+1}\varphi(\boldsymbol{r})}{\partial r_\alpha^{2n+1}} = -\int d\boldsymbol{r}\frac{\partial\varphi(\boldsymbol{r})}{\partial r_\alpha}\frac{\partial^{2n}\varphi(\boldsymbol{r})}{\partial r_\alpha^{2n}} = \dots$$

$$= -\int d\boldsymbol{r}\frac{\partial^{2n+1}\varphi(\boldsymbol{r})}{\partial r_\alpha^{2n+1}}\varphi(\boldsymbol{r}) = 0.$$

Here $n = 0, 1, \ldots$, and we made use of the fact that for localized state $\varphi(\boldsymbol{r})$ vanishes for $|\boldsymbol{r}| \to \infty$ and chose $\varphi(\boldsymbol{r})$ to be real (Landau and Lifshitz, 1977). Hence, for a localized electron, the spin-orbit interaction does not induce any spin splitting and spin precession.

In fact, the absence of the spin splitting of a localized electron state in the absence of an external magnetic field is a consequence of the time reversal symmetry, $t \to -t$. To prove this fact we note the following general transformation rule of the state with the angular momentum $\mathcal{J}$ and its z-component $\mathcal{J}_z$ at the time reversal operation $\hat{\mathcal{K}}$ (Edmonds, 1974):

$$\hat{\mathcal{K}}|\mathcal{J}, \mathcal{J}_z\rangle = (-1)^{\mathcal{J}+\mathcal{J}_z}|\mathcal{J}, -\mathcal{J}_z\rangle. \tag{2.56}$$

Particularly,

$$\hat{\mathcal{K}}|\uparrow\rangle = -|\downarrow\rangle, \quad \hat{\mathcal{K}}|\downarrow\rangle = |\uparrow\rangle, \tag{2.57}$$

therefore, the spin-up and spin-down states interchange at the time reversal. Moreover, it is impossible to form linear combinations $\alpha|\uparrow\rangle + \beta|\downarrow\rangle$ with coefficients α, β such that $|\alpha|^2 + |\beta|^2 = 1$, which are invariant at $t \to -t$. Hence, in absence of a magnetic field, localized electron states are at least two-fold degenerate. This effect is also known as the *Kramers degeneracy*. We note that for free electrons the time reversal operation implies, in addition to the spinor transformations described by Eq. (2.57), the reversal of the electron wavevector $\boldsymbol{k}$, ensuring the odd-in-$\boldsymbol{k}$ form of the spin-orbit terms (2.15).

We may conclude, therefore, that the role of the spin-orbit interaction is strongly suppressed for localized charge carriers. Hence, in semiconductors with nonzero nuclear spins of the host lattice, localized electron spin dynamics is mainly driven by the hyperfine interaction with lattice nuclear spins. Moreover, the hyperfine interaction is strongly enhanced for localized electrons: The localization increases the probability to find the electron at a given nucleus, $\propto |\varphi_c(\boldsymbol{R})|^2$, thus increasing the coupling strength in Eq. (2.22). Even if nuclear spin system is disordered, that is the average spin polarization of nuclei, $\langle \boldsymbol{I} \rangle$, is absent, random statistical fluctuations of nuclear spins within the electron localization volume give rise to the fluctuation of the Overhauser field acting on the electron spin:

$$\delta \boldsymbol{B}_N \sim \frac{A}{g\mu_B N_n} \delta \boldsymbol{I}. \tag{2.58}$$

Here, $\delta \boldsymbol{I}$ is the fluctuation of the total nuclear spin polarization in the electron localization volume, and N_n is the number of nuclei in this volume. The factor N_n in the denominator of Eq. (2.58) accounts for the spread of electron density over the nuclei. Assuming that these nuclear spins are independent, for the mean square fluctuation $\langle (\delta \boldsymbol{I})^2 \rangle$ we have

$$\langle(\delta \boldsymbol{I})^2\rangle = I(I+1)N_n. \tag{2.59}$$

Hence, unpolarized nuclei provide the fluctuation of the Overhauser field and corresponding Overhauser splitting, $\hbar\delta_e$, of the electron spin sublevels that can be estimated as

$$\sqrt{\langle(\delta \boldsymbol{B}_N)^2\rangle} \sim \frac{A\sqrt{I(I+1)}}{g\mu_B\sqrt{N_n}}, \quad \hbar\delta_e \sim \frac{A\sqrt{I(I+1)}}{\sqrt{N_n}}. \tag{2.60}$$

As expected, the fluctuating field and splitting are larger the smaller the number of nuclei N_n is that are interacting with the localized electron. Consequently, the stronger the localization, the larger the fluctuating fields are experienced by electron spin. In typical situations, the number of nuclei interacting with electron range from $10^4 \ldots 10^6$ depending on the material, type of localized state, localization radius. Hence, for GaAs-based structures in accordance with Eqs (2.51) and (2.60), the fluctuating Overhauser field ranges from several mT to several tens of mT and result in the spin precession frequencies δ_e in the range of $0.1 \ldots 1$ ns^{-1}.

These fluctuating nuclear fields provide the major source of localized electron spin decoherence. In order to illustrate this process we consider an ensemble of isolated quantum dots, each of these contains a single electron. The nuclear spin fluctuations in quantum dots, $\delta\boldsymbol{I}$, are random and are not correlated. For simplicity, we consider these fluctuations as frozen; that is, time independent. Let us assume that at $t = 0$ all electron spins are aligned along the z-axis. Due to spin precession as time goes by, the electron spins in quantum dots precess around the corresponding nuclear spin fluctuations. Since the magnitudes and directions of $\delta\boldsymbol{I}$ are random, the electron spin in the ensemble decays. The timescale of the decay $\sim \delta_e^{-1}$ is several nanoseconds. However, the spin decay is not complete, because the electron spin component along the nuclear field is conserved. As a result, $1/3$ of the initial electron spin polarization is conserved in the course of the spin precession in random nuclear fields provided that fluctuations $\delta\boldsymbol{I}$ are distributed isotropically. One can readily check that the approximation of frozen nuclear fields is indeed valid on a nanosecond timescale, because the nuclear spin dynamics is much slower. The electron spin dynamics at longer time scales is more complex since nuclear spin dynamics become important mainly due to action of electron spin on nuclear spins and also because of dipole–dipole interactions between the nuclei and quadrupole effects. Hence, on timescales exceeding tens of nanoseconds, electron and nuclear spin dynamics cannot be, as a rule, separated, which results in the ultraslow decay of electron spin polarization. The models of electron spin decoherence by nuclei in quantum dots were suggested in (Merkulov *et al.*, 2002; Khaetskii *et al.*, 2002). An additional aspect of electron spin dynamics in quantum dot ensembles and, importantly, in ensembles of donor-bound electrons, is the possibility of an electron changeing its localization site; for example, via photon-assisted hopping. In this case another parameter, the electron correlation time at a given site, τ_c, comes into play. Thus, different regimes of long and short correlation times as compared with the electron spin precession period in the field of nuclear fluctuation can be realized (Meier and Zakharchenya, 1984; Kalevich *et al.*, 2017). The details of the dynamics in the interacting electron and nuclear systems are presented in Chapters 7, 9, and 10.

This analysis demonstrates that for systems with localized charge carriers, the carrier spin dynamics are qualitatively different from that of free electrons. Due to the enhanced role of the hyperfine interaction, electron and nuclear spin dynamics are strongly intertwined. The coupling of electron and nuclear spins can be diminished in a magnetic field, in which case the electron–phonon interaction starts to serve as a source of electron spin relaxation. In this case, the electron spin flip is enabled by the spin-orbit interaction and the energy needed for transition between spin sublevels is given to or taken from the phonon. The electron spin relaxation rate in a magnetic field is thus proportional to a sufficiently high power of Zeeman splitting, $(g\mu_B B)^\nu$, $\nu > 5$, and millisecond-long electron spin relaxation can be achieved (Khaetskii and Nazarov, 2001; Woods *et al.*, 2002). The dynamics of electron spins in the regime where the hyperfine interaction becomes unimportant is presented and analyzed in Chapter 8.

It is also noteworthy that the dynamics of nuclear spins is also modified in structures with carrier localization. First, as was noted back in 1950s (Bloembergen, 1954), the enhancement of the hyperfine interaction due to electron localization increases the transfer rate of the nuclear spin to the electron, see Eq. (2.52). Second, the strain in quantum dot structures effectively introduces the quadrupole splittings of nuclear spin levels, see Section 2.3. The quadrupole splittings may compete with the dipole–dipole interaction between neighboring nuclei. As a result, nuclear spin decoherence and diffusion proceed differently in quantum dot structures compared with, for example, bulk semiconductors.

To conclude this section we list the features of the systems with localized charge carriers that are important for spin dynamics:

- Suppression of the spin-orbit interaction.
- Enhancement of the hyperfine interaction.
- Long spin decoherence and relaxation times of charge carriers.
- Intertwined dynamics of electron and nuclear spins.
- The possibility to address single charge carrier spins.

These effects are discussed in more detail in later chapters of the book.

3

Spin Resonance

3.1 Basics of spin resonance

The response of spin systems to static and alternating magnetic fields is at the heart of many techniques studying and controling spin dynamics. The static magnetic field results in the splitting of spin sublevels and the alternating field causes the transitions between split sublevels. Under resonant condition where the frequency of the oscillating field matches the splitting between the spin sublevels divided by the Planck constant, the alternating field can be efficiently absorbed. This model is visualized in Figure 2.5 and presented in Section 2.5. It is an illustration of the spin or paramagnetic resonance phenomenon and allows evaluation of the Zeeman splitting and estimation of the spin decoherence time. Since the discovery of nuclear magnetic resonance (NMR) by Isidor Rabi in 1938 (Rabi *et al.*, 1938) and electron spin resonance by Evgeny Zavoisky in 1944 (Zavoisky, 1945), magnetic resonance methods have became extremely popular in spin dynamics studies, see (Baranov *et al.*, 2017) for a recent review, and theoretical models of magnetic resonance effects have been elaborated on. We start with a concise description of the spin density matrix approach to magnetic resonance in a noninteracting system; further, we discuss NMR in the case where dipole–dipole interaction and quadrupole splittings in a nuclear system are present, but hyperfine coupling with electron spins can be treated as a perturbation.

The basic properties of the magnetic resonance theory can be understood with a spin-$1/2$ model. This model can be directly applied to the electron spin resonance or to nuclear resonance in an ensemble of spin-$1/2$ nuclei. We first present this model and briefly address its generalization for a case of higher spins relevant to nuclear spin systems in semiconductors. Let us consider a single nuclear spin $I = 1/2$ (or an ensemble of identical but noninteracting spins) subject to a sum of the static magnetic field $\boldsymbol{B}$ and time-dependent magnetic field $\tilde{\boldsymbol{B}}(t)$. Let $\boldsymbol{B} \parallel z$-axis and assume that $\tilde{\boldsymbol{B}} \perp \boldsymbol{B}$, see Figure 3.1. The static field produces equilibrium (or quasi-equilibrium) spin polarization, $\langle \boldsymbol{I} \rangle \neq 0$, see Eqs. (2.47) and (2.50):

Electron and Nuclear Spin Dynamics in Semiconductor Nanostructures. M. M. Glazov, Oxford University Press (2018). © M. M. Glazov. DOI: 10.1093/oso/9780198807308.001.0001

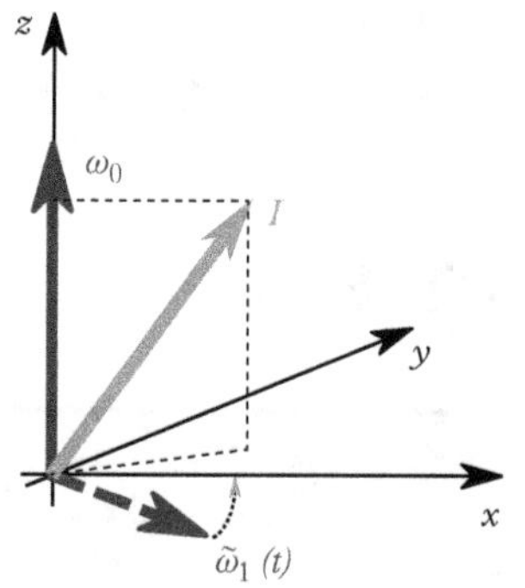

Figure 3.1 *Illustration of spin precession frequencies* $\boldsymbol{\omega}_0$ *associated with the static field* $\boldsymbol{B}$ *(solid dark shaded arrow) and* $\tilde{\boldsymbol{\omega}}_1(t)$ *associated with the alternating field* $\tilde{\boldsymbol{B}}$ *(dashed arrow). The solid light arrow shows the average spin* $\boldsymbol{I}$*. The thin dotted arrow shows the direction of rotation of the alternating field, see text for details.*

$$\langle \boldsymbol{I} \rangle = \frac{\boldsymbol{B}}{2B} \tanh\left(\frac{g_n \mu_N B}{2k_B T}\right). \tag{3.1}$$

We recall that the g_n is the nuclear g-factor, μ_N is the nuclear magneton, and $k_B T$ is the temperature measured in energy units. We assume that the alternating field is small enough $\tilde{B} \ll B$ and neglect the spin polarization induced by $\tilde{\boldsymbol{B}}$. Next, we show that even such a weak field can significantly affect the spin polarization in the system.

Following the general approach outlined in Section 2.4 we use the spin density matrix method and present a kinetic equation for the average spin of nuclei $\boldsymbol{I} = \mathrm{Sp}\{\rho \hat{\boldsymbol{I}}\}$, where ρ is the spin density matrix of the system, in the form [cf. Eq. (2.32)]

$$\frac{d\boldsymbol{I}}{dt} + \boldsymbol{I} \times \boldsymbol{\omega}_0 + \boldsymbol{I} \times \tilde{\boldsymbol{\omega}}_1(t) + \boldsymbol{Q}\{\boldsymbol{I}\} = 0. \tag{3.2}$$

Here, $\boldsymbol{\omega}_0 = -g_n \mu_N \boldsymbol{B}/\hbar = -\gamma_I \boldsymbol{B}$ is the nuclear spin precession frequency in the static field, $\tilde{\boldsymbol{\omega}}_1(t) = -g_n \mu_N \tilde{\boldsymbol{B}}(t)/\hbar = -\gamma_I \tilde{\boldsymbol{B}}(t)$ is the nuclear spin precession frequency in the alternating field (Figure 3.1), $\boldsymbol{Q}\{\boldsymbol{I}\}$ is the operator describing the relaxation and decoherence of the spin. For nuclei, these processes are caused by the hyperfine interaction with electron spin, dipole–dipole interaction with neighboring nuclei, and quadrupole interaction with fluctuating electric fields caused, for example, by charge fluctuations.

Here we abstain from discussion about particular relaxation mechanisms and present the phenomenological expression for the collision integral. At this level, the form of $\boldsymbol{Q}\{\boldsymbol{I}\}$ can be established from the requirements of the relaxation operator to be linear in $\boldsymbol{I}$, describe relaxation of $\boldsymbol{I}$ to its equilibrium value $\langle \boldsymbol{I} \rangle$, Eq. (3.1), and to be consistent with the system symmetry. For external magnetic fields exceeding local ones, $B \gg B_L$, the symmetry of crystalline lattice plays a much less important role compared with the symmetry reduction induced by the static field. As a result, the relaxation operator takes the commonly accepted form

$$Q_z\{\boldsymbol{I}\} = \frac{I_z - \langle I_z(\boldsymbol{B})\rangle}{T_1}, \quad Q_x\{\boldsymbol{I}\} = \frac{I_x}{T_2}, \quad Q_y\{\boldsymbol{I}\} = \frac{I_y}{T_2}. \tag{3.3}$$

Equation (3.3) demonstrates that the component of the spin parallel to the static field relaxes toward the equilibrium value with the characteristic time T_1 and the components in the plane perpendicular to the external field relax to zero with a time constant of T_2. Correspondingly, the times T_1 and T_2 are termed the *longitudinal* and *transversal* spin relaxation times, respectively. In terms of the 2×2 spin density matrix ρ, the T_2 time describes the relaxation of the off-diagonal density matrix elements, $\rho_{\pm 1/2, \mp 1/2}$ responsible for the coherence between the spin-up, $I_z = +1/2$, and spin-down, $I_z = -1/2$, sublevels. The T_1-time is responsible of the relaxation of the occupancy difference $\rho_{1/2,1/2} - \rho_{-1/2,-1/2}$ toward its equilibrium value. The pseudovector kinetic equation (3.2) with the relaxation operator in the form of Eq. (3.3) is equivalent to the set of three scalar equations for the spin components I_α, $\alpha = x, y, z$. This set is often referred to as the *Bloch equations* after the seminal work of Felix Bloch (Bloch, 1946) where these equations were introduced. More advanced forms of the relaxation operator in Eq. (3.2) applicable also for $I > 1/2$ are discussed in (Goldman, 2001; Ibarra-Sierra *et al.*, 2017).

It is noteworthy, that equations (3.2) and (3.3) can be equally applied to the electrons in order to describe the electron spin resonance, in which case the nuclear spin $\boldsymbol{I}$ is replaced by the electron spin $\boldsymbol{S}$, see Eq. (2.33), the frequencies $\boldsymbol{\omega}_0$ and $\tilde{\boldsymbol{\omega}}_1$ are associated with the electron precession frequencies in the magnetic field and the relaxation times T_1 and T_2 play the roles of the electron longitudinal and transversal relaxation times. Moreover, the simple form of Bloch equations (3.2) is valid for $I \neq 1/2$ provided that quadrupole and dipole–dipole interactions are unimportant.

In order to solve the Bloch equations (3.2) we assume that the alternating field $\tilde{\boldsymbol{B}}(t)$ is monochromatic [cf. Eq. (2.25)],

$$\tilde{\boldsymbol{B}}(t) = \boldsymbol{e} B_1 e^{-\mathrm{i}\omega t} + \text{c.c.}, \tag{3.4}$$

where ω is the frequency of the electromagnetic wave and $\boldsymbol{e} \parallel (xy)$ is its polarization vector. We introduce the basic polarization vectors for the σ^+ and σ^- circular polarization as

$$\boldsymbol{o}_+ = -\frac{\boldsymbol{o}_x + \mathrm{i}\boldsymbol{o}_y}{\sqrt{2}}, \quad \boldsymbol{o}_- = \frac{\boldsymbol{o}_x - \mathrm{i}\boldsymbol{o}_y}{\sqrt{2}}, \tag{3.5}$$

where $\boldsymbol{o}_\alpha$ ($\alpha = x, y, z$) are the unit vectors of Cartesian axes and we assume for simplicity that the incident field is right circularly polarized $\boldsymbol{e} = \boldsymbol{o}_+$, that is, $P_c = 1$

in Eq. (2.26a). Under these assumptions the vector $\tilde{\boldsymbol{\omega}}_1(t)$ rotates clockwise (if one looks along the positive direction of the z-axis) in the (xy) plane, see Figure 3.1. Hence, it is instructive to transform to the coordinate frame that rotates around the z-axis synchronously with the alternating field. In this reference frame the alternating field $\tilde{\boldsymbol{B}}_1(t)$ becomes static. However, the transformation of the frame is accompanied by the change of the time derivatives, namely:

$$\frac{d\boldsymbol{I}}{dt} \to \frac{d\boldsymbol{I}}{dt} + \omega \boldsymbol{o}_z \times \boldsymbol{I}.$$

Note that, for left circularly polarized radiation, one has to change the sign from a plus to minus in the rule of transformation to the synchronously rotating frame. Since $\boldsymbol{\omega}_0 \parallel z$, the static field $\boldsymbol{B} \parallel z$ does not change, the form of collision integrals also remains the same as in Eq. (3.3). Finally, we have

$$\frac{d\boldsymbol{I}}{dt} + \boldsymbol{I} \times (\boldsymbol{\omega}_0 - \omega \boldsymbol{o}_z) + \boldsymbol{I} \times \tilde{\boldsymbol{\omega}}_1 + \boldsymbol{Q}\{\boldsymbol{I}\} = 0. \tag{3.6}$$

Here we omitted the time argument of $\tilde{\boldsymbol{\omega}}_1$ because in the rotating frame the alternating field becomes static. It is worth noting that time dependence of the fields in Eq. (3.6) vanished after the transformation to the rotating frame of reference. Hence, in the steady-state regime the time derivative of the $\boldsymbol{I}$ can be omitted and in the rotating reference frame the Bloch equations are reduced to a set of linear algebraic equations:

$$\frac{I_+}{T_2} - \mathrm{i} I_+(\omega_0 - \omega) + \mathrm{i}\tilde{\omega}_{1,+} I_z = 0, \tag{3.7a}$$

$$\frac{I_-}{T_2} + \mathrm{i} I_-(\omega_0 - \omega) - \mathrm{i}\tilde{\omega}_{1,-} I_z = 0, \tag{3.7b}$$

$$\frac{I_z - \langle I_z \rangle}{T_1} + \mathrm{i}(I_+ \tilde{\omega}_{1,-} - I_- \tilde{\omega}_{1,+}) = 0. \tag{3.7c}$$

Here we introduced the following combinations

$$I_\pm = \frac{I_x \pm \mathrm{i} I_y}{\sqrt{2}}, \quad \tilde{\omega}_{1,\pm} = \frac{\tilde{\omega}_{1,x} \pm \mathrm{i}\tilde{\omega}_{1,y}}{\sqrt{2}}.$$

Note that I_+ and I_- as well as $\tilde{\omega}_{1,+}$ and $\tilde{\omega}_{1,-}$ are related by the complex conjugation operation.

Although Eqs. (3.7) can be readily solved, it is instructive to proceed step-by-step and, at first, to present the solution in the first order in $\tilde{\boldsymbol{\omega}}_1$. In this limit we have $I_z = \langle I_z \rangle$ and

$$I_\pm = \mp \mathrm{i} \langle I_z \rangle \frac{\tilde{\omega}_{1,\pm} T_2}{1 \mp \mathrm{i}(\omega_0 - \omega) T_2}. \tag{3.8}$$

Equation (3.8) demonstrates the appearance of the in-plane spin components in response to the oscillating magnetic field. As expected the effect of the $\tilde{\boldsymbol{\omega}}$ is largest at the resonance conditions where

$$\omega = \omega_0. \tag{3.9}$$

The resonance condition means that the alternating field rotates synchronously with the spin precession in the static field $\boldsymbol{B}_0$. In this case the denominator in Eq. (3.8) vanishes. The only field that acts on the spin in the rotating reference frame is, under condition (3.9), $\tilde{\boldsymbol{\omega}}$. The appearance of the in-plane components with the magnitude $I_\pm = \pm \mathrm{i}\langle I_z\rangle \tilde{\omega}_{1,\pm} T_2$ can be interpreted (in the rotating frame) as a result of the spin rotation with the precession frequency $\tilde{\boldsymbol{\omega}}_1$ during the decoherence time T_2 from its initial orientation $\boldsymbol{I} = \langle I\rangle \parallel z$. The deviation from the resonance condition (3.9) breaks the synchronicity of the spin precession in the static field and the rotation of the alternating field and results in suppression of the in-plane spin components.

It follows from Eq. (3.7c) that the variation of the z spin component takes place in the second order in $\tilde{\omega}_1$:

$$I_z - \langle I_z\rangle = -\langle I_z\rangle \frac{\tilde{\omega}_1^2 T_1 T_2}{1 + (\omega_0 - \omega)^2 T_2^2}. \tag{3.10}$$

Again, as expected, the variation is largest at the resonance where Eq. (3.9) holds and the denominator vanishes. Equations (3.7c) and (3.10) allow one to evaluate the absorbed power, $\mathcal{P}$: To that end we note that the spin flip from $I_z = 1/2$ to $I_z = -1/2$ state is accompanied by the absorption of the photon with energy $\hbar\omega$. In the steady state the energy pumped into the spin system is dissipated at the same rate due to the spin-flip processes, hence $\mathcal{P} = \hbar\omega |I_z - \langle I_z\rangle|/T_1$ or

$$\mathcal{P} = \hbar\omega |\langle I_z\rangle| \frac{\tilde{\omega}_1^2 T_2}{1 + (\omega_0 - \omega)^2 T_2^2}. \tag{3.11}$$

In the described model the absorbed power as a function of detuning between the resonance $\omega - \omega_0$ is described by the Lorentzian function with a half-width at a half-maximum (HWHM) being $1/T_2$. This is consistent with qualitative expectations presented in Figure 2.5: The transitions between spin levels are most efficient where the frequency of the incident radiation matches the splitting between the levels. In this respect, we note that Eq. (3.11) can be derived (in the limit of $|\omega_0 - \omega| T_2 \ll 1$) by virtue of Fermi's golden rule relating the absorbed power with the transition rate: $\mathcal{P} = \hbar\omega W_{-1/2,+1/2}$. Indeed, the rate of transitions from $I_z = +1/2$ to $I_z = -1/2$ spin level is given by

$$W_{-1/2,+1/2} = \frac{2\pi}{\hbar^2} \frac{\hbar^2 \tilde{\omega}_1^2}{4} \Delta_{T_2}(\omega - \omega_0) 2\langle I_z\rangle. \tag{3.12}$$

In this expression $(\hbar\tilde{\omega}_1)^2/4$ is the squared modulus of the perturbation matrix element, $\hbar(\tilde{\boldsymbol{\omega}}_1 \cdot \hat{\boldsymbol{I}})$, $2\langle I_z\rangle$ is the difference of occupancies of the ground and excited spin levels, and we introduced the broadened Dirac δ-function as

$$\Delta_T(x) = \frac{1}{\pi}\frac{T}{1+x^2T^2}. \tag{3.13}$$

It is worth noting that the transition rate induced by linearly polarized [in the (xy) plane] *ac* field is half the size, because in the studied configuration only one circular component of the radiation is absorbed. The z-component of the alternating field does not cause, in this model, any transitions between the $I_z = \pm 1/2$ levels.

The derivation of the absorbed power presented here is valid in the lowest order in the parameter $\tilde{\omega}_1^2 T_1 T_2$. Otherwise, a variation of the z component of the nuclear spin induced by the alternating field, Eq. (3.10), is not small compared with its mean value $\langle I_z\rangle$ and, thus, cannot be neglected. Let us now briefly address the case of strong perturbation, where $\tilde{\omega}_1^2 T_1 T_2$ can be arbitrarily large while, as before, $\tilde{\omega}_1 \ll \omega_0$. The modification of the expressions (3.8) and (3.10) is straightforward (Abragam, 2002). Here we present the expression for the absorbed power that immediately follows from the solution of Eqs. (3.7) and generalizes Eq. (3.11) for the case of arbitrary $\tilde{\omega}_1^2 T_1 T_2$:

$$\mathcal{P} = \hbar\omega\langle I_z\rangle \frac{\tilde{\omega}_1^2 T_2}{1+(\omega_0-\omega)^2T_2^2+\tilde{\omega}_1^2 T_1 T_2}. \tag{3.14}$$

It follows from Eq. (3.14) that with an increase of the *ac* field strength $\tilde{\omega}_1$ the absorbed power first increases $\propto \tilde{\omega}_1^2$, that is proportionally to the intensity of the alternating field, while for $\tilde{\omega}_1^2 T_1 T_2 \gg 1$ the absorbed power saturates at a value of

$$\mathcal{P}_{sat} = \frac{\hbar\omega\langle I_z\rangle}{T_1}. \tag{3.15}$$

The saturation power is the smaller the larger is the longitudinal relaxation time T_1. This expression has a simple physical meaning, it demonstrates that the maximum energy dissipation rate is limited by the $1/T_1$. Hence, studies of the magnetic resonance spectra as a function of the incident radiation intensity make it possible to extract both the transversal relaxation time T_2 and the longitudinal relaxation time T_1.

The model outlined here provides a description of the simplest regimes of magnetic resonance where the ensemble of noninteracting spins subject to the static field interacts with the monochromatic radiation. Various aspects of magnetic resonance phenomena, including time-resolved methods and spin echo techniques are presented in various references. We refer to (Slichter, 1990; Abragam, 2002) for further reading on this subject. Next, we turn to the questions on the origin of T_2 and discuss in Sections 3.2 and 3.4 the mechanisms related to dipole–dipole interaction between the nuclei and quadrupole splittings, respectively. We also address the manifestations of NMR in optical response of semiconductors and semiconductor nanostructures.

3.2 Magnetic resonance in ensembles of interacting nuclei

It is noteworthy that the absorption power spectrum (3.11) is broadened in spin ensembles if the spin precession frequency in the static field ω_0 fluctuates. Such fluctuations are possible in inhomogeneous ensembles where each spin experiences its own local magnetic field formed, for example, due to (i) interaction with other spins in the ensemble, (ii) inhomogeneity of the external field $\boldsymbol{B}$, or (iii) spatial fluctuations of g-factors. The latter are relevant for the electron ensembles (where the theory developed in Section 3.1 is also applicable) due to strong sensitivity of the electron gyromagnetic ratio in semiconductors to the confinement (Ivchenko, 2005). In solids, dipole–dipole interaction between nuclei is an apparent source of NMR broadening.

The interaction between the nuclei has a magnetic nature and can be simply interpreted as the interaction of the magnetic moment of a given nucleus with the magnetic field produced by the surrounding nuclei. Let us consider two nuclei with the moments $\boldsymbol{\mu}_I$ and $\boldsymbol{\mu}_{I'}$. The magnetic field produced by the nucleus with the moment $\boldsymbol{\mu}_{I'}$ at the point with the position vector $\boldsymbol{r}$ (reckoned from the nucleus) reads

$$\boldsymbol{B} = \mathrm{rot}\frac{\boldsymbol{\mu}_{I'} \times \boldsymbol{r}}{r^3}, \tag{3.16}$$

and the interaction energy of the nuclei has a form $-\boldsymbol{\mu}_I \cdot \boldsymbol{B}$. As a result the dipole–dipole interaction Hamiltonian takes the form [cf. Eq. (2.23)]:

$$\mathcal{H}_{int} = \frac{1}{2}\sum_{i\neq j}\frac{\mu_{I_i}\mu_{I_j}}{I_i I_j}\left(\frac{\hat{\boldsymbol{I}}_i \cdot \hat{\boldsymbol{I}}_j}{r_{ij}^3} - 3\frac{(\hat{\boldsymbol{I}}_i \cdot \boldsymbol{r}_{ij}) \cdot (\hat{\boldsymbol{I}}_j \cdot \boldsymbol{r}_{ij})}{r_{ij}^5}\right). \tag{3.17}$$

Here the summation takes place over all nuclei in the lattice, subscripts i and j enumerate nuclei, $\hat{\boldsymbol{I}}_i$ the spin of ith nucleus, $\boldsymbol{r}_{ij} = \boldsymbol{r}_i - \boldsymbol{r}_j$ is the vector that connects the nuclei, $\boldsymbol{r}_i$ denotes the positions of nuclei. The prefactor 1/2 in Eq. (3.17) ensures that the interaction between nucleus i and nucleus j is accounted for only once. Our goal is to analyze the effect of the interaction Hamiltonian (3.17) on the nuclear spin dynamics and, in particular, on NMR.

The interacting nuclear system in the static magnetic field $\boldsymbol{B} \parallel z$ is described by the Hamiltonian

$$\mathcal{H} = \mathcal{H}_{B,N} + \mathcal{H}_{int} = -\sum_i \frac{\mu_{I_i}}{I_i} \hat{\boldsymbol{I}}_i \cdot \boldsymbol{B} + \mathcal{H}_{int}. \tag{3.18}$$

We introduce the eigenvalues, E_n, and the eigenvectors, $|n\rangle$, of the Hamiltonian (3.18) with the index n enumerating all states of the multi-nuclear system. Let us calculate the absorbed power of the alternating magnetic field $\tilde{\boldsymbol{B}}$, Eq. (3.4). Here, for simplicity, we assume that the field is linearly polarized with the polarization vector $\boldsymbol{e} \parallel x$. The perturbation Hamiltonian reads

$$-\sum_i \frac{\mu_{I_i}}{I_i} \hat{I}_{i,x} \tilde{B}_1 e^{-\mathrm{i}\omega t} + \text{c.c.}.$$

The rate of quantum transitions between the states of the system accompanied by the absorption of the photon with energy $\hbar\omega$ is found from Fermi's golden rule (3.12) with the result

$$w(\omega) = \sum_{nn'} \left| \langle n' | \hat{M}_x | n \rangle \right|^2 \delta(E_{n'} - E_n - \hbar\omega). \tag{3.19}$$

Here we omitted the numerical coefficients and introduced the components of the total spin of the nuclei as

$$M_x = \sum_i \hat{I}_{i,x}, \quad M_y = \sum_i \hat{I}_{i,y}, \quad M_z = \sum_i \hat{I}_{i,z}. \tag{3.20}$$

Hereafter, we assume that all nuclei are of the same type, the general case is briefly discussed at the end of this section.

In absence of interactions, Eq. (3.19) is equivalent to Eq. (3.12) at $T_2 \to \infty$. Hence, as a function of ω the transition rate has a δ-function peak at $\omega = \omega_0 = |\mu_I B/(\hbar I)| = |\gamma_I B|$. Accounting for interactions, the transition rate is a superposition of the peaks corresponding to the distances $E_n - E_{n'}$ between the many nuclear states. In the macroscopic crystalline system with $\sim 10^{23}$ nuclei the inter-level spacings are so small that the energies E_n form an almost continuous spectrum. Therefore, the broadening of the resonance is expected. However, the direct diagonalization of the Hamiltonian (3.18) for the macroscopic number of nuclei is hardly possible, therefore, statistical methods should be used to study the role of dipole–dipole interactions between nuclei. In fact, it is natural to consider the transition rate $w(\omega)$ as a certain "distribution" function with a peak at $\omega = \omega_0$ and characterize it by the set of moments as suggested by Van Vleck (Van Vleck, 1948):

$$\mathcal{M}_m = \frac{\int d\omega (\omega - \omega_0)^m w(\omega)}{\int d\omega\, w(\omega)}, \quad m = 0, 1, 2, \ldots. \tag{3.21}$$

The integration in Eq. (3.21) ranges from $\omega \to -\infty$ to $+\infty$. The 0-th moment $\mathcal{M}_0 = 1$, the first one gives the shift of the peak due to the dipole–dipole interaction, the second one gives the variance of the distribution, and so on. The knowledge of the moments allows one to characterize the shape of $w(\omega)$ and, correspondingly, to estimate the width of the resonance. As we demonstrate next, the moments can be calculated without performing the diagonalization of the Hamiltonian (3.18).

Before we turn to the evaluation of moments in Eq. (3.21) we need to specify the situation in question. First, we rewrite the Hamiltonian (3.17) in the form

$$\mathcal{H}_{int} = -\frac{\mu_I^2}{I^2}\sqrt{\frac{6\pi}{5}}\sum_{i\neq j}\frac{1}{r_{ij}^3}\sum_{\mu=-2}^{2} Y_{2\mu}(\theta_{ij},\varphi_{ij})\mathcal{T}_\mu\{\hat{\boldsymbol{I}}_i,\hat{\boldsymbol{I}}_j\}, \tag{3.22}$$

where $Y_{l\mu}(\theta,\varphi)$ are the spherical harmonics with the angular momentum l and its z-component μ, $\mathcal{T}_\mu\{\hat{\boldsymbol{I}},\hat{\boldsymbol{J}}\}$ are irreducible second-rank tensors defined as (Abragam and Goldman, 1982)

$$\mathcal{T}_0\{\hat{\boldsymbol{I}},\hat{\boldsymbol{J}}\} = \frac{1}{\sqrt{6}}\left[3\hat{I}_z\hat{\mathcal{J}}_z - (\hat{\boldsymbol{I}}\cdot\hat{\boldsymbol{J}})\right], \tag{3.23a}$$

$$\mathcal{T}_{\pm 1}\{\hat{\boldsymbol{I}},\hat{\boldsymbol{J}}\} = \mp(I_\pm\mathcal{J}_z + \mathcal{J}_z I_\pm), \tag{3.23b}$$

$$\mathcal{T}_{\pm 2}\{\hat{\boldsymbol{I}},\hat{\boldsymbol{J}}\} = \frac{1}{2}I_\pm\mathcal{J}_\pm, \tag{3.23c}$$

$I_\pm = I_x \pm \mathrm{i}I_y$, and angles θ_{ij}, φ_{ij} are the polar and azimuthal angles of the vector $\boldsymbol{r}_{ij}$ in the spherical coordinate system with $z \parallel \boldsymbol{B}$. The form of the Hamiltonian (3.22) is convenient to clarify the origin of various perturbations induced by the dipole–dipole interaction of nuclei. Particularly, the term with $\mu = 0$, also known as a truncated or secular part of the dipole–dipole interaction,

$$\mathcal{H}'_{int} = -\frac{\mu_I^2}{I^2}\sqrt{\frac{6\pi}{5}}\sum_{i\neq j}\frac{1}{r_{ij}^3}Y_{20}(\theta_{ij},\varphi_{ij})\mathcal{T}_0\{\hat{\boldsymbol{I}}_i,\hat{\boldsymbol{I}}_j\}, \tag{3.24}$$

commutes with the Zeeman Hamiltonian $\propto \sum_i \hat{I}_{i,z}B$. The remaining terms with $|\mu| = 1$ and $|\mu| = 2$ change the total spin of a pair of interacting nuclei by 1 and 2, respectively. In what follows, we assume that the external static magnetic field is sufficiently large so that $\hbar\omega_0$ exceeds by far the splittings induced by the dipole–dipole interaction between nuclei, that is, B exceeds by far the local field B_L. In the noninteracting system the nuclear states are classified by the z-component of the total spin, M_z, Eq. (3.20). Hence, the truncated Hamiltonian (3.24) provides first-order corrections to the energies of the noninteracting nuclei in a magnetic field. It results in the broadening of the NMR peak at $\omega = \omega_0$. The remaining part, $\mathcal{H}_{int} - \mathcal{H}'_{int}$ mixes the states with M_z and $M_z \pm 1$ or $M_z \pm 2$ giving rise to the second-order corrections to the energies. Moreover, since the Hamiltonian $\mathcal{H}_{int} - \mathcal{H}'_{int}$ does not conserve the z-component of spin, the transitions between the mixed states with the momentum component change by 0 or 2 become possible. Hence, additional peaks at $\omega = 0$, $\omega = 2\omega_0$, ...appear in the NMR spectrum. The absorption spectrum, Eq. (3.19), consisting of the main NMR peak and side peaks is schematically shown in Figure 3.2. Since for a sufficiently large external field the broadening of the peaks is much smaller than the distance between the peaks, one has to exclude the contribution of the weak peaks in magnetic resonance absorption. Therefore, in order to find the broadening of the main NMR peak at $\omega = \omega_0$ one has to evaluate the moments of the spectrum, Eq. (3.21), taking into account only the secular part of the Hamiltonian (Van Vleck, 1948; Abragam and Goldman, 1982; Slichter, 1990; Abragam, 2002).

Now let us evaluate the moments in Eq. (3.21). First, we calculate the normalization integral in the denominator of $\mathcal{M}_m$, Eq. (3.21). Integration over ω removes the δ-function and the remaining sum can be readily transformed to the trace of $\hat{M}_x^2$ operator using the completeness relation:

$$\int d\omega\, w(\omega) = \sum_n \langle n|\hat{M}_x^2|n\rangle = \mathrm{Sp}\{\hat{M}_x^2\} = \frac{I(I+1)}{3} N(2I+1)^N, \tag{3.25}$$

where N is the number of nuclei in the normalization volume. It is easy to show that the first moment, $\mathcal{M}_1$ vanishes; that is, the NMR line is symmetric with respect to ω_0. This is because for a given pair of states n and n' there is a "symmetric" pair $\tilde{n}$ and $\tilde{n}'$ where all spins are reversed. Since $\mathcal{H}_{int}$ is even in nuclear spins and the Zeeman Hamiltonian is odd in $\boldsymbol{I}_i$, the following equality holds:

$$E_n - E_{n'} - \hbar\omega_0 = -(E_{\tilde{n}'} - E_{\tilde{n}} - \hbar\omega_0).$$

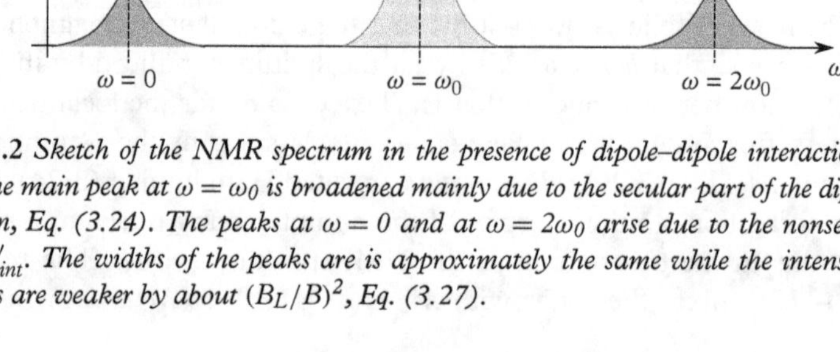

Figure 3.2 *Sketch of the NMR spectrum in the presence of dipole–dipole interaction between nuclei. The main peak at $\omega = \omega_0$ is broadened mainly due to the secular part of the dipole–dipole interaction, Eq. (3.24). The peaks at $\omega = 0$ and at $\omega = 2\omega_0$ arise due to the nonsecular part, $\mathcal{H}_{int} - \mathcal{H}'_{int}$. The widths of the peaks are is approximately the same while the intensities of the side peaks are weaker by about $(B_L/B)^2$, Eq. (3.27).*

Hence the pairs (n, n') and $(\tilde{n}', \tilde{n})$ provide same contributions to the absorption spectrum being symmetric with respect to ω_0; that is, $\mathcal{M}_1 = 0$. The straightforward calculation of $\mathcal{M}_1$ with the same result is presented, for example, in (Slichter, 1990).

Next, let us turn to the second moment, $\mathcal{M}_2$, which provides an estimate of the NMR peak width. Similar to the calculation (3.25), removing the δ-function and taking into account that $\mathcal{H}|n\rangle = E_n|n\rangle$, one can reduce the numerator in $\mathcal{M}_2$ to the trace of commutator squared:

$$\sum_{nn'} (E_{n'} - E_n - \hbar\omega_0)^2 \langle n|\hat{M}_x|n'\rangle\langle n'|\hat{M}_x|n\rangle$$
$$= \mathrm{Sp}\{[\mathcal{H}_{B,N} + \mathcal{H}'_{int}, \hat{M}_x]^2\} - (\hbar\omega_0)^2\,\mathrm{Sp}\{\hat{M}_x^2\}.$$

Here we excluded from the Hamiltonian the nonsecular part for the reasons presented previously and took into account that $\mathcal{M}_1 = 0$. Finally (Van Vleck, 1948; Slichter, 1990; Abragam, 2002),

$$\mathcal{M}_2 = \frac{3}{4}\frac{\mu_I^4}{I^4} I(I+1) \sum_{j \neq i} \frac{(3\cos^2\theta_{ij} - 1)^2}{r_{ij}^6}. \tag{3.26}$$

Here the summation is carried out over all nuclei j (at a fixed i). The moment $\mathcal{M}_2$ roughly defines the line broadening, while in order to study the line shape all remaining moments should be calculated. Although this is, in principle, possible, see for example (Abragam, 2002), already for $\mathcal{M}_4$ the results turn out to be quite cumbersome. Here, we note that the analysis based on the ratio of the second and fourth moment demonstrates that the line shape deviates from both Lorentzian and Gaussian functions.

It is noteworthy that for a system that consists of different types of nuclear spins, $\hat{\boldsymbol{I}}$ and $\hat{\boldsymbol{J}}$ for example, the contribution to the second moment caused by the interaction between different types of nuclei should be calculated in a large enough magnetic field, taking into account the additional terms $\propto \hat{I}_z\hat{J}_z$ in the dipole–dipole Hamiltonian only. This is because all remaining terms in the dipole–dipole Hamiltonian do not commute with the Zeeman term and, hence, contribute to the side peaks.

The sum in Eq. (3.26) rapidly converges, therefore, for a crude estimate one can take into account the contributions of the nearest neighbours only. In terms of the linewidth it corresponds to $\lesssim 1$ KHz for GaAs. Usually, to characterize the strength of the dipole–dipole interaction the local field, B_L, is formally introduced as

$$B_L^2 = \frac{\mathrm{Sp}\{\mathcal{H}_{int}^2\}}{(\hbar\gamma_I)^2\,\mathrm{Sp}\{\hat{M}_x^2\}}. \tag{3.27}$$

We recall that $\gamma_I = \mu_I/\hbar I$ is the gyromagnetic ratio for ith nucleus, and the trace in Eq. (3.27) is calculated with the *total* interaction Hamiltonian, including both secular and nonsecular parts. For a system consisting of different nuclear spins the denominator is replaced by the weighted average over all spin species. In GaAs the local field was calculated in (Paget *et al.*, 1977) with the result

$$B_L \approx 0.15 \text{ mT}.$$

It has a similar order of magnitude in other III-V semiconductors. The associated decoherence time scale is in the order of $T_2 = \hbar/(|g_n|\mu_N B_L) \sim 10^{-4}$ sec. It is worth noting that dipole–dipole interaction provides the lowest bound to the nuclear spin decoherence rate and, correspondingly, to the width of the NMR spectrum.

The method of moments can be also used to calculate the weights of contributions to NMR spectra at combined frequencies, for example, at twice the Larmor frequency of the nucleus, $\omega = 2\omega_0$, see Figure 3.2, or, if different nuclei are present, at sum (or difference) Larmor frequencies. In this case one has to take into account the perturbation related to the $\mu = \pm 1$ terms in Eq. (3.22). These contributions to NMR are due to the first-order corrections to the wavefunctions of nuclear spin systems in the small parameter $B_L/B \ll 1$ because the strength of admixture of states with $M_z \pm 1$ to M_z is proportional to the local fields acting on the nucleus. As a result, the peaks at the combined frequencies are smaller compared with the main peak at $\omega = \omega_0$ by about $(B_L/B)^2$. The microscopic calculation for bulk III-V crystals were performed in (Kalevich *et al.*, 1982) and support this qualitative analysis. In a weak magnetic field, $B \lesssim B_L$, the peaks merge. In this limit one has to use the complete interaction Hamiltonian, $\mathcal{H}_{int}$, rather than its secular part $\mathcal{H}'_{int}$ to evaluate the moments of the magnetic resonance spectrum.

3.3 Optically detected NMR

In classical NMR the absorption or transmission of the radio-frequency radiation interacting with the sample is measured. In semiconductors the nuclear and electron spin systems are coupled via the hyperfine interaction and electronic spin system effectively interacts with light, see Chapter 2. It opens up possibilities for detecting NMR and, hence, for addressing the nuclear spin dynamics by studying the response at optical frequencies.

The basic physical idea behind the optical detection of NMR (ODNMR) can be formulated as follows: We consider a semiconductor in a static magnetic field $\boldsymbol{B}$. First, let us prepare the nuclear spin system in a state with low enough spin temperature, $|\Theta_N|$. In other words, let the average static nuclear polarization $\langle \boldsymbol{I} \rangle$ be significant in a sense that it provides observable effect on the electron spin dynamics, for example, the induced Overhauser field produces measurable splitting of electronic levels, variation in the electron spin precession frequency, or it just stabilizes the electron spin polarization affecting the electron spin relaxation time. Second, let us apply the radio-frequency radiation that causes transitions between the nuclear spin states; that is, NMR. As we demonstrated before, see Eq. (3.10), the absorption of radio-frequency quanta tend to equalize the occupancies of the nearest nuclear spin sublevels and, correspondingly, to suppress the nuclear spin polarization $\langle \boldsymbol{I} \rangle$. In terms of nuclear spin temperature the "warm-up" of the nuclear spin system occurs. As a result, the Overhauser field acting on the electron spin changes. This variation of the Overhauser field is detected optically, for instance, via the variation of the circular polarization degree of luminescence, see Section 2.4 for details. This is the basics of ODNMR.

The optical detection of NMR in semiconductors has been demonstrated in pioneering work (Ekimov and Safarov, 1972) where the bulk solid solutions $Ga_{0.8}Al_{0.2}As$ were used. An experiment on the optical orientation of electron spins in the longitudinal magnetic field (Faraday geometry) has been performed in the presence of radio-frequency radiation of varied frequency. The dips in the circular polarization of luminescence, P_c^{lum}, as a function of frequency of the applied radio-frequency field, were detected. Their positions matched the expected NMR frequencies for the relevant isotopes of ^{69}Ga, ^{71}Ga, ^{75}As, and ^{27}Al. The optical detection scheme of the NMR offered several advantages, including increased sensitivity [the signal to noise ratio of 100:1 has been reported already in the first study (Ekimov and Safarov, 1972)] and the possibility to address small sample volumes, which lifts constraints on high homogeneity of the external magnetic field. The latter was an important experimental obstruction for magnetic resonance detection.

The ODNMR method in semiconductors has gained high popularity (Paget, 1982; Meier and Zakharchenya, 1984; Dyakonov, 2017) and experiments have also been performed on quantum well (Kalevich *et al.*, 1990) and quantum dot (Gammon *et al.*, 1997; Poggio and Awschalom, 2005) structures. Various experimental schemes have been used, including, for example, oblique field geometry, where the NMR reveals itself as sharp features on the Hanle curve describing electron spin depolarization by the transverse magnetic field component. These features occur where the external field is compensated for by the nuclear one (D'yakonov *et al.*, 1975). Instead of NMR detection via the luminescence polarization one can use the time-resolved Kerr or Faraday effects to observe the variation of electron spin precession frequency, see for example (Poggio and Awschalom, 2005; Chen *et al.*, 2011*b*) or, if electron Zeeman sublevels can be resolved, one can detect the variation of the Overhauser field directly (Gammon *et al.*, 1997). Quite recently, NMR has been observed in the Faraday rotation signal induced by nuclear magnetization in bulk GaAs placed in a microcavity (Giri *et al.*, 2013).

Figure 3.3 shows the results of ODNMR on a bulk Si-doped GaAs sample the donor density at approximately 5×10^{16} cm^3 obtained in (Chen *et al.*, 2011*b*). The authors used used the pump-probe experiment based on time-resolved

Kerr rotation to monitor the electron spin precession around the static external magnetic field $\boldsymbol{B} = \boldsymbol{B}_{\text{ext}}$. The short, circularly polarized pulsed created electron spin polarization and induced its precession around the magnetic field. The detection of Larmor precession has been carried out by the linearly polarized pulse as described in Section 2.4 and Eq. (2.41). Additionally, the σ^+ or σ^- optical pumping took place in order to cool down the nuclear spin system and create an Overhauser field $\boldsymbol{B}_N \parallel \boldsymbol{B}_{\text{ext}}$, Eq. (3.16), which produced the shift of the electron Larmor frequency as

$$\Omega_L = \Omega_B + \Omega_N = \frac{g\mu_B}{\hbar}(B_{\text{ext}} + B_N), \tag{3.28}$$

where g is the electron g-factor. Depending on the sign of helicity of the pumping light the nuclear field B_N was either parallel or antiparallel to the external field resulting in the increase or decrease of the electron Larmor frequency, respectively, as shown by circles and squares in Figure 3.3. For reference, linearly polarized excitation has been also carried out, in which case no build-up of the nuclear

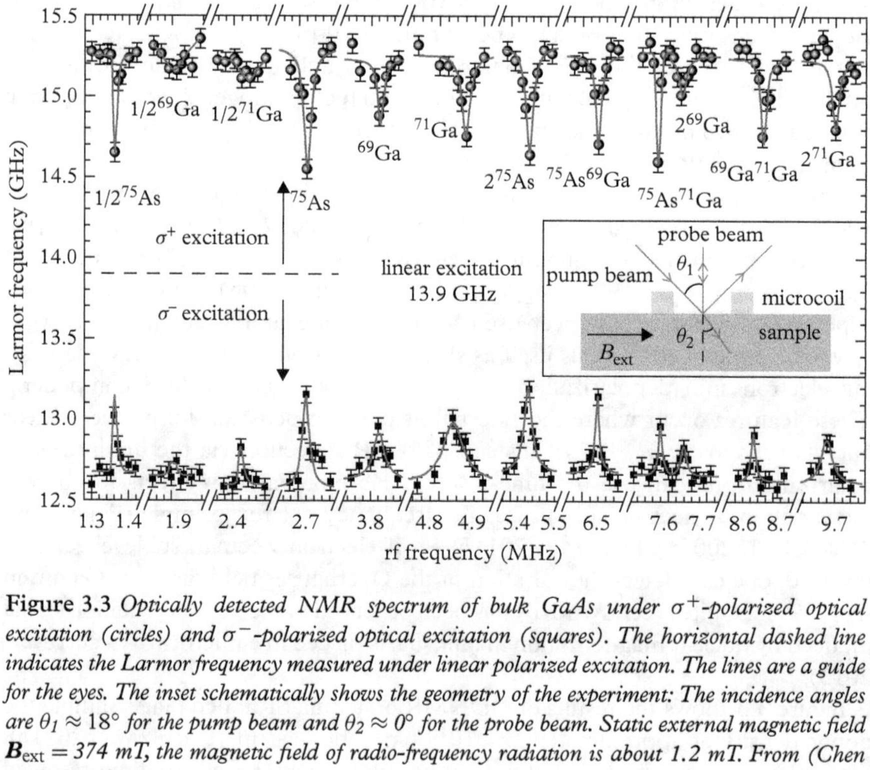

Figure 3.3 *Optically detected NMR spectrum of bulk GaAs under σ^+-polarized optical excitation (circles) and σ^--polarized optical excitation (squares). The horizontal dashed line indicates the Larmor frequency measured under linear polarized excitation. The lines are a guide for the eyes. The inset schematically shows the geometry of the experiment: The incidence angles are $\theta_1 \approx 18°$ for the pump beam and $\theta_2 \approx 0°$ for the probe beam. Static external magnetic field $\boldsymbol{B}_{\text{ext}} = 374$ mT, the magnetic field of radio-frequency radiation is about 1.2 mT. From (Chen et al., 2011b).*

spin polarization was observed, see the dashed line in Figure 3.3. The application of the radio-frequency radiation from the on-chip microscale current loop – see (Chen *et al.*, 2011*b*,) for technical details – resulted in the appearance of resonant features in the electron Larmor frequency related to NMR: If the radio frequency coincides with one of nuclear resonance frequency in the sample, then the corresponding contribution to the nuclear spin polarization $\langle \boldsymbol{I} \rangle$ decreases and the electron spin precession frequency either decreases or increases (depending on the initial direction of the nuclear field) tending to approach its value for the unpolarized situation.

The analysis performed in the (Chen *et al.*, 2011*b*) has made it possible to identify four types of NMR feature marked in Figure 3.3:

1. Fundamental frequencies of ^{75}As, ^{69}Ga, and ^{71}Ga isotopes of host lattice nuclei;
2. Two-spin (combined) resonances involving the same isotope with the frequencies being twice the nuclear precession frequency of a given isotope, for example, 2 ^{75}As;
3. Two-spin resonances involving different isotopes;
4. Half-harmonic resonances with the frequencies being half of the NMR frequency of a given isotope, for example, $1/2\,^{69}$Ga.

The origin of the combined NMR frequencies [also observed, e.g., in (Kalevich *et al.*, 1982; Buratto *et al.*, 1993; Strand *et al.*, 2005)] is presented Section 3.2, it is related to the dipole–dipole interaction between the nuclei. It is also worth noting that quadrupole splittings induced by donors may also lead to the resonance at the double frequency of the isotopes (Paget *et al.*, 2008; Chen *et al.*, 2011*a*).

The observation (Berkovits and Safarov, 1977; Zakharchenya *et al.*, 1977; Strand *et al.*, 2005) of half-harmonic features at

$$2\omega = \omega_N, \tag{3.29}$$

where ω_N is the nuclear Larmor frequency, demonstrates that two radio-frequency quanta get absorbed in the course of the transition between nuclear spin levels. This is a nonlinear effect (Berkovits and Safarov, 1977; Eles and Michal, 2010) and, as one can verify from the symmetry arguments, is absent in the geometry with alternating magnetic field being perpendicular to the static field, $\tilde{\boldsymbol{B}} \perp \boldsymbol{B}$, see also Eq. (3.14). However, if the radio-frequency field is tilted with respect to the plane perpendicular to $\boldsymbol{B}$ the two-photon processes become possible: In a quantum-mechanical description these transitions involve a virtual intermediate state coinciding either with the initial or with the final state. The composite matrix element describing such transitions of a single nucleus from the state with $I_z = +1/2$ to the state with $I_z = -1/2$ in the case where radio field is linearly polarized in the (xz) plane can be written as (Landau and Lifshitz, 1977)

$$M^{(2)}_{-1/2,+1/2} = \frac{M_{-1/2,+1/2}M_{+1/2,+1/2}}{\hbar\omega} + \frac{M_{-1/2,-1/2}M_{-1/2,+1/2}}{\hbar\omega - \hbar\omega_N} \tag{3.30}$$
$$= \frac{M_{-1/2,+1/2}}{\hbar\omega}\left(M_{+1/2,+1/2} - M_{-1/2,-1/2}\right) = \frac{\hbar\tilde{\omega}^2}{4\omega}\sin 2\alpha.$$

We remind that $\tilde{\omega}$ is the nuclear Larmor frequency associated with the radio-frequency field, we introduced the angle α between the polarization plane of $\tilde{\boldsymbol{B}}$ and the static field direction, and, finally, we made use of Eq. (3.29) in order to relate the numerators in the first and second summands. Equation (3.30) demonstrates that the alternating field should have both longitudinal and transverse components with respect to the static field in order to induce two-quantum transitions. The rate of these transitions can be calculated via Fermi's golden rule, see Eq. (3.12), with the replacement of $(\hbar\tilde{\omega}/2)^2$ by $|M^{(2)}_{-1/2,+1/2}|^2$. This result can be also obtained by means of classical spin precession equations, see (Eles and Michal, 2010) for a review. It is worth stressing that quadrupole effects can also give rise to the two-photon transitions mixing the states with different I_z (for nuclei with $I > 1/2$), however, the authors of (Chen *et al.*, 2011*b*) claim that for two-photon transitions presented in Figure 3.3 the dominant mechanism is related to the tilt of $\tilde{\boldsymbol{B}}$. Other subharmonics and fractional frequencies, for example, $2\omega_N/3$ and $3\omega_N/2$ can be also observed (Berkovits and Safarov, 1977; Strand *et al.*, 2005).

The development of nanotechnology and optical spectroscopy of semiconductors made it possible, by the end of the 1990s, to address optically single charge carriers localized in quantum dots (Brunner *et al.*, 1994; Hess *et al.*, 1994; Gammon *et al.*, 1996). In the pioneering work by (Gammon *et al.*, 1997), ODNMR was observed in single quantum dots formed as a result of atomic fluctuations at the interfaces in GaAs/$Al_{0.3}Ga_{0.7}As$ quantum wells. In the photoluminescence spectra of the studied quantum well system excited through a $\sim 1\ \mu$m aperture one observes narrow ($\sim 10\ \mu$eV) lines related to the recombination of electron-hole pairs, also termed zero-dimensional excitons, localized at the quantum dots, see Figure 3.4(a). In the presence of magnetic field $\boldsymbol{B}$ applied in Faraday geometry, the authors of (Gammon *et al.*, 1997) were able to polarize nuclear spins in the quantum dot, see (Brown *et al.*, 1996) and Chapter 5 for details. The nuclear spin polarization results in the Overhauser field acting on the electron and hole spin. This Overhauser field provides additional contribution to the Zeeman splitting of the excitonic levels. Application of the transverse oscillating magnetic field at the NMR frequency induces the transitions between the nuclear spin levels and, as we discussed before, suppresses the nuclear spin polarization. As a result, the effective Zeeman splitting decreases or increases (depending on the relative sign the external field z-component and the radiation helicity, cf. Figure 3.3) provided that the frequency of the alternating magnetic field equals to NMR frequency. Such dips and peaks in the dependence of the Zeeman splitting on the frequency of the magnetic field oscillations have indeed been observed. Figure 3.4(b) shows exemplary dependence of excitonic Zeeman splitting in the frequency region close to the ^{75}As resonance. Two set of points demonstrate the evolution of the

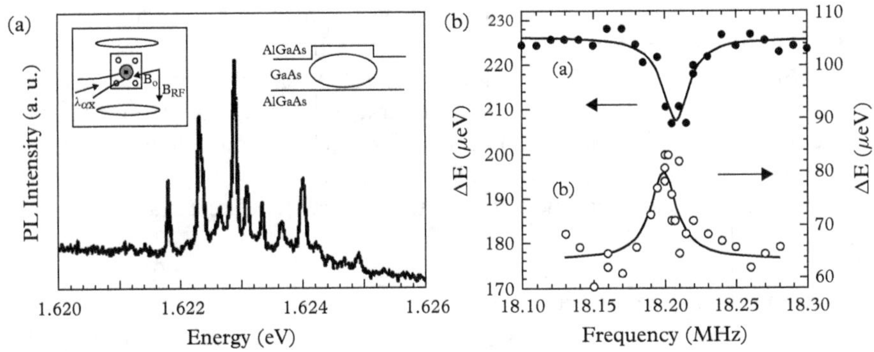

Figure 3.4 *(a) Low temperature ($T = 6$ K) PL spectrum from a 4.5-nm quantum well excited through a 1.5 μm aperture. Sharp, discrete lines are related to localized excitons (at atomic fluctuations of interfaces, right inset) recombination. The sketch of the sample and setup is given in the left inset. (b) Changes in excitonic Zeeman splitting from a single quantum dot in an external magnetic field of $\mathbf{B} = 2.5$ T as the function of frequency of the alternating magnetic field [B_{RF} in the left inset to the panel (a)] in the vicinity of ^{75}As resonance. The sample was excited with σ^+ (curve a) and σ^- (curve b) polarized light. The solid lines are Lorentzian fits to the data. From (Brown et al., 1998).*

Zeeman splitting with nuclei initially polarized by σ^+ (filled circles) and σ^- (open circles). Distinct NMR is seen. Compared with previous works on NMR in bulk materials and quantum well structures, enhancement of the sensitivity by five orders of magnitude has been reported (Gammon *et al.*, 1997). Interestingly, typical widths of the resonances in this experiment are about 10 KHz, while it is noted in (Gammon *et al.*, 1997) that, for some quantum dots, the NMR lines are much broader. The origin of the broadening is, most likely related, with the quadrupole splittings of nuclear spin levels, the inherent feature of quantum dot structures, which we discuss in Section 3.4.

3.4 Role of strain and quadrupole splittings

Semiconductor nanostructures are, as a rule, strained, because an elastic strain is inevitable at combination of semiconductors with different lattice constants. Moreover, the strain plays an important role in quantum dot formation, particularly in such prominent quantum dot systems as InGaAs/GaAs quantum dots (Bimberg *et al.*, 1999). In semiconductors, strain produces inhomogeneous electric fields, quantified by the second-rank tensor of electric field gradients:

$$\mathcal{V}_{\alpha\beta} \equiv \frac{\partial^2 \mathcal{V}}{\partial r_\alpha \partial r_\beta} = \sum_{\gamma\delta} S_{\alpha\beta\gamma\delta}\varepsilon_{\gamma\delta}. \tag{3.31}$$

Here $\mathcal{V}$ is the potential energy associated with the field, $\varepsilon_{\gamma\delta} = \varepsilon_{\delta\gamma}$ are the components of the strain tensor and $S_{\alpha\beta\gamma\delta}$ is the tensor of gradient-elastic constants (Shulman *et al.*, 1957; Sundfors, 1969*a*, 1974). In accordance with the definition in Eq. (3.31) the electric field gradients tensor is symmetric with respect to the subscripts permutation: $\mathcal{V}_{\alpha\beta} = \mathcal{V}_{\beta\alpha}$. Hence, the following relations hold: $S_{\alpha\beta\gamma\delta} = S_{\beta\alpha\gamma\delta} = S_{\alpha\beta\delta\gamma}$. The rank-four tensor $S_{\alpha\beta\gamma\delta}$ has the same linearly independent coefficients as the stiffness tensor, which relates strain and stress tensors (Landau *et al.*, 1999).

The fact that the shape of a nucleus with $I > 1/2$ is different from spherical gives rise to the nonzero electric quadrupole moment of the nucleus and, correspondingly, to the interaction of the nuclear spin with electric field gradients in the form [cf. Eq. (2.24)]:

$$\mathcal{H}_Q = \frac{|e|Q}{6I(2I-1)} \sum_{\alpha\beta} \mathcal{V}_{\alpha\beta} \left[\frac{3}{2}(\hat{I}_\alpha \hat{I}_\beta + \hat{I}_\beta \hat{I}_\alpha) - I(I+1) \right]. \tag{3.32}$$

We recall that Q is the nuclear quadrupole moment, Eq. 2.1). Additionally, in solid solutions such as $Ga_xAl_{1-x}As$ the substitutional disorder produces electric field gradients owing to different electronic structure of the solid solution constituents. It follows from Eq. (3.32) that the trace $\sum_\alpha \mathcal{V}_{\alpha\alpha}$ does not contribute to the quadrupole splittings (Harrison and Sagalyn, 1962). Hence, one can impose an additional constraint by setting the trace of the tensor $\mathcal{V}_{\alpha\beta}$ to zero, which imposes an additional relation between the components of the fourth rank tensor $S_{\alpha\beta\gamma\delta}$ in Eq. (3.31). As a result, for cubic semiconductors such as GaAs or InAs there are only two independent components, denoted usually as S_{11} and S_{44} (given in Table 3.1) allowing relation of the strain and electric field gradients tensors as

$$\mathcal{V}_{zz} = S_{11} \left(\varepsilon_{zz} - \frac{\varepsilon_{xx} + \varepsilon_{yy}}{2} \right), \quad \mathcal{V}_{xy} = 2S_{44}\varepsilon_{xy}. \tag{3.33}$$

Table 3.1 *Components of gradient elastic constants tensor* $S_{\alpha\beta\gamma\delta}$ *(in units of* 10^{15} *statcoulomb×cm*$^{-3}$*) for several prominent semiconductor compounds. The sign of* S_{44} *is fixed to be positive. From (Sundfors, 1974).*

Compound	Isotope	S_{11}	S_{44}
GaAs	^{69}Ga	−9.1	9.2
	^{75}As	13.2	26.5
InAs	^{115}In	−16.7	10.0
	^{75}As	13.0	25.2
InP	^{115}In	−19.7	8.7

Here we use the cubic axes $x \parallel [100], y \parallel [010]$, and $z \parallel [001]$, the relations between the other components of $\mathcal{V}_{\alpha\beta}$ and $\varepsilon_{\alpha\beta}$ can be easily obtained from Eq. (3.33) by making cyclic permutations.

Just like any symmetric two-rank tensor the electric field gradients tensor $\mathcal{V}_{\alpha\beta}$ can be transformed to its principal axes x', y', z', where the off-diagonal components of the tensor vanish. Let $\mathcal{V}_{z'z'}$ be the component of the electric field gradients tensor with the largest absolute value. We introduce the field gradient parameter, $q = \mathcal{V}_{z'z'}/|e|$, and the asymmetry parameter $\eta = (\mathcal{V}_{x'x'} - \mathcal{V}_{y'y'})/\mathcal{V}_{z'z'}$. The latter parameter characterizes the degree of biaxiality of the electric field gradients. In terms of η and q, the quadrupole Hamiltonian (3.32) takes a particularly simple form

$$\mathcal{H}_Q = \frac{e^2 qQ}{4I(2I-1)}\left[3\hat{I}_{z'}^2 - I(I+1) + \eta\left(\hat{I}_{x'}^2 - \hat{I}_{y'}^2\right)\right]. \tag{3.34}$$

Here, $\hat{I}_{x'}, \hat{I}_{y'}$, and $\hat{I}_{z'}$ are the nuclear spin matrices in the (x', y', z') reference frame.

In nanostructures the strain and, correspondingly, the electric field gradients are coordinate dependent; therefore, the set of axes (x', y', z') and the parameters q, η, can depend on the nucleus position within the system. One can readily check that for half-integer I the eigenstates of (3.34) are doubly degenerate. Indeed, the interaction of spin with electric field does not break the time reversal symmetry. In weak and moderate magnetic fields where the nuclear Zeeman splitting, $|\hbar\gamma_I B|$ is significantly smaller compared with the typical quadrupole energy (we will clarify its meaning in the next paragraph)

$$E_Q = \frac{3}{2}\frac{e^2 qQ}{I(2I-1)}, \tag{3.35}$$

the Kramers doublets related to each two-fold degenerate pair of quadrupole-split nuclear levels can be treated independently. Each doublet can be considered as an effective pseudospin-1/2 state, however, with the anisotropic g-factor (D'yakonov *et al.*, 1979; Meier and Zakharchenya, 1984). Note that for the integer-spin nuclei the Hamiltonian (3.34) can fully lift the degeneracy of the spin states.

To analyze the spectrum further, let us consider an illustrative case of uniaxial symmetry of the electric field gradients tensor, where $\eta = 0$. In this situation the nuclear spin states form doublets with a definite absolute value of z'-component of the nuclear spin. The splitting between the doublet with $|I_{z'}| = 3/2$ and the doublet with $|I_{z'}| = 1/2$ is given by E_Q in Eq. (3.35). The next doublet with $|I_{z'}| = 5/2$ is split from $\pm 3/2$ states by $2E_Q$, and so on. This set of non equidistant spin doublets is illustrated in Figure 3.5(a). The estimates show that in typical InGaAs quantum dot structures quadrupole splittings can be very significant: For several percent of strain the quadrupole splitting expressed in frequency units may reach from ~ 100 KHz to ~ 1 MHz (Bulutay, 2012), which exceeds by far the nuclear spin precession rates in the local fields.

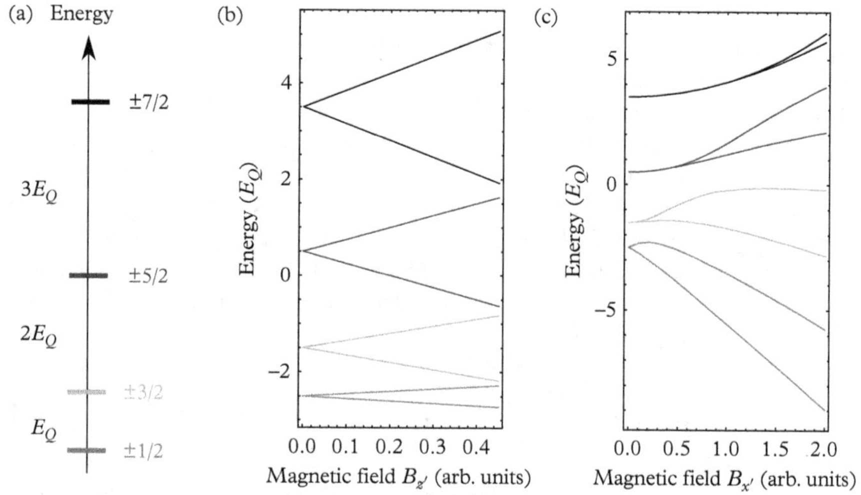

Figure 3.5 *(a) Schematics of nuclear spin states in the presence of uniaxial electric field gradients, $\eta = 0$ in Eq. (3.34), $I = 7/2$. (b) Energies of nuclear spin in the presence of uniaxial quadrupole splittings as a function of magnetic field **B** applied along the main axis, $\boldsymbol{B} \parallel z'$. (c) The same as (b) but for $\boldsymbol{B} \parallel x'$. The shading in panels (b) and (c) corresponds to the shading in panel (a) and denotes the spin components of the states in the limit of $B \to 0$.*

Furthermore, in the presence of external magnetic field the nuclear energy spectrum and, correspondingly, the magnetic resonance frequencies are determined by an interplay of the quadrupole effects and Zeeman splitting induced by the magnetic field. Depending on the magnetic field orientation with respect to the quadrupole axes (x', y', z') the Zeeman splittings of nuclear spin states differ strongly. Panels (b) and (c) of Figure 3.5 show the positions of the nuclear spin sublevels in the presence of the uniaxial quadrupole splitting and external magnetic field, $\boldsymbol{B}$, applied along the quadrupole axis z' [panel (b)] and perpendicular to it [panel (c)] calculated after Eqs. (3.34) and (2.21). Figure 3.5(b) shows linear in-field Zeeman splitting with the effective g-factor being larger, the larger $|I_{z'}|$ is. Indeed, in this configuration, $\boldsymbol{B} \parallel z'$ and the Zeeman and quadrupole effects are additive. By contrast, for $\boldsymbol{B} \perp z'$, as can be seen from Figure 3.5(c), the magnetic field induced splitting of nuclear spin states is, in general, a nonlinear function of the field. The linear Zeeman effect is observed only for relatively weak fields on the $\pm 1/2$ doublet, see the lower curves in Figure 3.5(c). This is also expected from the symmetry arguments because the transversal to z'-axis magnetic field changes, in the first order, the z' spin component by ± 1. Correspondingly, it mixes and splits $|I_{z'}| = 1/2$ doublet, while for the states with larger $|I_{z'}|$, it admixes the neighboring doublets only, which are distant in energy. Hence, the splitting of spin sublevels with $|I_{z'}| > 1/2$ takes place in higher-order perturbation theory resulting in the $\propto B_{x'}^3$ splitting for the $3/2$ doublet, and so on. In terms of pseudospin-$1/2$ states with anisotropic g-factors the studied situation of the uniaxial symmetry corresponds to the extreme anisotropy of g-factors for $|I_{z'}| > 1/2$ pairs: The in-plane components of the effective g-factor tensor vanish, $g_{x'x'} = g_{y'y'} = 0$. In contrast, the longitudinal components of g-factor increases linearly with an increase of $|I_{z'}|$: $g_{z'z'} = 2|I_{z'}|g_n$, where g_n is the nuclear g-factor in absence of quadrupole coupling. The account for the biaxiality of the electric field gradients and, hence, of the quadrupole interaction produces an even richer phenomenology. For weak biaxiality, $|\eta| \ll 1$, the symmetry reduction manifests itself as avoided crossings between Zeeman sublevels stemming from the $|I_{z'}|$ and $|I_{z'}|+1$ manifolds in the $\boldsymbol{B} \parallel z'$ geometry.

Due to quadrupole splittings of nuclear spin sublevels, the NMR spectrum changes. For example, already at $\boldsymbol{B} \to 0$ the states, for example, $I_{z'} = +1/2$ and $I_{z'} = +3/2$ are split and, hence, at zero field the transitions occur at a nonzero frequency. The general approach to calculate the NMR spectrum is as follows. First, the distribution of strain in the quantum dot, $\varepsilon_{\alpha\beta}(\boldsymbol{r})$, should be found by means, for example, of elasticity theory or atomistic modeling. Second, for each nucleus the spin eigenstates and eigenenergies should be found by diagonalizing the Hamiltonian including the external (static) magnetic field and quadrupole interactions,

$$\mathcal{H} = -\frac{\mu_I}{I}\hat{\boldsymbol{I}} \cdot \boldsymbol{B} + \mathcal{H}_Q.$$

The numerical diagonalization of this Hamiltonian is quite straightforward due to the limited number of nuclear spin states, typically, four for Ga or As isotopes and ten for an In isotope. Third, the radio frequency absorption spectrum of a given nucleus can be then calculated by means of Fermi's golden rule, see Eq. (3.19), or by directly solving the density matrix equations if the alternating magnetic field should be taken nonperturbatively. Note that, in the described model the local fields are disregarded since, as we mentioned before typical quadrupole splittings in quantum dots exceed by far the local fields. The effect of local fields could be phenomenologically included in the broadening of the δ-function in Fermi's golden rule, see Eq. (3.13). Finally, the NMR spectrum should be averaged over the isotope positions, which probe different values of the strain components, $\varepsilon_{\alpha\beta}(\boldsymbol{r})$, and over relevant isotopes.

The specific calculations performed in (Bulutay, 2012) for InAs and $In_{0.7}Ga_{0.3}As$ quantum dots grown on a (001) plane demonstrate that in such structures the in-plane strain is compressive $\varepsilon_\perp = (\varepsilon_{xx} + \varepsilon_{yy})/2 < 0$, while $\varepsilon_{zz} > 0$, which corresponds to the tensile strain. The parameters of the strain tensor, $\varepsilon_B = \varepsilon_{zz} - \varepsilon_\perp$ and of the shear strain $\varepsilon_S = |\varepsilon_{xy}| + |\varepsilon_{yz}| + |\varepsilon_{xz}|$ have been introduced. According to (Bulutay, 2012) both strain parameters are of importance in InGaAs quantum dots: in this particular calculations $\varepsilon_B \approx 1\%$ and $\varepsilon_S \approx 0.5\%$. The calculated NMR spectra have a complex form and a multitude of peaks corresponding to various transitions between, generally speaking, all nuclear spin sublevels. The positions of the peaks and their magnetic field dependence are strongly dependent on the strain distribution in the quantum dot. This has opened up the possibility of using NMR spectroscopy to study both the chemical composition and the strain distribution in quantum dots (Chekhovich *et al.*, 2012). In this work optically detected NMR has been used to study InP/GaInP and InGaAs/GaAs quantum dots. In order to improve the NMR sensitivity the authors of (Chekhovich *et al.*, 2012) have suggested *inverse* NMR spectroscopy. In contrast to the standard NMR where relatively narrow-band radio frequency radiation is used to induce the transitions between the nuclear Zeeman sublevels and, ultimately, to equalize their occupancies, in inverse spectroscopy the broadband radio frequency spectrum with a gap at a frequency ν with a width w_{gap} is used. In this method, all transitions but one (resonant with the gap) are saturated and a stronger, compared with the standard NMR method, variation of nuclear polarization occurs. The NMR spectra measured in such a way at a rather strong external magnetic field, $B \approx 5.3$ T, demonstrate strong "central" transitions $-1/2 \leftrightarrow +1/2$ of relevant isotopes and broad bands (stretching for ± 7 MHz with respect to the central transition) related to the "satellite" transitions, corresponding to quadrupole interaction shifted $|I_z| > 1/2$ spin sublevels. The effective inhomogeneous broadening results from the strain distribution over the quantum dot volume. The satellite transitions involving $+1/2 \leftrightarrow +3/2$ $(-1/2 \leftrightarrow -3/2)$ spin states can be resolved for the ^{115}In isotope, which has allowed the authors of (Chekhovich *et al.*, 2012) to evaluate the quadrupole splitting frequency $E_Q/h \approx 1.15$ MHz (1.4 MHz) for InP (InGaAs) quantum dots. Corresponding strain parameter $|\varepsilon_B|$ has been evaluated to be in the order of 2%...3% depending on the material.

Further theoretical and experimental studies of quadrupole splittings of nuclear spin states and their relation with the strain effects in InGaAs/GaAs quantum dot ensembles have been performed in (Bulutay *et al.*, 2014; Kuznetsova *et al.*, 2014; Sokolov *et al.*, 2016). Figure 3.6 demonstrates the results of the analysis of the optically detected NMR spectra performed in (Kuznetsova *et al.*, 2014) for two samples annealed at higher (sample #1) and lower (sample #2) temperatures. Points demonstrate the experimentally obtained NMR peak positions while solid lines correspond to the calculated peaks. The authors of (Kuznetsova *et al.*, 2014) have neglected the in-plane anisotropy of the strain, $\eta = 0$ in Eq. (3.34). The only fitting parameter was the strain parameter ϵ_B defined in this work as $\epsilon_B = 2\varepsilon_{zz} - (\varepsilon_{xx} + \varepsilon_{yy})$. The results obtained show that, despite the large broadening of the resonances in the studied structure, most of them can be reliably identified and their behavior as a function of magnetic field can be described by a simple model. The values of the strain parameter ϵ_B obtained from experiments on the samples under study are very close to those found from data fitting. These results have been corroborated by the high-resolution transmission electron microscopy analysis of individual InGaAs/GaAs quantum dots performed in (Sokolov *et al.*, 2016). In this work the authors extract the strain distribution from the microscopy image under additional assumptions $\varepsilon_{xx} = \varepsilon_{yy}$ and $\varepsilon_{xy} = 0$. In this way three components of the strain tensor ε_{zz}, ε_{xx}, and ε_{zz} have been determined and the NMR spectra have

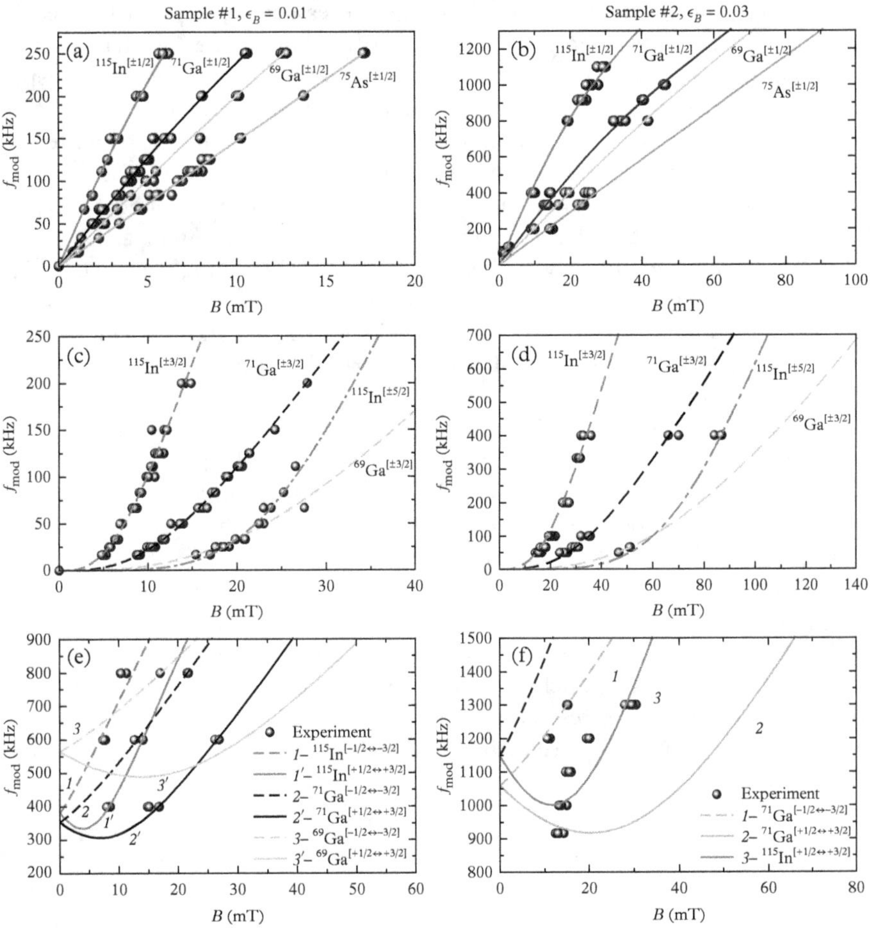

Figure 3.6 *Magnetic-field dependence of the nuclear spin transition frequencies: $-1/2 \leftrightarrow +1/2$ (a, b), $-3/2 \leftrightarrow +3/2$, and $-5/2 \leftrightarrow +5/2$ (for In nuclei only) (c, d), and $\pm 1/2 \leftrightarrow \pm 3/2$ (e, f). Points are the positions of the resonant peaks obtained from the experiments on two samples of InGaAs/GaAs quantum dots annealed at different temperatures. Lines are the calculations. From (Kuznetsova et al., 2014).*

been reconstructed. Ultimately, the important role of shear strain in formation of anticrossings in NMR spectra has been studied.

To summarize, NMR is a very powerful tool to address nuclear spin systems. Also, it can provide important information about strain and composition of quantum dot structures due to quadrupole effects. In semiconductors, ODNMR is often used where the variations of the nuclear spin states occupancies induced by the radio frequency radiation (oscillating magnetic field) are detected via the response at optical frequencies, that is, variation of the photoluminescence polarization, change of the electron Larmor precession frequency (detected in the pump-probe experiment), or directly via monitoring the Overhauser field acting on the electron and hole spins in single quantum dots. The nuclear spin system can be affected by quadrupole interactions due to electric field gradients induced, for example, by elastic strain. It opens up possibilities to excite the transitions between the nuclear spin sublevels by acoustic waves and observe *nuclear acoustic resonance* (Bolef and Menes, 1959; Sundfors, 1969*a*, *b*, 1974). Moreover, as demonstrated in (Kalevich, 1986; Kikkawa and Awschalom, 2000) NMR can be induced optically. In this case the role of alternating magnetic field is played by the Knight field exerted on nuclear spins by a spin-polarized electron. Due to modulation of the intensity or ellipticity of the excitation radiation at the NMR frequency or by using the train of polarized pulses with an appropriate repetition rate, the components of electron spin oscillate with time at the NMR frequency and produce the oscillating effective field acting on the nuclear spins. An interesting proposal has been put forward in (Aronov and Lyanda-Geller, 1989), where the modulation of the electron spin, and, correspondingly, the Knight field can be achieved by modulating the electric current flowing through a system with linear coupling between the electron wavevector and spin, Eq. (2.15). In this case the alternating current induces the alternating electron spin polarization which, in turn, causes NMR.

Hence, the hyperfine interaction between the electron and nuclear spins not only allows one to detect nuclear polarization and, correspondingly, NMR, but also allows to induce the resonant transitions between nuclear spin states. The hyperfine interaction in bulk semiconductors and nanostructures is addressed in Chapter 4.

4

Hyperfine Interaction of Electron and Nuclear Spins

4.1 Contact interaction of electron and nuclear spins

Hyperfine interaction between nuclear and charge carriers spins in semiconductors and semiconductor nanosystems is one of the key interactions in spin physics. We have already briefly reviewed the consequences in Chapter 2, such as production of the effective Overhauser and Knight fields acting on the electron and nuclei spins from their counterparts, Eq. (2.22), and the role of hyperfine coupling in dynamic polarization of nuclear spins, nuclear spin relaxation, and the depolarization of localized charge carriers. Moreover, the hyperfine interaction allows one to address optically nuclear spin resonance, see Section 3.3. This chapter aims to present the microscopic theory of the hyperfine interaction in semiconductors and an overview of the relevant experimental data.

In this section, we present the general model and address in detail the hyperfine interaction of the conduction electron spins with nuclear spins in GaAs-like semiconductors. In order to introduce the hyperfine interaction we recall that the nucleus can be considered as a point-like magnetic moment $\boldsymbol{\mu}_I = \mu_I \boldsymbol{I}/I$, which produces the coordinate-dependent magnetic field $\boldsymbol{B} = \boldsymbol{B}(\boldsymbol{r})$ in the form [cf. Eq. (3.16)]

$$\boldsymbol{B} = \operatorname{rot} \frac{\boldsymbol{\mu}_I \times \boldsymbol{r}}{r^3}, \tag{4.1}$$

where $\boldsymbol{r}$ is reckoned from the nucleus. It is convenient to introduce the vector potential $\boldsymbol{A}$ associated with the field as

$$\boldsymbol{A} = \frac{\boldsymbol{\mu}_I \times \boldsymbol{r}}{r^3} = \operatorname{rot} \frac{\boldsymbol{\mu}_I}{r}. \tag{4.2}$$

The magnetic field, in agreement with Eq. (4.1), is related to its vector potential by the general expression $\boldsymbol{B} = \operatorname{rot} \boldsymbol{A}$. The vector potential and the magnetic field affect the electron orbital and spin degrees of freedom, just like any vector potential and magnetic field irrespective of its source. However, in derivation of the effective

Electron and Nuclear Spin Dynamics in Semiconductor Nanostructures. M. M. Glazov, Oxford University Press (2018). © M. M. Glazov. DOI: 10.1093/oso/9780198807308.001.0001

interaction Hamiltonian care should be taken because both the field and the vector potential diverge at the nucleus position, such that at $r \to 0$ the vector potential $A \propto r^{-2}$ and $B \propto r^{-3}$. At the same time, the electron wavefunctions do not necessary vanish at the nuclear positions. Therefore, the magnetic field and vector potential should be included in the free-electron Schrödinger equation with periodic potential, Eq. (2.2), rather than in the effective $\boldsymbol{k}\cdot\boldsymbol{p}$ Hamiltonian. Treating the interaction of the electron with the magnetic field, Eq. (4.1), and with the vector potential, Eq. (4.2), as a small perturbation we derive an effective Hamiltonian that describes the hyperfine interaction of the electron in bulk material with the nucleus. Further, this effective Hamiltonian can be used to evaluate the hyperfine structure of electronic states in semiconductor nanosystems.

The details of derivation are as follows. In the presence of a spin-polarized nucleus the electron Hamiltonian takes the form [cf. Eq. (2.2)]

$$\mathcal{H} = \frac{1}{2m_0}\left(\boldsymbol{p} - \frac{e}{c}\boldsymbol{A}\right)^2 + \frac{1}{2}g_0\mu_B\boldsymbol{\sigma}\cdot\boldsymbol{B} + V(\boldsymbol{r}) + \mathcal{H}_{so}. \tag{4.3}$$

We recall that m_0 is the free electron mass, $\mu_B = |e|\hbar/(2m_0c)$ is the Bohr magneton, $g_0 = 2$ is the free electron g-factor, and $V(\boldsymbol{r})$ is the periodic potential. The inclusion of the vector potential $\boldsymbol{A}$ into the spin-orbit term, $\mathcal{H}_{so}$, is not needed. Indeed, it gives rise to parametrically smaller terms, where the relativistic effects are accounted for twice: once in the nuclear field and second time in the spin-orbit interaction. One can readily check this making use of the explicit form $\mathcal{H}_{so}$ in Eq. (2.11), see also Appendix B. Hence, the Hamiltonian (4.3) can be recast as the sum $\mathcal{H}_0 + \mathcal{H}_{hf}$, where $\mathcal{H}_0$ given by Eq. (2.2) (or Eq. (4.3) with $\boldsymbol{B} \equiv 0, \boldsymbol{A} \equiv 0$) describes the electron Bloch states in the absence of a nuclear field and the linear in $\boldsymbol{\mu}_I$ perturbation

$$\mathcal{H}_{hf} = -\frac{e}{2m_0c}(\boldsymbol{p}\cdot\boldsymbol{A} + \boldsymbol{A}\cdot\boldsymbol{p}) + \frac{1}{2}g_0\mu_B\boldsymbol{\sigma}\cdot\boldsymbol{B}. \tag{4.4}$$

Quadratic in $\boldsymbol{A}$ term is disregarded, its contribution is minor and plays no role in semiconductors. The contribution to $\mathcal{H}_{hf}$ containing the vector potential can be transformed to a particularly simple form, $2\mu_B(\boldsymbol{\mu}_I\cdot\boldsymbol{L})/r^3$, where $\boldsymbol{L} = [\boldsymbol{r}\times\boldsymbol{p}]/\hbar$ is the electron angular momentum. The term with the Zeeman effect induced by the nuclear magnetic field can be simplified by virtue of equality

$$\begin{aligned}\boldsymbol{\sigma}\cdot \text{rot rot}\frac{\boldsymbol{\mu}}{r} &= \boldsymbol{\sigma}\cdot\left[\nabla\times\left[\nabla\times\frac{\boldsymbol{\mu}}{r}\right]\right] = [(\boldsymbol{\sigma}\cdot\nabla)(\boldsymbol{\mu}\cdot\nabla) - (\boldsymbol{\sigma}\cdot\boldsymbol{\mu})\Delta]\frac{1}{r}\\ &= \left[(\boldsymbol{\sigma}\cdot\nabla)(\boldsymbol{\mu}\cdot\nabla) - \frac{1}{3}(\boldsymbol{\sigma}\cdot\boldsymbol{\mu})\Delta\right]\frac{1}{r} - \frac{2}{3}(\boldsymbol{\sigma}\cdot\boldsymbol{\mu})\Delta\frac{1}{r},\end{aligned}$$

where the singular at $r \to 0$ part, $-(2/3)(\boldsymbol{\sigma}\cdot\boldsymbol{\mu})\Delta r^{-1}$ is separated from the regular one. Finally, making use of definitions $\hat{\boldsymbol{\mu}}_I = \mu_I\hat{\boldsymbol{I}}/I$ and $\hat{\boldsymbol{s}} = \hat{\boldsymbol{\sigma}}/2$ we arrive at

$$\mathcal{H}_{hf} = \frac{2\mu_B\mu_I}{I}\hat{\boldsymbol{I}}\cdot\left[\frac{\hat{\boldsymbol{L}}}{r^3} - \frac{\hat{\boldsymbol{s}}}{r^3} + 3\frac{\boldsymbol{r}(\boldsymbol{r}\cdot\hat{\boldsymbol{s}})}{r^5} + \frac{8\pi}{3}\hat{\boldsymbol{s}}\delta(\boldsymbol{r})\right]. \tag{4.5}$$

Here $\delta(\boldsymbol{r})$ is the Dirac δ-function. We stress again that the operator in Eq. (4.5) acts on the full Bloch function of the electron in crystal, Eq. (3.2), which includes both the periodic Bloch amplitude and the smooth envelope function. In what follows, we transform Eq. (4.5) to a more convenient form to study the hyperfine coupling in semiconductors by separating the factors due to the Bloch amplitude and envelope.

All terms in Eq. (4.5) allow for transparent physical interpretation. The contribution $\propto \hat{\boldsymbol{I}}\cdot\hat{\boldsymbol{L}}$ describes the interaction of nuclear magnetic moment with electric current associated with electron orbital motion. This term can be derived from Biot–Savart law that relates current density and induced magnetic field. The contribution

$$\propto \frac{\hat{\boldsymbol{I}}\cdot\hat{\boldsymbol{s}}}{r^3} - 3\frac{(\hat{\boldsymbol{I}}\cdot\boldsymbol{r})(\hat{\boldsymbol{s}}\cdot\boldsymbol{r})}{r^5},$$

describes the dipole–dipole interaction of electron and nuclear spins. This is the interaction of two magnetic moments separated by the distance $\boldsymbol{r}$. The last term in Eq. (4.5),

$$\propto (\hat{\boldsymbol{I}}\cdot\hat{\boldsymbol{s}})\delta(\boldsymbol{r})$$

is also a result of the magnetic interaction between two spins, it accounts for the possibilities for the spatial distribution of magnetic moments to overlap. This contribution has a contact form and it is proportional to the probability of finding the electron at the nucleus. The contact interaction is not a novel type of coupling, it can be derived by a purely electrodynamic approach by considering the electromagnetic interaction of magnetic moments. To illustrate it we assume that the electron magnetic moment is distributed in space with the density $-2\mu_B\boldsymbol{s}|\Psi(\boldsymbol{r})|^2$, where $\Psi(\boldsymbol{r})$ is the orbital wavefunction of the electron that does not significantly vary on the scale of the nucleus. On the other hand, formally, nuclear magnetic moment $\boldsymbol{\mu}_I$ can be associated with fictitious current density $\boldsymbol{j}_I$ as $\boldsymbol{\mu}_I = \int d\boldsymbol{r}[\boldsymbol{r}\times\boldsymbol{j}_I]$. Making use of the Maxwell equation $\mathrm{rot}\boldsymbol{B} = (4\pi/c)\boldsymbol{j}$ one has $\int_v d\boldsymbol{r}\boldsymbol{B} = -(8\pi/3)\boldsymbol{\mu}_I$, where the integration takes place over a small volume v surrounding a given nucleus. Ultimately, we can evaluate the interaction energy as

$$2\mu_B\int_v d\boldsymbol{r}|\Psi(\boldsymbol{r})|^2(\boldsymbol{s}\cdot\boldsymbol{B}) \approx 2\mu_B|\Psi(0)|^2\boldsymbol{s}\cdot\int_v d\boldsymbol{r}\boldsymbol{B} = 2\mu_B|\Psi(0)|^2\frac{8\pi}{3}(\boldsymbol{s}\cdot\boldsymbol{\mu}_i), \tag{4.6}$$

in agreement with the last term of Eq. (4.5).

The hyperfine interaction between the electron and nucleus was first derived by Enrico Fermi from relativistic arguments (Fermi, 1930). The contact term is known as the *Fermi contact interaction*. The derivation along the similar lines to these can be found in many references, see for example (Abragam, 2002). Derivation of the dipole–dipole and contact terms in the framework of classical electrodynamics can be found in (Jackson, 1962).

Let us now analyze the particular form of the hyperfine interaction for conduction band electron in a bulk zinc-blende lattice semiconductor with a band structure similar to that of GaAs. At the Γ-point the conduction band state is doubly degenerate and transforms according to the Γ_6 spinor representation of the T_d point symmetry group, see Appendix A. The orbital part of the Bloch amplitude $\mathcal{S}(\boldsymbol{r})$ is invariant under all transformations of T_d point group and does not vanish at any host lattice nucleus. Hence, for the conduction band electron, the hyperfine interaction is dominated by the contact part. Exactly at the Γ-point all remaining terms in Eq. (4.5) vanish for symmetry reasons. Away from the Γ-point the dipole–dipole and orbital terms may contribute, but their effect is minor because it is controlled by the small parameter, ka_0, where k is the magnitude of the electron wavevector and a_0 is the lattice constant. Hence, to derive the hyperfine interaction for the conduction band electron with the host lattice nuclei it is sufficient to represent the orbital part of the wavefunction in the form

$$\Psi(\boldsymbol{r}) = \varphi_c(\boldsymbol{r})\mathcal{S}(\boldsymbol{r}),$$

where $\varphi_c(\boldsymbol{r})$ is the conduction band envelope function, and retain only the contact part of the hyperfine interaction. As a result, the effective Fermi contact interaction Hamiltonian takes the form [cf. Eq. (2.22)]

$$\mathcal{H}_{hf} = Av_0(\hat{\boldsymbol{I}} \cdot \hat{\boldsymbol{s}})|\varphi_c(\boldsymbol{R})|^2, \tag{4.7}$$

where $\boldsymbol{R}$ is the nucleus position and the parameter A is given by

$$A = \frac{16\pi \mu_B \mu_I}{3Iv_0}|\mathcal{S}(\boldsymbol{R})|^2. \tag{4.8}$$

Here, v_0 is the primitive cell volume ($v_0 = a_0^3/4$ for a zinc-blende lattice structure), the Bloch amplitude is assumed to be normalized within the primitive cell as $\int_{v_0} |\mathcal{S}(\boldsymbol{r})|^2 = v_0$ (Meier and Zakharchenya, 1984). Equation (4.7) can be recast in the alternative form

$$\mathcal{H}_{hf} = Av_0(\hat{\boldsymbol{I}} \cdot \hat{\boldsymbol{s}})\delta(\boldsymbol{r}), \tag{4.9}$$

where $\delta(\boldsymbol{r})$ acts on the smooth envelope only. Both forms, Eq. (4.7) and (4.9), can be readily obtained from the symmetry arguments: The only combination of the electron spin, $\hat{\boldsymbol{s}}$, and the nuclear spin, $\hat{\boldsymbol{I}}$, operators being invariant in the T_d point group relevant for zinc-blende semiconductors is their scalar product.

The constant A, see Eq. (4.8), has units of energy. The product AI gives the Overhauser splitting of the electron spin states if all nuclei within the localization volume were fully polarized. If the Bloch amplitude were constant over the primitive cell volume then $|\mathcal{S}(\boldsymbol{R})|^2 = 1$ and typical values of the constant A would be in the order of $10^{-3} \ldots 10^{-2}$ μeV. However, in semiconductor crystals the Bloch amplitudes are strongly inhomogeneous and have sharp extrema around nuclei, which results in $10^3 \ldots 10^4$ enhancement of the hyperfine coupling constants yielding the values of A about $10 \ldots 100$ μeV.

High values of hyperfine coupling constants were first observed in nuclear magnetic resonance experiments performed on InSb (Shulman *et al.*, 1955; Gueron, 1964). In particular, (Gueron, 1964) reported the results of the direct measurements of the Overhauser shifts in indium antimonide: The nuclei were first polarized in a strong magnetic field. Then the magnetic field was reduced fast enough to preserve nuclear polarization [the time of experiment was smaller than the nuclear spin relaxation time $T_1 \sim 1$ hour, Eq. (2.52)]. Second, electron spin splitting was observed by means of the electron spin resonance technique and the isotopes of In and Sb were depolarized by the radio frequency field. In this way, one can detect the variation of the Overhauser field induced by each of the isotopes. Since the initial nuclear polarization was known from the experimental conditions it was possible by making use of Eqs. (4.7) and (4.8) to estimate the $|\mathcal{S}(\boldsymbol{R})|^2$. A similar technique has been used to evaluate hyperfine interaction in InP (Gotschy *et al.*, 1989). The authors of (Paget *et al.*, 1977) have rescaled the hyperfine coupling constants measured in InSb to GaAs assuming that the hyperfine interaction in crystals scales like the hyperfine interaction in isolated atoms. In CdS and CdTe the electron density at the Cd sites has been evaluated from the Knight field measurements: In the experiments performed in (Adams *et al.*, 1971; Look and Moore, 1972), see also (Nakamura *et al.*, 1979), the nuclear spin relaxation rate due to hyperfine interaction with free electrons in semiconductors, that is, due to the Knight field fluctuations, was measured. Very recently, in (Chekhovich *et al.*, 2017) the authors were able to evaluate accurately the nuclear spin polarization degree in GaAs/AlGaAs quantum dots under conditions of optically detected nuclear magnetic resonance (ODNMR) and measure directly the hyperfine coupling constants of electron spin with Ga and As isotopes. These data are in good agreement with the estimate made in Ref. (Paget *et al.*, 1977). The estimates of the Bloch amplitude at the nuclei and of the hyperfine interaction constants in several semiconductors are summarized in Table 4.1.

Interestingly, the order of magnitude estimates of hyperfine coupling constants can be obtained from atomic values (Morton and Preston, 1978; Koh and Miller, 1985). Depending on the model and level of complexity of the calculation, the atomic hyperfine interaction constant for a given isotope such as In or Ga is higher than that in semiconductors by roughly 20–50%. Quite crude estimates, but still typically valid within the order of magnitude accuracy can be obtained by using simplest possible approximation, namely, using the hydrogenic wavefunctions

Table 4.1 *Bloch amplitudes at nucleus positions and hyperfine interaction constants for conduction band electrons in zinc-blende lattice semiconductors. From (Gueron, 1964; Paget et al., 1977; Nakamura et al., 1979; Gotschy et al., 1989; Testelin et al., 2008; Syperek et al., 2011; Chekhovich et al., 2017). The isotope-average values of* $|S(\boldsymbol{R})|^2$ *are presented for In, Sb, Cd, and Te isotopes.*

Compound	a_0 Å	Isotope	$\|S(\boldsymbol{R})\|^2$	A μeV
GaAs[a]	5.65	^{69}Ga	2.61×10^3	38
		^{71}Ga	2.61×10^3	48
		^{75}As	4.42×10^3	46
GaAs[b]	5.65	^{69}Ga	2.96×10^3	43.1
		^{71}Ga	2.96×10^3	54.8
		^{75}As	4.17×10^3	43.5
InSb	6.48	^{113,115}In	6.35×10^3	56
		^{121}Sb	10.8×10^3	105
		^{123}Sb	10.8×10^3	57
InP	5.87	^{113,115}In	3.9×10^3	45
		^{31}P	1.6×10^3	36
CdTe	6.48	^{111,113}Cd	3.6×10^3	−31
		^{123,125}Te	3.6×10^3	−45
CdS	6.08	^{111,113}Cd	3.8×10^3	−33

[a] From (Paget *et al.*, 1977)
[b] From (Chekhovich *et al.*, 2017)

with effectively screened Coulomb interaction (Clementi and Raimondi, 1963). The results of first-principles calculations of hyperfine coupling constants in some semiconductors are given in (Van de Walle and Blöchl, 1993). It is worth stressing that further experiments and detailed first-principles calculations of Bloch functions in semiconductors are needed, in particular, to improve the accuracy of the values of the hyperfine coupling constants.

4.2 Dipole–dipole interaction of hole and nuclear spins

Let us now turn to the analysis of the hyperfine interaction in the valence band. We recall that in GaAs and similar semiconductors the orbital Bloch amplitudes of the valence band $\mathcal{X}(\boldsymbol{r})$, $\mathcal{Y}(\boldsymbol{r})$, and $\mathcal{Z}(\boldsymbol{r})$ transform according to the irreducible representation F_2, that is, as the coordinates x, y, and z, respectively, see Figure 2.2, Section 2.3, and Appendix A for details. These functions vanish at nuclei and, therefore, the contact part of the hyperfine interaction described by the δ-function term in Eq. (4.5) vanishes. The resulting interaction of the valence band hole spins and the nuclear spins is thus described by the remaining terms in the Hamiltonian (4.5):

$$\mathcal{H}_{hf} = \frac{2\mu_B\mu_I}{I}\hat{\boldsymbol{I}}\cdot\left[\frac{\hat{\boldsymbol{L}}}{r^3} - \frac{\hat{\boldsymbol{s}}}{r^3} + 3\frac{\boldsymbol{r}(\boldsymbol{r}\cdot\hat{\boldsymbol{s}})}{r^5}\right] = \frac{2\mu_B\mu_I}{I}(\hat{V}_1 - \hat{V}_2). \tag{4.10}$$

Here we introduced two operators

$$\hat{V}_1 = \frac{\hat{\boldsymbol{I}}\cdot\hat{\boldsymbol{L}}}{r^3}, \quad \hat{V}_2 = P_{\alpha\beta}(\boldsymbol{r})\hat{s}_\alpha\hat{I}_\beta, \tag{4.11}$$

with $P_{\alpha\beta}(\boldsymbol{r}) = \delta_{\alpha\beta}/r^3 - r_\alpha r_\beta/r^5$, $\delta_{\alpha\beta}$ is the Kronecker symbol. In Eq. (4.11) the summation over the repeated Cartesian indices $\alpha, \beta = x, y, z$ is assumed. We focus ourselves on the topmost valence band that transforms according to the Γ_8 irreducible representation of the T_d point symmetry group, that is, its Bloch amplitudes (3.2) behave, at all point-group transformations, as the angular momentum 3/2 wavefunctions. In the following, we outline the derivation from Eq. (4.10) an effective Hamiltonian of the hyperfine interaction in the Γ_8 valence band (Obata, 1963; Gryncharova and Perel', 1977).

We introduce the following notations: Let $\boldsymbol{\mathcal{R}}(\boldsymbol{r}) = [\mathcal{X}(\boldsymbol{r}), \mathcal{Y}(\boldsymbol{r}), \mathcal{Z}(\boldsymbol{r})]$ be the vector of the valence band orbital functions. We denote its Cartesian components as $\mathcal{R}_\alpha(\boldsymbol{r})$; for example, $\mathcal{R}_x(\boldsymbol{r}) = \mathcal{X}(\boldsymbol{r})$. In accordance with Eqs. (3.2), (2.12) the valence band Bloch functions can be written as linear combinations of the basic functions $\exp(\mathrm{i}\boldsymbol{k}\boldsymbol{r})\mathcal{R}_\alpha(\boldsymbol{r})|\uparrow\rangle$, $\exp(\mathrm{i}\boldsymbol{k}\boldsymbol{r})\mathcal{R}_\alpha(\boldsymbol{r})|\downarrow\rangle$, where $\boldsymbol{k}$ is the wavevector. The hyperfine interaction is thus described by the set of matrix elements of Eq. (4.10) calculated with the basic functions with the wavevectors $\boldsymbol{k}$ and $\boldsymbol{k}'$. Let $\boldsymbol{q} = \boldsymbol{k} - \boldsymbol{k}'$ and

$$Q_{\alpha\beta\gamma\delta} = \int d\boldsymbol{r} e^{\mathrm{i}\boldsymbol{q}\boldsymbol{r}}\mathcal{R}_\alpha^*(\boldsymbol{r})\mathcal{R}_\beta(\boldsymbol{r})P_{\gamma\delta}(\boldsymbol{r}), \tag{4.12a}$$

$$T_{\alpha\beta\gamma} = -\mathrm{i}\int d\boldsymbol{r} e^{\mathrm{i}\boldsymbol{q}\boldsymbol{r}}\mathcal{R}_\alpha^*(\boldsymbol{r})\frac{[\boldsymbol{r}\times\nabla]_\gamma}{r^3}\mathcal{R}_\beta(\boldsymbol{r}). \tag{4.12b}$$

The integrals $T_{\alpha\beta\gamma}$ and $Q_{\alpha\beta\gamma\delta}$ involving orbital parts of the Bloch functions determine the orbital part of the matrix elements of the operators $\hat{V}_1$ and $\hat{V}_2$, respectively. The analysis presented in (Gryncharova and Perel', 1977) demonstrates that each of the integrals can be recast as a sum of two contributions: The *short-range* one is contributed by the part of the Bloch function in the vicinity of the nucleus and the *long-range* one is related to the interaction at the large distances, exceeding by far the lattice constant. Mathematically, one can introduce a sphere of a radius R_0 that satisfies the following inequalities

$$a_0 \ll R_0 \ll 1/q. \tag{4.13}$$

The integrals in (4.12) can be separated into the integral over $r < R_0$, where one can replace $\exp(\mathrm{i}\boldsymbol{qr})$ by unity, and the integral over $r > R_0$, where one can first integrate the product of the Bloch amplitudes over the primitive cell and then sum over the cells. Correspondingly, the expression for $\mathcal{Q}_{\alpha\beta\gamma\delta}$ takes the form

$$\mathcal{Q}_{\alpha\beta\gamma\delta} = \mathcal{A}_{\alpha\beta\gamma\delta} + \mathcal{B}_{\alpha\beta\gamma\delta}, \tag{4.14}$$

where the short-range part

$$\mathcal{A}_{\alpha\beta\gamma\delta} = \int_{r<R_0} d\boldsymbol{r}\mathcal{R}_\alpha^*(\boldsymbol{r})\mathcal{R}_\beta(\boldsymbol{r})P_{\gamma\delta}(\boldsymbol{r}), \tag{4.15}$$

and the long-range part

$$\mathcal{B}_{\alpha\beta\gamma\delta} = \frac{\delta_{\alpha\beta}}{v_0}\int d\boldsymbol{r}e^{\mathrm{i}\boldsymbol{qr}}P_{\gamma\delta}(\boldsymbol{r}) \propto \frac{\delta_{\alpha\beta}}{v_0}\frac{q^2\delta_{\gamma\delta} - 3q_\gamma q_\delta}{q^2}, \tag{4.16}$$

are introduced. Similar to the case of the conduction band, the valence band wavefunctions are not uniform in the primitive cell but rather concentrated in the vicinity of nucleus, although, of course, $\mathcal{R}_\alpha(0) = 0$. Hence, the long-range part $\mathcal{B}_{\alpha\beta\gamma\delta}$ is small as compared with the short-range part $\mathcal{A}_{\alpha\beta\gamma\delta}$. Formally, this is because the condition

$$\int_{v_0} d\boldsymbol{r}\frac{|\mathcal{R}_\alpha(\boldsymbol{r})|^2}{r^3} \equiv \left\langle\frac{1}{r^3}\right\rangle \gg \frac{1}{v_0},$$

usually holds (Gryncharova and Perel', 1977; Fischer *et al.*, 2008; Testelin *et al.*, 2009). Similar decomposition into the long-range and short-range parts takes place for the integrals $\mathcal{T}_{\alpha\beta\gamma} = \mathrm{i}\mathcal{D}_{\alpha\beta\gamma} + \mathrm{i}\mathcal{E}_{\alpha\beta\gamma}$. Here we present only the dominant, short-range, contribution,

$$\mathcal{D}_{\alpha\beta\gamma} = -\int_{r<R_0} d\boldsymbol{r}\mathcal{R}_\alpha^*(\boldsymbol{r})\frac{[\boldsymbol{r}\times\nabla]_\gamma}{r^3}\mathcal{R}_\beta(\boldsymbol{r}). \tag{4.17}$$

In Eqs. (4.16) and (4.17) the integration is carried out over a sphere of the radius R_0 that satisfies the condition (4.13). However, since the short-range contribution is dominated by the Bloch functions in the vicinity of the nucleus one can restrict integration in Eqs. (4.16) and (4.17) over a primitive cell. Note that in spherical approximation where the cubic symmetry of GaAs semiconductor is ignored, the rank-four tensor $A_{\alpha\beta\gamma\delta}$ reduces to a single independent parameter A_0 as $A_{\alpha\beta\gamma\delta} = A_0\delta_{\alpha\beta}\delta_{\gamma\delta} - (3A_0/2)(\delta_{\alpha\gamma}\delta_{\beta\delta} + \delta_{\alpha\delta}\delta_{\beta\gamma})$. Under the same approximation, $D_{\alpha\beta\gamma} = \epsilon_{\alpha\beta\gamma}D_0$, where D_0 is a constant and $\epsilon_{\alpha\beta\gamma}$ is the unit antisymmetric tensor also known as the Levi–Civita symbol.

Phenomenologically, in cubic semiconductors the effective Hamiltonian of the short-range dipole–dipole interaction of the Γ_8 valence band electron and nuclear spins takes the form

$$\mathcal{H}_{hf,v} = \frac{Av_0}{2}\delta(\boldsymbol{r})\left(M_1\hat{\boldsymbol{I}}\cdot\hat{\boldsymbol{J}} + M_2\hat{\boldsymbol{I}}\cdot\hat{\boldsymbol{J}}^3\right). \tag{4.18}$$

Here, A is the conduction band hyperfine interaction constant given in Eq. (4.8), introduced here to provide the correct dimensionality, $\delta(\boldsymbol{r})$ is the δ-function acting on the smooth envelope, $\hat{\boldsymbol{J}} = (\hat{J}_x, \hat{J}_y, \hat{J}_z)$ are the angular momentum 3/2 basic matrices that act in the space of $|\Gamma_8, J_z\rangle$ ($J_z = -3/2, \ldots 3/2$) Bloch amplitudes, and the notation $\hat{\boldsymbol{J}}^3 = (\hat{J}_x^3, \hat{J}_y^3, \hat{J}_z^3)$ is introduced. Furthermore, in Eq. (4.18) we use cubic frame of axis with $x \parallel [100]$, $y \parallel [010]$ and $z \parallel [001]$, and introduce two dimensionless parameters M_1 and M_2 that describe relative strengths of the isotropic, $\propto (\hat{\boldsymbol{I}}\cdot\hat{\boldsymbol{J}})$, and anisotropic, $\propto (\hat{\boldsymbol{I}}\cdot\hat{\boldsymbol{J}}^3)$, contributions in Eq. (4.18). Such a form of dipole–dipole interaction is consistent with the symmetry requirements because in the T_d point group both $\hat{\boldsymbol{J}}$ and cubic combinations $\hat{\boldsymbol{J}}^3$ form the basis of the same irreducible representation, F_1. The form of Eq. (4.18) is thus similar to the expression describing the Zeeman effect in the valence band (Bir and Pikus, 1974).

The products AM_1 and AM_2 can be expressed via linearly independent components of the tensors $A_{\alpha\beta\gamma\delta}$ and $D_{\alpha\beta\gamma}$ (Gryncharova and Perel', 1977). Usually, just like for the conduction band, for estimations of the valence band hyperfine constants, atomic orbitals are used instead of Bloch amplitudes. A rough estimate for M_1 follows from the values of $\langle 1/r^3\rangle$ for p-shell electrons in corresponding atoms, see for example (Morton and Preston, 1978; Koh and Miller, 1985). Such estimations performed in (Gryncharova and Perel', 1977; Fischer *et al.*, 2008; Testelin *et al.*, 2009) using p-atomic orbitals give $M_2 \equiv 0$ and $M_1 \sim 0.1$, that is, the hyperfine interaction in the valence band is about an order of magnitude weaker compared with the conduction band. In the model where only p-atomic orbitals are accounted for, the dipole–dipole interaction of the valence band electron and the nuclear spins in cubic semiconductors is isotropic: Symmetry-wise, such a situation is equivalent to the case of an isolated atom.

The anisotropic contribution may be derived taking into account that the valence band of GaAs-like semiconductors is formed not only by p-type atomic orbitals but also by d-type ones as demonstrated in (Chekhovich *et al.*, 2013). In the T_d point symmetry group, both the set of cubic axes x, y, z and the set of quadratic combinations yz, xz, xy form equivalent bases of the irreducible representation F_2, see Appendix A. In order to calculate hyperfine interaction the Bloch amplitudes

$\mathcal{X}(\mathbf{r})$, $\mathcal{Y}(\mathbf{r})$, and $\mathcal{Z}(\mathbf{r})$ in the vicinity of nucleus can be approximated as linear combinations of hydrogenic wavefunctions with defined orbital momentum l and satisfying the symmetry of the F_2 representation as

$$\mathcal{X}(\mathbf{r}) = \sum_l \alpha_l X_l(\theta,\phi) R_l(r), \tag{4.19a}$$

$$\mathcal{Y}(\mathbf{r}) = \sum_l \alpha_l Y_l(\theta,\phi) R_l(r), \tag{4.19b}$$

$$\mathcal{Z}(\mathbf{r}) = \sum_l \alpha_l Z_l(\theta,\phi) R_l(r), \tag{4.19c}$$

where $l = 1$ or 2 for the p- and d-states, respectively, α_l represents the admixture coefficients,

$$\sum_l |\alpha_l|^2 = 1,$$

and in Eqs. (4.19) each term has been factorized into the real radial part $R_l(r)$ and the angular part $X_l(\theta,\phi)$, $Y_l(\theta,\phi)$, $Z_l(\theta,\phi)$ corresponding to orbital momentum l. In the canonical basis, the following real linear combinations of spherical harmonics Y_{lm} transform according to the F_2 representation and correspond to the p orbitals

$$X_p(\theta,\phi) = -\frac{1}{\sqrt{2}}[Y_{1,+1}(\theta,\phi) - Y_{1,-1}(\theta,\phi)], \tag{4.20a}$$

$$Y_p(\theta,\phi) = \frac{\mathrm{i}}{\sqrt{2}}[Y_{1,+1}(\theta,\phi) + Y_{1,-1}(\theta,\phi)], \tag{4.20b}$$

$$Z_p(\theta,\phi) = Y_{1,0}(\theta,\phi), \tag{4.20c}$$

and d orbitals

$$X_d(\theta,\phi) = \frac{\mathrm{i}}{\sqrt{2}}[Y_{2,+1}(\theta,\phi) + Y_{2,-1}(\theta,\phi)], \tag{4.21a}$$

$$Y_d(\theta,\phi) = -\frac{1}{\sqrt{2}}[Y_{2,+1}(\theta,\phi) - Y_{2,-1}(\theta,\phi)], \tag{4.21b}$$

$$Z_d(\theta,\phi) = -\frac{\mathrm{i}}{\sqrt{2}}[Y_{2,+2}(\theta,\phi) - Y_{2,-2}(\theta,\phi)], \tag{4.21c}$$

respectively. The calculations performed in Ref. (Chekhovich *et al.*, 2013) demonstrate that the parameters M_1 and M_2 in Eq. (4.18) take the form

$$M_1 = \frac{8}{5}|\alpha_p|^2 M_p - \frac{39}{7}|\alpha_d|^2 M_d, \quad M_2 = \frac{12}{7}|\alpha_d|^2 M_d, \tag{4.22}$$

with

$$M_{p,d} = \frac{4\pi}{|\mathcal{S}(\boldsymbol{R})|^2} \int_0^\infty \frac{R_{1,2}^2(r)}{r} dr. \tag{4.23}$$

We recall that $\boldsymbol{R}$ is the nucleus position and $\mathcal{S}(\boldsymbol{R})$ is the conduction band Bloch amplitude at the nucleus. Equations (4.22) and (4.23) clearly show that the account of both p- and d-orbitals gives rise to the cubic anisotropy term in the hyperfine interaction of the nuclear spins and the valence band electron spins. The account for the d-atomic orbitals also renormalizes the isotropic hyperfine interaction constant.

It is important to note that the same form of the hyperfine Hamiltonian, Eq. (4.18), is valid both in the electron and the hole representation. This is because at the transformation from the electron to the hole representation one has to (i) perform the time reversal operation $\hat{\mathcal{K}}$ Eq. (2.56) to the electron Bloch functions, and (ii) invert the energies of the states (Bir and Pikus, 1974; Ivchenko, 2005). As a result, both energy [in the left-hand side of Eq. (4.18)] and the electron spin operator [in the right-hand side of Eq. (4.18)] change signs. The proportionality constants, therefore, should remain unchanged. In what follows, we use the Hamiltonian (4.18) to describe the hyperfine interaction of the hole and nuclear spins in nanosystems.

4.3 Hyperfine interaction in nanosystems

Size quantization and strain effects in quantum wells and quantum dot structures lift the degeneracy of the heavy- and light-hole states of the four-fold degenerate valence band at the Γ point of bulk semiconductors. Depending on the growth direction and the shape of heteropotential, various point symmetries of nanosystems can be realized. In a popular case of quantum dots grown on a $z \parallel (001)$ plane the point symmetry is D_{2d} (if the heteropotential has a $z \to -z$ mirror reflection plane) or C_{2v}, the latter symmetry is typical for most of quantum dot structures (Bimberg *et al.*, 1999; Bester and Zunger, 2005; Ivchenko, 2005). The ground heavy-hole subband and the ground conduction band electron subband transform in the same way, namely, according to the Γ_6 spinor representation in the D_{2d} point symmetry group or according to the Γ_5 spinor representation in the C_{2v} point symmetry group, see Appendix A. It follows from the symmetry arguments that the hyperfine interaction for the conduction band electron with a given nucleus in a system described by the C_{2v} point symmetry can be presented in the form:

$$\mathcal{H}_{hf,c} = \frac{|\varphi_c(\boldsymbol{R})|^2 v_0}{2} (A_\parallel \hat{I}_z \hat{\sigma}_z + A_{x_1} \hat{I}_{x_1} \hat{\sigma}_{x_1} + A_{y_1} \hat{I}_{y_1} \hat{\sigma}_{y_1}). \tag{4.24}$$

Here, $\varphi_c(\boldsymbol{R})$ is the conduction band envelope function at the position of the nucleus; $\sigma_{x_1},\dots$ are the electron Pauli matrices; $A_\parallel, A_{x_1}$, and A_{y_1} are the constants; and we use the in-plane axes $x_1 \parallel [1\bar{1}0]$ and $y_1 \parallel [110]$, which are the main axes to the C_{2v} point group. Note that in the D_{2d} point symmetry group the constants A_{x_1} and A_{y_1} are equal due to the presence of the mirror-rotational axis $S_4 \parallel z$. Equation (4.24) extends Eq. (4.7) to account for the microscopic symmetry of the quantum well or quantum dot structure.

However, microscopically, the difference of constants $A_\parallel$, A_{x_1}, and A_{y_1} which gives rise to the hyperfine coupling anisotropy appears with account for the $\boldsymbol{k}\cdot\boldsymbol{p}$ mixing of the conduction and valence band states. The resulting difference of the parameters $A_\parallel$, A_{x_1}, and A_{y_1} is negligible for typical quantum wells and quantum dots. Therefore, the isotropic form of the hyperfine interaction, $A_\parallel = A_{x_1} = A_{y_1} \equiv A$, Eq. (4.8), is generally used in order to describe the hyperfine coupling in the conduction band for low-dimensional systems. Note that the situation is more involved in multi-valley semiconductor systems, see (Shchepetilnikov *et al.*, 2016) for details.

Here we need to make a general remark about the form of phenomenological Hamiltonian (4.24) and similar Hamiltonians for the valence band holes. In fact, the symmetry of a crystalline environment around a given nucleus is, generally, lower than the symmetry of the quantum well or quantum dot. This is because the position of the nucleus and the origin of the point group transformations may not coincide. Hereafter, we disregard the effects related to such a symmetry reduction, although these effects deserve further study.

The situation is qualitatively different for the heavy holes in nanosystems. The pair of $|\Gamma_8, \pm 3/2\rangle$ Bloch amplitudes of the heavy-hole states can be mapped onto the pseudospin 1/2 states. It is convenient to associate the $\mathcal{J}_z = +3/2$ heavy hole state with the pseudospin $+1/2$ and the $-3/2$ heavy hole state with the pseudospin $-1/2$, and introduce the set of hole Pauli matrices $\sigma_{x_1}^{(h)}, \sigma_{y_1}^{(h)}, \sigma_z^{(h)}$. The general expression for the hyperfine interaction for heavy hole has the same form as Eq. (4.24)

$$\mathcal{H}_{hf,hh} = \frac{|\varphi_{hh}(\boldsymbol{R})|^2 v_0}{2}(C_\parallel \hat{I}_z \hat{\sigma}_z^{(h)} + C_{x_1}\hat{I}_{x_1}\hat{\sigma}_{x_1}^{(h)} + C_{y_1}\hat{I}_{y_1}\hat{\sigma}_{y_1}^{(h)}). \tag{4.25}$$

Here, $\varphi_{hh}(\boldsymbol{R})$ is the heavy hole envelope function at the nucleus position and $C_\parallel$, C_{x_1}, and C_{y_1} are three independent parameters. In the axial approximation, where the crystalline symmetry is disregarded and only the z-axis is considered to be non-equivalent to the in-plane axes, $C_{x_1} = C_{x_2} = 0$. The hyperfine interaction in this model has the Ising form

$$\mathcal{H}_{hf,hh} = \frac{|\varphi_{hh}(\boldsymbol{R})|^2 v_0}{2} C_\parallel \hat{I}_z \hat{\sigma}_z^{(h)}.$$

The expression for $C_\parallel$ can be derived from Eq. (4.18) with $M_2 = 0$ with the result

$$C_\parallel = \frac{3AM_1}{2}. \tag{4.26}$$

The analysis performed in (Fischer *et al.*, 2008) demonstrates that the account of the $\boldsymbol{k}\cdot\boldsymbol{p}$ admixture of the conduction band states to the valence band states provides minor, $\sim 1\%$ corrections to the constant $C_\parallel$ in Eq. (4.26). The cubic anisotropy term $\propto M_2$ in Eq. (4.18) as well as the heavy-light hole mixing give rise to the parameters C_{x_1} and C_{x_2} responsible for the coupling of the in-plane hole and nuclear spin components. Allowance for both M_1 and M_2 in Eq. (4.18) corresponds to the D_{2d} point symmetry of the nanosystem and yields

$$C_\parallel = \frac{3AM_1}{2} + \frac{27AM_2}{8}, \quad C_{x_1} = -C_{y_1} = \frac{3AM_2}{8}. \tag{4.27}$$

Note that the negative relative sign of C_{x_1} and C_{y_1} results from the particular choice of the pseudospin representation, in which case the $+3/2$ heavy-hole state (pseudospin $+1/2$) transforms like $-1/2$ conduction band electron state and the $-3/2$ heavy-hole (pseudospin $-1/2$) transforms like a $+1/2$ conduction band. This is similar to the description of Zeeman and spin-orbit effects for heavy holes in quantum wells (Kusrayev *et al.*, 1999; Marie *et al.*, 1999; Durnev *et al.*, 2014). The account of the hole mixing induced, for example, by the elastic strain in the quantum dots or by the anisotropy of chemical bonds at the interfaces makes C_{x_1} and C_{y_1} independent. In what follows, for simplicity we disregard the difference of the light- and heavy-hole envelope functions and introducing the common envelope $\varphi_v(\boldsymbol{R})$ for both heavy- and light-holes. As a result we can recast the hole wavefunctions in the form

$$|hh,+3/2\rangle = \frac{\varphi_v(\boldsymbol{r})}{\sqrt{1+\beta^2}}\left(|\Gamma_8,3/2\rangle + \beta|\Gamma_8,-1/2\rangle\right), \tag{4.28a}$$

$$|hh,-3/2\rangle = \frac{\varphi_v(\boldsymbol{r})}{\sqrt{1+\beta^2}}\left(|\Gamma_8,-3/2\rangle + \beta|\Gamma_8,+1/2\rangle\right), \tag{4.28b}$$

where β is the admixture parameter, it is assumed to be real for simplicity, see (Ivchenko *et al.*, 1996; Testelin *et al.*, 2009). It follows from Eq. (4.18) in the first order in β that

$$C_{x_1} = C_{y_1} = \frac{\beta}{\sqrt{3}} C_\parallel. \tag{4.29}$$

Typically, $|C_{x_1}|, |C_{y_1}| \sim 0.1 C_\parallel$ (Testelin *et al.*, 2009; Chekhovich *et al.*, 2013)

It follows from the symmetry analysis that the hyperfine interaction for the light-hole doublet $\Gamma_8, \pm 1/2$, is described by an expression similar to Eq. (4.25). Interestingly, for the light holes the hyperfine interaction in quantum dots is already strongly anisotropic in the axial approximation. Making use of the Hamiltonian (4.18) with $M_2 = 0$ one has for light holes $C_{x_1} = C_{y_1} = 2C_{\parallel}$ (Testelin *et al.*, 2009). The analysis of the cubic anisotropy effects in the dipole–dipole interaction of the light hole and nuclei has been carried out in (Chekhovich *et al.*, 2013).

The situation becomes particularly interesting in quantum dots grown on a (111) surface, with the most prominent example being GaAs/AlGaAs quantum dots grown by the droplet epitaxy (Mano *et al.*, 2010; Sallen *et al.*, 2011). Such quantum dots possess the trigonal, C_{3v}, point symmetry, which does not distinguish the states whose angular momentum z-components differ by multiples of three, see Appendix A. The same point symmetry is realized in quantum dots grown by organometallic chemical vapor deposition in prepatterned GaAs(111) substrates (Karlsson *et al.*, 2010; Dupertuis *et al.*, 2011). The trigonal symmetry results in peculiarities in the hyperfine interaction of the heavy hole and nuclear spins. In this symmetry the electron Kramers doublet transforms according to the two-dimensional irreducible representation Γ_4, while the heavy-hole doublet transforms according to the reducible representation $\Gamma_5 \oplus \Gamma_6$ (Sallen *et al.*, 2011; Durnev *et al.*, 2013), see Appendix A for details.

In order to analyze the hyperfine coupling in (111)-grown quantum dots in more detail we use the axes frame $x_2 \parallel [11\bar{2}]$, $y_2 \parallel [\bar{1}10]$, $z_2 \parallel [111]$ and express the effective Hamiltonians of hyperfine interaction via the corresponding electron spin-1/2 and hole-pseudospin-1/2 operators represented by the sets of the Pauli matrices in this coordinate frame. The contact hyperfine interaction of the conduction band electron and nuclear spin is given by the effective Hamiltonian

$$\mathcal{H}_{hf,e} = \frac{|\varphi_c(\boldsymbol{R})|^2 v_0}{2} \left[A_{\parallel} \hat{\sigma}_{z_2} \hat{I}_{z_2} + A_{\perp} (\hat{\sigma}_{x_2} \hat{I}_{x_2} + \hat{\sigma}_{y_2} \hat{I}_{y_2}) \right], \tag{4.30}$$

with two independent parameters $A_{\parallel}$ and $A_{\perp}$. Just as for (001)-grown quantum dots with C_{2v} point symmetry described before, one may expect that the anisotropy of the hyperfine interaction in (111) quantum dots is minor and $A_{\parallel} \approx A_{\perp} = A$ in Eq. (4.8). The light-hole hyperfine interaction is described by the Hamiltonian of the same form as Eq. (4.30) because the light hole states transform according to the same two-dimensional irreducible representation, Γ_4.

For heavy holes the situation is different due to the fact that the $|\pm 3/2\rangle$ doublet transforms according to the reducible representation $\Gamma_5 \oplus \Gamma_6$. Hence, the basic matrices $\sigma_{z_2}^{(h)}$ and $\sigma_{x_2}^{(h)}$ transform according to the Γ_2 irreducible representation of the C_{3v} point group, while the matrix $\sigma_{y_2}^{(h)}$ is invariant under all point-group transformations and is odd at time reversal. As a result, the effective heavy-hole hyperfine coupling Hamiltonian, $\mathcal{H}_{hf,h}$, assumes the form

$$\mathcal{H}_{hf,h} = \frac{|\varphi_{hh}(\boldsymbol{R})|^2 v_0}{2} (C_1 \hat{\sigma}_{z_2}^{(h)} + C_2 \hat{\sigma}_{x_2}^{(h)}) \hat{I}_{z_2}, \tag{4.31}$$

with two constants C_1 and C_2. This is because the direct product $(\Gamma_5 \oplus \Gamma_6) \otimes (\Gamma_5 \oplus \Gamma_6)^* = 2\Gamma_1 + 2\Gamma_2$ contains twice the irreducible representation Γ_2 relevant for $\hat{I}_{z2}$ and does not contain any Γ_3 representation relevant for the in-plane nuclear spin components, $\hat{I}_{x2}, \hat{I}_{y2}$. This form of Hamiltonian is similar to the effective Hamiltonian describing the Zeeman effect for heavy holes in trigonal quantum dots (Sallen *et al.*, 2011). In the axial approximation, the constant C_2 vanishes and the Ising form of the hyperfine interaction is recovered. The parameter C_2 gives rise to the *non-collinear* interaction of z_2 component of nuclear spin and the in-plane, x_2, component of heavy hole spin. Microscopically, both parameters C_1 and C_2 can be derived from Eq. (4.18) describing the hyperfine interaction in the bulk material after transformation to the (x_2, y_2, z_2) axes frame and averaging over the heavy hole wavefunction with the result (Vidal *et al.*, 2016):

$$C_1 = \frac{1}{2}\left(3AM_1 + \frac{23}{4}AM_2\right), \quad C_2 = -\frac{\sqrt{2}}{2}AM_2. \tag{4.32}$$

As it is expected, the $C_2 \propto M_2$ vanishes if cubic anisotropy contribution in the bulk Hamiltonian (4.18) is disregarded. It has been shown in (Vidal *et al.*, 2016) that heavy-light hole mixing caused by the trigonal shape of the quantum dot potential and valence band warping terms in the Γ_8 band Luttinger Hamiltonian give rise to the renormalization of C_1 and C_2 and yield nonzero C_2, even if M_2 is set to zero in Eq. (4.18).

4.4 Experiments and theory

Experimentally, the Overhauser splittings of electronic states in GaAs, InAs, and similar quantum dots in the order of ~ 10 μeV have been directly observed in the experiments on dynamical nuclear polarization, see Chapter 3 and (Brown *et al.*, 1996; Gammon *et al.*, 1997; Urbaszek *et al.*, 2007) for details. Combined with the estimates of the nuclear polarization efficiency of about $\sim$ 50–80% in these experiments (Urbaszek *et al.*, 2007, 2013; Chekhovich *et al.*, 2017), see also Chapter 5 where the processes of dynamical nuclear polarization are discussed in detail – this supports the estimates of the conduction band hyperfine coupling constant $A \sim 100$ μeV summarized in Table 4.1. Similar estimates are supported by the experimental data on the electron spin decoherence in quantum dots discussed in Chapter 7. For such semi-quantitative estimates the valence band hyperfine interaction, being an order of magnitude weaker than in the conduction band, see Section 4.2, can be neglected.

In order to address quantitatively the hyperfine interaction of the hole and nuclear spin in individual quantum dots more sophisticated techniques were developed (Fallahi *et al.*, 2010; Chekhovich *et al.*, 2011, 2013; Vidal *et al.*, 2016). In these studies, the ratio of the hole and electron hyperfine interaction constants have been evaluated experimentally. Next, we describe in some detail the results of some studies (Chekhovich *et al.*, 2013; Vidal *et al.*, 2016) performed on (001) and (111) grown quantum dots, respectively.

The experiments reported in (Chekhovich *et al.*, 2013) have been carried out on InGaAs/GaAs, InP/GaInP, and GaAs/AlGaAs single quantum dots. The determination of the ratio of the electron to hole hyperfine interaction constants has been made by detecting the photoluminescence of the electron-hole pairs localized in the dots under the conditions of dynamic nuclear polarization, see Figure 4.1 (a). The experimental protocol is shown in Figure 4.1 (b), it consists of three steps. In a first step, the dynamical nuclear polarization has been achieved in the dots by means of pumping by the circularly polarized light for several seconds in the longitudinal magnetic field $\boldsymbol{B} \parallel z$, see Section 2.5 and Chapter 5 for details. The resulting nuclear polarization is opposite for σ^+ and σ^- excitation. Under the assumption that the nuclear spin polarization is uniform within the quantum dot volume, the Overhauser shifts for the electron, ΔE_e, and the hole, ΔE_h, can be represented in the following simple form

$$\Delta E_{e,h} = \sum_k \Delta E^k_{e,h}, \quad \Delta E^k_e = A^k_{\parallel}\langle I_{z,k}\rangle, \quad \Delta E^k_h = C^k_{\parallel}\langle I_{z,k}\rangle. \tag{4.33}$$

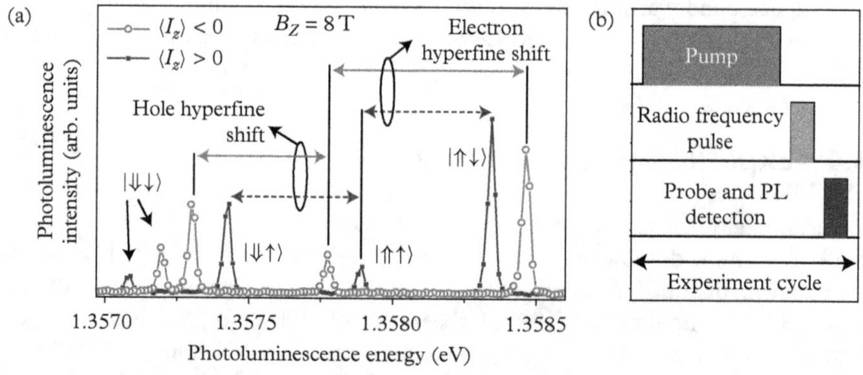

Figure 4.1 *(a) Photoluminescence spectra of a single neutral InGaAs/GaAs quantum dot in a magnetic field $B_z \approx 8$ T. For low-power optical excitation, four photoluminescence lines are observed in each spectrum corresponding to all possible combinations of the electron spin states ($\uparrow, \downarrow$) and hole states ($\Uparrow, \Downarrow$) forming bright and dark excitons, the latter have a small (< 0.01) admixture of bright states making dark states visible in photoluminescence spectra Two spectra are shown corresponding to negative (open symbols) and positive (solid symbols) nuclear spin polarization, $\langle I_z\rangle$, induced on the dot by pumping with σ^+ and σ^- polarized light, respectively. (b) Timing diagram of the pump-probe experiment used in the measurements of the hole hyperfine constants: nuclear spins are polarized by a high-power optical pump pulse. Afterward, a radio frequency oscillating magnetic field is switched on to achieve isotope-selective depolarization of nuclear spins. Finally, the sample is excited with a low-power probe laser pulse, during which the photoluminescence spectrum of both bright and dark excitons is measured. From (Chekhovich et al., 2013).*

Here, the subscript k enumerates isotopes, for example, ^{69}Ga and ^{71}Ga, $\langle I_{z,k}\rangle$ is the average polarization of the isotope, see Eq. (2.48), and the hyperfine interaction constants of kth isotope. $A_{\parallel}^{k}$ and $C_{\parallel}^{k}$, are introduced in Eqs. (4.24) and (4.25), respectively. In the second step of experiments, a radio frequency oscillating magnetic field is switched on to cause the nuclear magnetic resonance and achieve isotope-selective depolarization of nuclear spins as described in Chapter 3. The radio frequency pulse duration varies between 0.15 s and 35 s depending on the material. Finally, the sample is excited with a low-power short, ~0.3 s probe laser pulse, during which photoluminescence spectrum is measured, as shown in Figure 4.1(a). Note that in all experiments the durations of the radio frequency and probe pulses are much smaller than the natural decay time of nuclear polarization, therefore, the polarization of all isotopes but one (resonant with the radio frequency field) is preserved.

The possibility to determine experimentally both electron and hole Overhauser shifts in the studied quantum dots relies on the fact that in such structures all four possible spin states of the electron-hole pair are visible. In accordance with the selection rules, see Section 2.4, the σ^+ polarized photon promotes the electron from the $|\Gamma_8,-3/2\rangle$ state to the $|\Gamma_6,-1/2\rangle$ state thus creating the exciton in the state $(-1/2_e,+3/2_h)$ denoted as $\downarrow\Uparrow$, where the single arrow denotes the electron spin and the double arrow denotes the heavy-hole spin. Note that since the electron and hole representations are related by the time reversal the hole angular moment is opposite to that of the unoccupied valence band state. Similarly, the σ^- polarized photon creates the electron-hole pair in the state $(+1/2_e,-3/2_h)$ denoted as $\uparrow\Downarrow$. The recombination of these bright states gives rise to the dominant lines in the quantum dot photoluminescence spectrum in Figure 4.1(a). Two remaining exciton states $\uparrow\Uparrow$ and $\downarrow\Downarrow$ with parallel electron and hole spins have total angular moment z component of ± 2 and are nominally dark. Due to practically unavoidable symmetry reduction in real quantum dot samples the dark states mix with bright ones and gain small oscillator strength, producing weaker lines in Figure 4.1(a) (Zieliński *et al.*, 2015). As a result, the electron (hole) hyperfine shift of a chosen (kth) isotope can be deduced from a differential measurement: The spectral splitting between the electron-hole pairs with opposite electron (hole) spins, see Figure 4.1(a), is measured with a radio frequency pulse that depolarizes only the kth isotope and without any radio frequency pulse. The difference between these two splittings is equal to ΔE_e^k (ΔE_h^k).

Figure 4.2 presents an example of such measurements performed in GaAs [panel (a)] and InGaAs [panel (b)] quantum dots. Each point in the plot corresponds to the pair of values $\Delta E_e^k, \Delta E_h^k$ observed for different values of the initial nuclear spin polarization. Although in such experiments the magnitude of the initial nuclear spin polarization cannot be measured and, hence, the absolute values of the hyperfine constants $A_{\parallel}$, $C_{\parallel}$ are inaccessible, the ratio of the hole to the electron hyperfine coupling constants can be reliably detected from the slope of ΔE_h^k on ΔE_e^k dependence. The slopes are shown by straight lines in Figure 4.2. We introduce the ratio of the hyperfine coupling constants

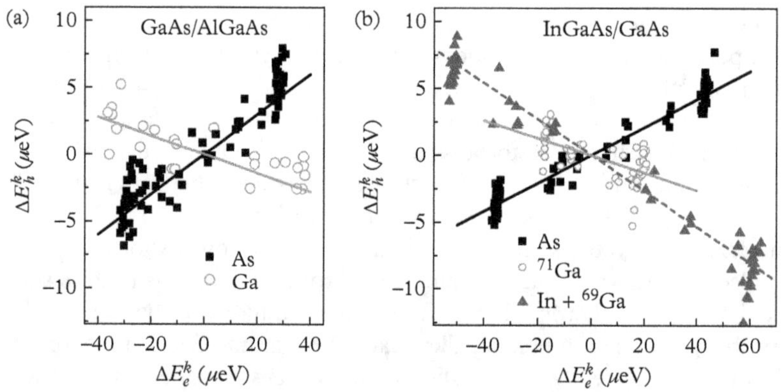

Figure 4.2 *Dependence of the hole hyperfine shift ΔE_h^k on the electron hyperfine shift ΔE_e^k for different isotopes in GaAs quantum dot (a) and InGaAs quantum dot (b). Solid lines show fitting: the slopes correspond to the relative hole-nuclear hyperfine constants, γ_k. The NMR resonances of ^{69}Ga and ^{115}In in InGaAs cannot be resolved, hence, the total hyperfine shift produced by these isotopes is presented. The dashed line in panel (b) is a guide to the eye. From (Chekhovich et al., 2013).*

$$\gamma_k = \frac{C_{\parallel,k}}{A_{\parallel,k}}. \tag{4.34}$$

Interestingly, although the absolute value of the ratio $|\gamma_k|$ is in the order of 10% for all studied elements, its magnitude and, in particularly, the sign vary strongly. Moreover, in all studied materials, cations (gallium, indium) have a negative hole hyperfine constant, whereas it is positive for anions (phosphorus, arsenic). The experimental results are summarized in Table 4.2. The results have been interpreted in (Chekhovich *et al.*, 2013) as a manifestation of the interplay between p- and d-type atomic orbitals in the valence band Bloch amplitudes. Indeed, assuming that $A_\parallel = A$ in Eq. (4.8) and making use of Eqs. (4.22) and (4.27), one arrives at the

$$\gamma_k = \frac{12}{5}|\alpha_p^k|^2 M_p^k - \frac{18}{7}|\alpha_d^k|^2 M_d^k. \tag{4.35}$$

It follows from Eq. (4.35) that p- and d-orbitals provide the contributions to the valence band hyperfine coupling constants of the opposite signs. This fact can be qualitatively understood from the comparison of the z-component of the angular momentum of p- and d-orbitals forming the Bloch functions of $|\Gamma_8, \pm 3/2\rangle$ states, see Eqs. (4.20) and (4.21). The estimates performed in (Chekhovich *et al.*, 2013) using hydrogenic wavefunctions with the effective screening radii from (Clementi and Raimondi, 1963) show that it is sufficient to include ~20% of the d-orbital admixture to the p-orbitals of Ga to account for the negative hole hyperfine

Table 4.2 *Experimentally measured ratio of the hole hyperfine constant to the electron one, γ_k, for different chemical elements k in GaAs and InGaAs quantum dots. The data are averaged over the isotopes. From (Chekhovich et al., 2013).*

Material/QD	γ^{Ga},%	γ^{In},%	γ^{As},%	γ^{P},%
InGaAs/GaAs	-5.0 ± 4.5	-15.0 ± 3.5	$+9.0\pm2$	-
GaAs/AlGaAs	-7.5 ± 3	—	$+16.0\pm3.5$	—
InP/GaInP	—	-12.5 ± 3	—	$+18.0\pm8$

constant measured experimentally. In fact, the substantial admixture of the d-orbitals in the Bloch functions of the valence band is expected in III-V semiconductors as demonstrated, for example, in (Chadi, 1976; Boguslawski and Gorczyca, 1994; Jancu *et al.*, 1998; Persson and Zunger, 2003). It is worth stressing, however, that the so far precise and reliable calculation of Bloch amplitudes in the vicinity of nucleus are not well developed and further experimental and theoretical studies are needed to clarify the role of d-orbitals in the hyperfine interaction in the valence band and, generally, the manifestations of manybody effects in the hyperfine interaction.

As we demonstrated before, the possibility to measure experimentally both electron and hole hyperfine shifts relies on the simultaneous observation of the bright and dark states of excitons in quantum dots. The quantum dots grown on (111) crystalline plane offer a unique opportunity to observe all four transitions with comparable intensities due to specific trigonal symmetry of the dots and magneto-induced heavy hole mixing (Sallen *et al.*, 2011; Durnev *et al.*, 2013). Indeed, as follows from the symmetry arguments and microscopic calculations, the Zeeman effect for heavy holes is governed by two g-factors, g_{h1} and g_{h2}, describing the diagonal and off-diagonal contributions to the effective Hamiltonian in the basis of the hole states with the spin projections of $\pm3/2$ onto the growth axis $z_2 \parallel [111]$:

$$\mathcal{H}_B = \frac{1}{2}\mu_B B_{z_2}\left(g_{h1}\hat{\sigma}_{z2}^{(h)} + g_{h2}\hat{\sigma}_{x2}^{(h)}\right). \tag{4.36}$$

As already mentioned, this form of the Zeeman Hamiltonian for heavy holes is similar to that of the hyperfine interaction, Eq. (4.31), because the symmetry of the perturbation, z_2-component of the magnetic field or z_2-component of the nuclear field, is the same. Typically, the diagonal, g_{h1}, and the off-diagonal, g_{h2}, components of the g-factor tensor are comparable in magnitude. Therefore, the eigenstates of the heavy hole in the magnetic field applied along the trigonal axis are the linear combinations of the $\pm3/2$ basic states with comparable weights. Hence, all four states of the electron-hole pair are optically active in the magnetic field applied along the trigonal axis, $\boldsymbol{B} \parallel [111]$.

In (Vidal *et al.*, 2016) the dynamic polarization of nuclei in single GaAs/AlGaAs (111) quantum dots embedded in the charge tunable device has been achieved. In this work, the longitudinal optical phonon-assisted excitation of a positively charged X^+-trion in quantum dots with resident holes has been employed and the Overhauser shifts for electrons and holes measured, as shown in Figure 4.3. The X^+-trion consists of two holes in the spin singlet state and an electron with unpaired spin that recombines leaving behind the heavy hole, as illustrated in Figure 4.3(e). In the longitudinal magnetic field both "bright" and "dark" transitions depicted by thick and thin arrows in Figure 4.3(e) are visible in accordance with Eq. (4.36). In the experiments performed in (Vidal *et al.*, 2016) the radio frequency fields were not applied, thus the overall hyperfine shifts induced by all present isotopes of Ga and As have only been measured. In analogy with Eq. (4.33) the electron Overhauser shift is quantified as $\Delta E_e = A_\| \langle I_{z_2} \rangle$, where $\langle I_z \rangle$ is the overall nuclear spin polarization and the constant $A_\|$ is introduced in Eq. (4.30). For the holes the situation is more involved, since both the hyperfine coupling, see Eq. (4.31), and Zeeman, see Eq. (4.36), Hamiltonians contain collinear and noncollinear terms. Thus depending on the relation between the Overhauser field and external field the effective hole splitting is a nonlinear function of B_{z_2} and $\langle I_{z_2} \rangle$.

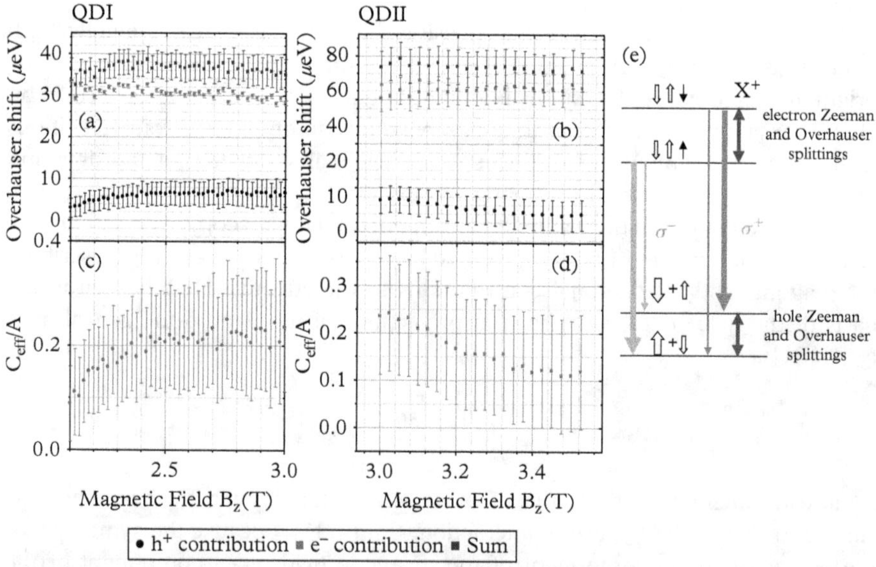

Figure 4.3 *(a) and (b) Overhauser shift for hole (black) and electron (light gray) and the sum of the two contributions (dark gray) for two different dots GaAs grown on (111)A plane. (c) and (d) The ratio of the effective hole hyperfine constant and electron hyperfine constant. (e) Initial and final states for radiative recombination of the X^+-trion in a longitudinal applied magnetic field* $\mathbf{B} \parallel$ *[111] in the presence of the dynamical nuclear polarization;* ↑ (↓) *and* ⇑ (⇓) *represent the electron and hole spins, respectively. From (Vidal et al., 2016).*

However, in experiments the external magnetic field is sufficiently strong, that is,

$$\sqrt{C_1^2+C_2^2}|\langle I_z\rangle| \ll g_h\mu_B|B_{z_2}|,$$

where $g_h=\sqrt{g_{h1}^2+g_{h2}^2}$. Therefore, the heavy-hole spin states are defined by the Zeeman effect, Eq. (4.36), and the hyperfine Hamiltonian (4.31) can be treated as a small perturbation with respect to Eq. (4.36). In this case, one has a linear Overhauser shift, $\Delta E_h = C_{\text{eff}}\langle I_{z2}\rangle$, with the effective hyperfine coupling constant

$$C_{\text{eff}} = \frac{g_{h1}}{g_h}C_1 + \frac{g_{h2}}{g_h}C_2. \tag{4.37}$$

It then follows that the heavy-hole hyperfine interaction in the dots under study is defined not only by the material constants C_1 and C_2 but also, through the effective g-factors g_{h1} and g_{h2}, by the dot size and shape. The ratio is $C_{\text{eff}}/A_{\parallel} = 0.2 \pm 0.1$ for typical dots studied in (Vidal *et al.*, 2016), see Figure 4.3(c,d). The analysis performed in this work shows that such a ratio is not in contradiction with the model of d-atomic orbitals contributions to the Bloch amplitudes of the valence band given the accuracy of the measurements and estimates presented before. Moreover, an additional orbital contribution to C_2 may play a role. To provide a complete description of the hyperfine interaction in trigonal quantum dots further experimental and theoretical studies are needed. On the experimental side, element-sensitive measurements on the (111) quantum dots could provide additional information on the signs and magnitudes for the ratio C_{eff}/A for Ga and As isotopes. This, together with the data in (Chekhovich *et al.*, 2013), could provide a more complete picture of the hyperfine coupling for holes and establish stronger restrictions on the model parameters. As for the theory, a simultaneous fitting of data on (001) and (111) quantum dots also taking into account the orbital contribution to C_2 is challenging. Of course, the first principles modeling of the electron and hole states in quantum dots is needed in order to quantify the hyperfine coupling of the conduction band electrons and valence band holes in semiconductors.

5

Dynamical Nuclear Polarization

5.1 General model of dynamical nuclear polarization

The smallness of a nuclear gyromagnetic ratio results in the practically negligible equilibrium thermal polarization of nuclei in experimentally available external magnetic fields. The hyperfine interaction of electron and nuclear spins makes it possible to transfer the electron spin polarization to the nuclear spin system and achieve the dynamical nuclear polarization. The effect of the dynamical nuclear polarization lies at the heart of nonmagnetic nuclear spin manipulation, as the electron spin dynamics can be controlled optically. We have already presented the minimal model of dynamical nuclear spin polarization based on the phenomenological rate equations in Section 2.5, see Eqs. (2.44) and (2.48). Next, we address various aspects of multifaceted phenomenon of the dynamical nuclear polarization and we start with the microscopic model of the nuclear spin polarization by electrons, which justifies and extends the approach of Section 2.5.

In fact, the process of dynamical nuclear polarization can be described within the kinetic equation for nuclei spin density matrix and the transition rates can be calculated, in a first approximation, within Fermi's golden rule approach. It is instructive, however, to evaluate the transition rates in somewhat different, although equivalent, way following (Dyakonov and Perel', 1974). Hereafter, we consider the simplest possible geometry of the optical orientation experiment, with the light propagating along the z-axis and the external magnetic field $\boldsymbol{B}$ applied in the Faraday configuration, $\boldsymbol{B} \parallel z$. The particular symmetry of the structure under study is not relevant here and we disregard for a moment the dipole–dipole interaction between the nuclei and the quadrupole splittings. We also assume that the nuclei are identical. Hence, just as in Section 2.5, it is possible to separate the density matrix of the whole ensemble of nuclei into the independent density matrices of individual nuclei and, moreover, parametrize the density matrix of a nucleus by the diagonal matrix elements, N_{I_z}, corresponding to the nuclei spin projection I_z onto the z-axis.

In what follows, we assume that the nuclear spin dynamics is controlled solely by a hyperfine interaction with electron spins. In accordance with Eq. (4.9) we present the hyperfine coupling Hamiltonian in the form

Electron and Nuclear Spin Dynamics in Semiconductor Nanostructures. M. M. Glazov, Oxford University Press (2018). DOI: 10.1093/oso/9780198807308.001.0001

$$\mathcal{H}_{hf} = Av_0 \sum_n (\hat{\boldsymbol{I}} \cdot \hat{\boldsymbol{s}}_n)\delta(\boldsymbol{r}_n - \boldsymbol{R}), \tag{5.1}$$

where the summation takes place over the electrons in the system, $\boldsymbol{r}_n$ is the position vector of the nth electron, $\hat{\boldsymbol{s}}_n$ is the nth electron spin operator, and $\boldsymbol{R}$ is the position of nucleus. We recall that A is the hyperfine coupling constant and v_0 is the primitive cell volume. The kinetic equation (2.31) for the nuclear spin density matrix ρ in the second order in the hyperfine interaction, Eq. (5.1), and neglecting the nuclear Zeeman splitting in the external field $\boldsymbol{B}$ can be recast in the following form (D'yakonov and Perel', 1975)

$$\frac{\partial \rho}{\partial t} = -\frac{1}{\hbar^2}\int_{-\infty}^{t} d\tau \,\mathrm{Sp}_e \left\{ [\mathcal{H}'_{hf}(t), [\mathcal{H}'_{hf}(\tau), \rho_e \otimes \rho]] \right\}. \tag{5.2}$$

Here, ρ_e is the electron density matrix, the trace Sp_e is taken over the electron variables and the electron dynamics is assumed to be independent of the nuclear one, therefore, the total density matrix of the system is written as a direct product, $\rho_e \otimes \rho$. The operator $\mathcal{H}'_{hf}(t)$ is the hyperfine interaction operator (5.1) taken in the interaction representation with respect to the electron Hamiltonian. Equation (5.2) is valid provided that the variations of the hyperfine fields acting on the nuclei are fast as compared with the nuclear spin dynamics. As already noted, for noninteracting nuclei, the density matrix can be separated in the product of independent density matrices for individual nuclei, $\rho^{(i)}$, which turn out to be the same. In what follows, we are interested in the diagonal elements of the single-nucleus spin density matrix, $N_{I_z} = \rho^{(i)}_{I_z,I_z}$, where I_z is the z component of the nuclear spin. As a result, the kinetic equation for the diagonal elements of the density matrix of the nucleus, N_{I_z}, reads

$$\frac{dN_{I_z}}{dt} = \mathcal{W}_{I_z,I_z-1}N_{I_z-1} + \mathcal{W}_{I_z,I_z+1}N_{I_z+1} - (\mathcal{W}_{I_z-1,I_z} + \mathcal{W}_{I_z+1,I_z})N_{I_z}. \tag{5.3}$$

Making use of Eq. (5.2) we introduce the rate of nuclear transitions from the level I_z to $I_z - 1$ in agreement with (Dyakonov and Perel', 1974; Abragam, 2002)

$$\mathcal{W}_{I_z-1,I_z} = \left(\frac{Av_0}{\hbar}\right)^2 (I + I_z)(I - I_z + 1)\,\mathrm{Re}\int_0^{\infty} d\tau \langle \hat{S}_-(0)\hat{S}_+(\tau)\rangle. \tag{5.4}$$

Here $\hat{\boldsymbol{S}} = \sum_n \hat{\boldsymbol{s}}_n \delta(\boldsymbol{r}_n - \boldsymbol{R})$ is the electron spin density operator taken in the interaction representation with the circular components defined as

$$\hat{S}_\pm = \frac{\hat{S}_x \pm \mathrm{i}\hat{S}_y}{\sqrt{2}}.$$

Angular brackets in Eq. (5.4) denote the averaging over the electron density matrix. Similarly, for the reverse transition $I_z - 1 \to I_z$ we have

$$\mathcal{W}_{I_z,I_z-1} = \left(\frac{Av_0}{\hbar}\right)^2 (I+I_z)(I-I_z+1)\,\mathrm{Re}\int_0^\infty d\tau \langle \hat{S}_+(0)\hat{S}_-(\tau)\rangle, \tag{5.5}$$

which differs from Eq. (5.4) by the order of $\hat{S}_+, \hat{S}_-$ operators. In what follows, we discuss the validity range of kinetic equation (5.3) with the transition rates in the form of (5.4) in more detail.

Equations (5.4) and (5.5) demonstrate that the problem of calculation of the nuclear spin flip rates reduces to the calculation of the electron spin correlation functions, $\langle \hat{S}_\pm(0)\hat{S}_\mp(\tau)\rangle$. Similar spin-spin correlators are discussed in detail in Chapter 10 in relation to spin noise spectroscopy. In this section, in order to study the dynamical nuclear polarization, we disregard effects of nuclei on electron spin dynamics. Hence, the electron spin components transversal to the external field simply precess around the field at the frequency $\Omega_B = g\mu_B B/\hbar$ [cf. Eq. (2.39)], where g is the electron g-factor, whose possible anisotropy is also disregarded for simplicity (Kalevich and Korenev, 1993; Kiselev, 1993). Under these assumptions, the electron spin operator correlator can be readily expressed via the correlator of the electron density operators,

$$\hat{\rho}(\boldsymbol{R},t) = \sum_n \delta(\boldsymbol{r}_n - \boldsymbol{R}),$$

as

$$\langle \hat{S}_-(0)\hat{S}_+(\tau)\rangle = \frac{1}{2}\left(\frac{1}{2} - S_z\right)\mathcal{K}_1(\boldsymbol{R},\Omega_B), \tag{5.6a}$$

$$\langle \hat{S}_+(0)\hat{S}_-(\tau)\rangle = \frac{1}{2}\left(\frac{1}{2} + S_z\right)\mathcal{K}_1(\boldsymbol{R},-\Omega_B), \tag{5.6b}$$

where S_z is the average value of the electron spin-z component, and

$$\mathcal{K}_1(\boldsymbol{R},\omega) = \mathrm{Re}\int_0^\infty d\tau e^{i\omega\tau}\langle \hat{\rho}(\boldsymbol{R},0)\hat{\rho}(\boldsymbol{R},\tau)\rangle. \tag{5.7}$$

It is convenient to express the spin correlators, instead of Eq. (5.7), via the symmetrized correlation function of the electron density

$$\mathcal{K}(\boldsymbol{R},\omega) = \frac{1}{2}\mathrm{Re}\int_0^\infty d\tau e^{i\omega\tau}\left[\langle \hat{\rho}(\boldsymbol{R},0)\hat{\rho}(\boldsymbol{R},\tau)\rangle + \langle \hat{\rho}(\boldsymbol{R},\tau)\hat{\rho}(\boldsymbol{R},0)\rangle\right], \tag{5.8}$$

making use of the equality (Dyakonov and Perel', 1974; Landau and Lifshitz, 2004)

$$\mathcal{K}_1(\boldsymbol{R},\pm\Omega_B) = (1 \mp 2S_T)\mathcal{K}(\boldsymbol{R},\Omega_B),$$

where S_T is the thermal electron spin polarization in the external magnetic field [cf. Eq. (2.47)]

$$S_T = -\frac{1}{2}\tanh\left(\frac{\hbar\Omega_B}{2k_BT}\right). \tag{5.9}$$

Finally, for the transition rates we obtain

$$\mathcal{W}_{I_z-1,I_z} = \left(\frac{Av_0}{2\hbar}\right)^2 \mathcal{K}(\boldsymbol{R},\Omega_B)(I+I_z)(I-I_z+1)(1+2S_T)(1-2S_z), \tag{5.10a}$$

$$\mathcal{W}_{I_z,I_z-1} = \left(\frac{Av_0}{2\hbar}\right)^2 \mathcal{K}(\boldsymbol{R},\Omega_B)(I+I_z)(I-I_z+1)(1-2S_T)(1+2S_z). \tag{5.10b}$$

Kinetic equation (5.3) with the transition rates in the form of Eqs. (5.10) allows one to study the nuclear spin dynamics and, in particular, dynamic nuclear polarization.

Taking into account that the transition rates $\mathcal{W}_{I_z,I_z'}$ introduced here differ from the rates $W_{I_z,I_z'}$ introduced in Section 2.5 by the electron spin distribution function only, one can indeed get convinced that the model derived here provides the microscopic justification of the phenomenological equations presented in Chapter 2. Particularly, as follows from Eq. (5.3) with $dN_{I_z}/dt = 0$, the ratio

$$\frac{\mathcal{W}_{I_z-1,I_z}}{\mathcal{W}_{I_z,I_z-1}} = \frac{(1+2S_T)(1-2S_z)}{(1-2S_T)(1+2S_z)},$$

is equivalent to the steady-state ratio of the nuclear spin states occupancies, Eq. (2.46). The complete steady-state solution of Eq. (5.3) can be conveniently recast as

$$N_{I_z} = \frac{\exp\left(\frac{I_z x}{I}\right)}{\sum_{I_z}\exp\left(\frac{I_z x}{I}\right)}, \quad \text{with} \quad x = I\left(\frac{g\mu_B B}{k_BT} + \ln\frac{1+2S_z}{1-2S_z}\right), \tag{5.11}$$

and the average nuclear spin polarization reads

$$\langle I_z\rangle = \sum_{I_z} I_z N_z = I\mathcal{B}_I(x), \tag{5.12}$$

where $\mathcal{B}_I(x)$ is the Brillouin function. It is in agreement with Eq. (2.48) derived in Section 2.5 in the limit of small electron spin polarization. For small electron spin polarization $|S_z| \ll 1$ and a not too strong magnetic field $|g\mu_B B/(k_BT)| \ll 1$ the parameter x in Eq. (5.12) can be recast in agreement with Eqs. (5.9) and (5.11) as

$$x = -4I(S_T - S_z).$$

It is evident that, for electrons in thermal equilibrium with the lattice, $S_z = S_T$, the argument x of the Brillouin function vanishes and the dynamical nuclear polarization is absent, $\langle I_z \rangle = 0$.

Equations (5.11) and (5.12) describe the dynamical nuclear polarization or the *Overhauser effect* (Overhauser, 1953). Significant polarization of nuclear spins with $|\langle I_z \rangle|$ that exceeds by far the thermal equilibrium polarization of nuclei is achieved due to the hyperfine interaction of nuclear spins with electron spins, which are driven out of equilibrium. Note that under conditions of electron spin resonance, other mechanisms of dynamical nuclear polarizations, for example, *solid effect*, can take place where the flip-flop transitions are driven by microwave radiation (Abragam and Goldman, 1978; Abragam, 2002; Wenckebach, 2008).

It also follows from Eq. (5.11) that the rate equation for the dynamical nuclear polarization can be presented as

$$\frac{d\langle I_z \rangle}{dt} = -\frac{1}{T_1}\left[\langle I_z \rangle - (I(I+1) - \langle I_z^2 \rangle)\tanh\left(\frac{x}{2I}\right)\right], \tag{5.13}$$

where

$$\frac{1}{T_1} = \frac{1}{2}\left(\frac{Av_0}{\hbar}\right)^2 \mathcal{K}(\boldsymbol{R}, \Omega_B)(1 - 4S_z S_T), \tag{5.14}$$

is the nuclear spin relaxation rate. Correspondingly, T_1 is the longitudinal nuclear spin relaxation time due to the hyperfine coupling. The general equation (5.13) describes both nuclear spin polarization and relaxation depending on the relation between S_z and S_T. It is noteworthy that in Eq. (5.13) $\langle I_z^2 \rangle$ depends on $\langle I_z \rangle$. In actual calculations it might be more convenient to solve the set of kinetic equations (5.3) instead of solving Eq. (5.13), see also Sections 5.2 and 5.3 for discussion.

Let us now specify the form of the electron density correlation function in Eq. (5.8). This correlation function for a localized electron is usually taken in the simplest possible form

$$\mathcal{K}(\boldsymbol{r}, \omega) = F|\varphi_c(\boldsymbol{r})|^4 \frac{\tau_c}{1 + (\omega\tau_c)^2}. \tag{5.15}$$

Here, $\varphi_c(\boldsymbol{r})$ is the envelope function, F is the probability of finding the electron at the localization site, that is, the site occupancy, and τ_c is the electron correlation time. Equation (5.15) corresponds to the model where the electron remains in the localized state during the time interval τ with the probability $\exp(-\tau/\tau_c)$. The physical origin of the correlation time τ_c can be related, for example, to the processes of hopping between the localization sites, exchange scattering with free electrons (if any), photoexcitation processes, or inter level transitions in quantum dot structures (Dyakonov and Perel', 1974; Kavokin, 2008; Urbaszek *et al.*, 2013).

Qualitatively, $\hbar/\tau_c$ gives the broadening of the electron energy levels, which allows one to overcome the energy mismatch between the Zeeman splitting of the electron and of the nucleus: Indeed, for $\Omega_B\tau_c \gg 1$ the electron spin states are well-resolved and the nuclear spin flip transitions become suppressed as the spin-flip rates, Eqs. (5.10) and (5.14), vanish $\propto 1/(\Omega_B^2\tau_c)$.

The estimates show that, for example, for the nuclei interacting with donor-bound electrons in GaAs with $\tau_c = 100$ ps and $F = 10^{-3}$ the nuclear spin relaxation time in Eq. (5.14) is in the order of several seconds for a nucleus in the vicinity of the donor. The relaxation time exponentially increases with an increase in distance from the donor due to suppression of the $|\varphi_c(\boldsymbol{r})|^4$. This leads to a specific situation where nuclei get polarized faster in the vicinity of the donor and the nuclear spin polarization becomes inhomogeneous in the space. The dipole–dipole interaction between the neighboring nuclei, Eq. (2.23), enables the transfer of the nuclear polarization from a region with high polarization to a region with low polarization; that is, to nuclear spin diffusion. The interplay of nuclear spin polarization, relaxation, and diffusion has been studied in detail in (Paget, 1982), where nuclear spin diffusion has been evidenced. The inhomogeneity of the nuclear spin polarization is also very prominent in a ZnSe semiconductor with fluorine donors as evidenced in (Heisterkamp *et al.*, 2015*a*). In quantum dot structures the situation is more complex, since the elastic strain, if any, is generally inhomogeneous in space and it may produce sufficient coordinate-dependent quadrupole splittings of nuclear spin levels, see Eq. (2.24) and Section 3.4, which may block the nuclear spin diffusion due to dipole–dipole interaction (Latta *et al.*, 2011). Similarly, the gradients of the magnetic field, for example, the Knight field experienced by the nuclei can also suppress nuclear spin diffusion (Kalevich *et al.*, 2017), see also (Nikolaenko *et al.*, 2009; Sallen *et al.*, 2014). The concept of the nuclear spin diffusion is addressed in more detail in the end of Section 5.2.

To conclude the section, let us analyze the applicability of the model developed here. The possibility of using the second order perturbation theory in hyperfine coupling, that is, in the Knight field exerted by the spin-polarized electron on the nuclei, to calculate the transition rates, see Eqs. (5.4) and (5.5), is justified under condition that

$$\left(\frac{Av_0\varphi_c(\boldsymbol{r})\tau_c}{\hbar}\right)^2 \ll 1 \quad \text{or} \quad \left(\frac{Av_0\varphi_c(\boldsymbol{r})}{\hbar\Omega_B}\right)^2 \ll 1. \tag{5.16}$$

Equation (5.16) ensures that the nuclear spin rotation angle in the fluctuating Knight field induced by the electron during the electron spin correlation time is small. The electron spin correlation time can be roughly estimated as τ_c or $1/\Omega_B$, whichever is shorter. Typically, the applicability conditions of the model are fulfilled for donor-bound electrons in bulk semiconductors, and for quantum dot structures, particularly if quantum dots are empty on average or contain resident holes. In the latter case, the dynamical nuclear polarization takes place via hyperfine interaction with an electron. Note, however, that the weaker dipole–dipole interaction of hole spin and nuclear spins may also give rise to the dynamical nuclear polarization for acceptor-bound holes as demonstrated theoretically in (Kavokin and Koudinov, 2013). For quantum dots with resident electrons the situation can be more involved due to the long electron correlation time, τ_c (Merkulov *et al.*, 2010), especially if electron spin coherence is excited by the train of short optical pulses. Note, however, that photoexcitation processes can strongly reduce the correlation time (Urbaszek *et al.*, 2013). The specifics of nuclear spin polarization arising in the n-doped quantum dots under these conditions is addressed in Chapter 9.

It is also instructive to draw an analogy here between the nuclear spin flips caused by the hyperfine interaction with electron spin and the nuclear magnetic resonance where the spin flips are induced by the Zeeman interaction with an alternating magnetic field. In the NMR case discussed in Chapter 3 the transitions are induced by the monochromatic radio frequency field, hence, the transition rate contains the broadened δ-function, see Eq. (3.12), whose broadening accounts for the nuclear spin decoherence processes. Moreover, a local environment of the nucleus affects the NMR spectrum via dipole–dipole interaction or quadrupole splittings as discussed in Section 3.2. Here, in the regime of dynamical nuclear polarization, the transitions between the nuclear spin levels are induced by a noisy, nonmonochromatic Knight field with the power spectrum in the form [cf. Eqs. (5.4), (5.5), and (5.15)]

$$\mathcal{K}(\boldsymbol{R},\omega-\Omega) \propto \frac{1}{1+[(\omega-\Omega_B)\tau_c]^2}.$$

The typical frequency range of the spectrum exceeds by far the relevant frequencies of the nuclear magnetic resonance as well as the frequencies related to the local field B_L, see Eq. (3.27). This justifies the possibility to disregard, in many cases, the nuclear spin dynamics in deriving the expressions for the transition rates $\mathcal{W}_{I_z,I'_z}$. However, as mentioned previously, the dipole–dipole interactions between the nuclei redistribute the spin polarization around the nuclei ensemble. The role of dipole–dipole interactions between the nuclei is particularly important for small magnetic fields $B \lesssim B_L$, where B_L is the local field, see Eq. (3.27), or if a radio-frequency field is applied and nuclear magnetic resonance conditions are realized. In the latter case in the rotating frame of reference, as discussed in Section 3.1, the external field is compensated while local fields are of importance. It is thus important to develop a formalism that allows us to describe the dynamical polarization effect in the ensemble of interacting nuclei. This is possible making use of the concept of the nuclear spin temperature as described in the Section 5.2.

5.2 Nuclear spin temperature

The Hamiltonian of the nuclear spin system contains, besides the hyperfine interaction Hamiltonian (5.1), two more terms. One is responsible for the Zeeman effect in the external magnetic field, $\mathcal{H}_{B,N}$, Eq. (2.20), and another one accounts for the dipole–dipole interaction of nuclear spins, $\mathcal{H}_{dd}$, Eq. (2.23). These additional terms read

$$\mathcal{H}_N = -\sum_i \frac{\mu_{I_i}}{I_i}(\hat{\boldsymbol{I}}_i \cdot \boldsymbol{B}) + \frac{1}{2}\sum_{i\neq j} \frac{\mu_{I_i}\mu_{I_j}}{I_i I_j}\left(\frac{\hat{\boldsymbol{I}}_i \cdot \hat{\boldsymbol{I}}_j}{r_{ij}^3} - 3\frac{(\hat{\boldsymbol{I}}_i \cdot \boldsymbol{r}_{ij}) \cdot (\hat{\boldsymbol{I}}_j \cdot \boldsymbol{r}_{ij})}{r_{ij}^5}\right). \qquad (5.17)$$

Here the summation takes place over all nuclei in the lattice, subscripts i and j enumerate nuclei, $\boldsymbol{r}_{ij}$ is the vector connecting nuclei i and j, and, as in Chapter 3, the factor $1/2$ ensures that the interaction of two nuclei is taken into account only once. With account for the $\mathcal{H}_N$ the kinetic equation (5.2) acquires the form

$$\frac{\partial \rho}{\partial t} = -\frac{\mathrm{i}}{\hbar}[\mathcal{H}_N, \rho] - \frac{1}{\hbar^2}\int_{-\infty}^{t} d\tau\, \mathrm{Sp}_e\left\{[\mathcal{H}'_{hf}(t), [\mathcal{H}'_{hf}(\tau), \rho_e \otimes \rho]\right\}. \qquad (5.18)$$

The exact solution of Eq. (5.18) cannot be found in a general case of the interacting nuclear system due to the huge number of involved nuclei. However, the nuclear spin system is characterized by drastically different relaxation time scales as already noted in Chapter 3. The dipole–dipole interaction is relatively efficient providing the nuclear spin decoherence on the time scale of $T_2 \sim 10^{-4}$ s due to the action of the local fields, see Section 3.2. On the other hand, the longitudinal relaxation of the nuclear spin, that is, the spin-lattice relaxation, is controlled mainly by the hyperfine interaction with electron spins. It is, in accordance with Eqs. (2.52) and (5.14), is much slower with $T_1 \sim 1$ s or longer. Hence, during the T_2 time scale the nuclear polarization is redistributed over the ensemble of the nuclei and a quasi-equilibrium within the nuclear spin system is established. The nuclear spin system remains quite isolated from the "outside world" on the time scale $t \ll T_1$. One may, therefore, apply thermodynamical methods and describe the nuclear spin system as a subsystem the effective *nuclear spin temperature*, Θ_N (Landau and Lifshitz, 2000). The concept of nuclear spin temperature takes an important place in the field of nuclear spin dynamics. Various aspects of the thermodynamic approach to nuclear spin systems including its precise applicability conditions and the way to introduce Θ_N for the system with different isotopes are addressed in many references (Purcell and Pound, 1951; Abragam and Proctor, 1958; Provotorov, 1962; Abragam and Goldman, 1982; Landau and Lifshitz, 2000; Abragam, 2002).

In the nuclear spin temperature approach the nuclear spin density matrix is presented in the form

$$\rho = Z^{-1} \exp\left(-\frac{\mathcal{H}_N}{k_B \Theta_N}\right), \tag{5.19}$$

with the normalization constant

$$Z = \mathrm{Sp}\left\{\exp\left(-\frac{\mathcal{H}_N}{k_B \Theta_N}\right)\right\}.$$

Equation (5.19) corresponds to the equilibrium state of the interacting system of nuclei at a given temperature Θ_N. The nuclear spin temperature, Θ_N, is a parameter that depends on conditions of the orientation of the electron spin system and time. The inverse product $k_B\Theta_N$ is usually denoted as

$$\beta = \frac{1}{k_B \Theta_N}.$$

This quantity is the inverse spin temperature measured in the energy units. If the nuclear spin system is far from complete polarization, which is usually the case in many experimental situations, the density matrix (5.19) can be recast in a simplified form

$$\rho \propto \hat{1} - \frac{\mathcal{H}_N}{k_B \Theta_N} = \hat{1} - \beta \mathcal{H}_N, \tag{5.20}$$

where $\hat{1}$ denotes the identity matrix of the corresponding dimensionality. Equation (5.20) corresponds to the *high temperature* approximation. It is noteworthy that, in the case of noninteracting nuclei where the dipole–dipole interaction, $\mathcal{H}_{dd}$, is disregarded, the form of the density matrix, Eq. (5.19), is exact as discussed in Section 2.5, see Eqs. (2.49), (5.11), and (5.12).

One can substitute the nuclear spin density matrix in the form of Eq. (5.19) or Eq. (5.20) into the kinetic equation (5.18) and derive the equation for the nuclear spin temperature, Θ_N. We assume that the external field is applied along the z-axis and apply the high temperature approximation. Technically, the density matrix in the form of Eq. (5.20) with the time-dependent $\Theta_N \equiv \Theta_N(t)$ is substituted into Eq. (5.18), the resulting equation is multiplied by $\mathcal{H}_N$ and the trace over the nuclear states is taken (Abragam, 2002). Finally, the equation for the nuclear spin temperature reads (D'yakonov and Perel', 1975)

$$\frac{d}{dt}\frac{1}{\Theta_N} = -\frac{1}{T_1}\left(\frac{1}{\Theta_N} - \frac{1}{\Theta_0}\right). \tag{5.21}$$

Here

$$\frac{1}{\Theta_0} = \frac{4k_B I}{\mu_I}\frac{B}{B^2 + \xi B_L^2}\frac{S}{1 - 4SS_T}, \tag{5.22}$$

and

$$\frac{1}{T_1} = \frac{1}{2}\left(\frac{Av_0}{\hbar}\right)^2 \mathcal{K}(\boldsymbol{R}, \Omega_B)(1 - 4S_z S_T)\frac{B^2 + \xi B_L^2}{B^2 + B_L^2}, \tag{5.23}$$

the local magnetic field, B_L, is introduced in accordance with Eq. (3.27):

$$B_L^2 = \frac{\mathrm{Sp}\{\mathcal{H}_{int}^2\}}{(\hbar\gamma_I)^2\,\mathrm{Sp}\{\hat{M}_x^2\}}, \quad \mathrm{Sp}\{\hat{M}_x^2\} = \frac{I(I+1)}{3}N(2I+1)^N, \tag{5.24}$$

N is the number of nuclei in the normalization volume and γ_I is the nuclear gyromagnetic ratio. The factor ξ can be related to the correlation of the spins on neighboring nuclei produced by the hyperfine interaction with the electron. In the presented model, it depends solely on the degree of electron spin relaxation anisotropy. If an external magnetic field is relatively weak and the spin relaxation anisotropy related to the anisotropic spin-orbit interaction can be neglected, the factor $\xi = 3$ (D'yakonov and Perel', 1975; Abragam, 2002). Just as in Section 3.2, only one type of nuclei is assumed to be relevant in Eq. (5.24), the generalization for the arbitrary number of nuclear species is obvious, see for example (Paget *et al.*, 1977). For $B \gg B_L$ definitions (5.14) and (5.23) of T_1 agree with each other.

Equation (5.21) can be easily solved in the steady state regime with the result

$$\Theta_N = \Theta_0,$$

hence the physical meaning of Θ_0 in Eq. (5.22) is the steady-state nuclear spin temperature. The external magnetic field, in accordance with the expression for the nuclear spin density matrix (5.20), produces the nuclear spin polarization in the form

$$\langle I_z \rangle_0 = \frac{I+1}{3}\frac{\mu_I B}{k_B \Theta_0}. \tag{5.25}$$

For sufficiently strong external magnetic fields, $B \gg \xi B_L$, Eqs. (5.23) and (5.25) reduce to Eqs. (5.14) and (5.12), and Eq. (2.50) is derived here neglecting the dipole–dipole interaction of nuclear spins. Similarly, in this limit Eq. (5.21) is equivalent to the rate equation (5.13) for nuclear polarization. The analysis performed in (D'yakonov and Perel', 1975) demonstrates that the generalized Eqs. (5.21), (5.23), and (5.22) remain valid for the external fields $B \gtrsim B_L\sqrt{T_2/T_1}$, otherwise the nonequilibrium correction to the nuclear spin polarization $\sim T_2/T_1$ becomes comparable with the quasi-equilibrium one $\sim B^2/B_L^2$.

The concept of the nuclear spin temperature is schematically illustrated in the Figure 5.1 where the relative occupancies of the nuclear spin states in a sufficiently large magnetic field, $B \gg B_L$, are illustrated in the case of positive nuclear spin temperature, $\Theta_N > 0$, panel (a) and negative Θ_N, panel (b). The ratio

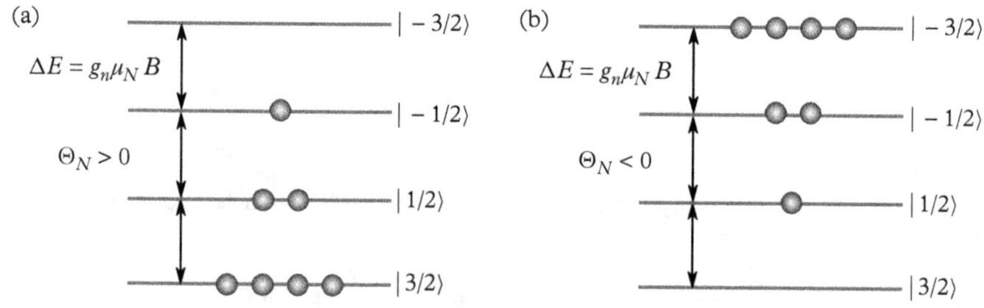

Figure 5.1 *Schematic illustration of the nuclear spin temperature concept. Gray lines show the states of the $I = 3/2$ nucleus in the magnetic field $\boldsymbol{B} \parallel z$. The balls denote relative occupancies of the nuclear spin states. (a) Positive spin temperature, $\Theta_N > 0$. (b) Negative spin temperature, $\Theta_N < 0$.*

of occupancies of the nuclear spin states with neighboring z spin components is given by the thermodynamic equilibrium formula [cf. Eq. (2.49)]

$$\frac{N_{I_z}}{N_{I_z-1}} = \exp\left(\frac{g_n\mu_N B}{k_B\Theta_N}\right). \tag{5.26}$$

However, depending on the sign of the Θ_N the occupation of the state with higher energy can be either smaller ($\Theta_N > 0$) or larger ($\Theta_N < 0$) compared to that of the state with lower energy. In the latter case, the distribution of the nuclear spins is "inverted." However, the description of the nuclear spin distribution with negative temperatures does not violate general laws of the thermodynamics, because the nuclear energy spectrum given by the eigenenergies of the Hamiltonian (5.17) is bounded from both above and below. Hence, the total energy of the state with any Θ_N is finite. Naturally, if Θ_N differs from the lattice temperature T (particularly, if $\Theta_N < 0$) and the excitation of the electron spin subsystem stops ($S = 0$ in Eq. (5.22)) the nuclear spin density matrix will eventually relax to its equilibrium form with $\Theta_N = T$. In this sense the nuclear spin system is out of thermodynamic equilibrium with respect to the crystal.

It is instructive to derive the the same results, Eqs. (5.21) and (5.22), from more general but phenomenological arguments based on the energy transfer balance between the electron and nuclear spin systems. Here we follow (Meier and Zakharchenya, 1984). The spin flux from the electron to the nuclei via the hyperfine interaction can be expressed as

$$\boldsymbol{q}_s = \frac{4I(I+1)}{3}\frac{\boldsymbol{S}}{T_{1s}}. \tag{5.27}$$

Here, and in what follows, we neglected the thermal spin orientation of electrons, S_T. Generally, in the right-hand side of Eq. (5.27) only the non-equilibrium spin polarization, related, for example, with the optical orientation of electrons, should be included. The time T_{1s} is defined in accordance with (Meier and Zakharchenya, 1984; Abragam, 2002):

$$\frac{T_{1s}}{T_1} = \frac{B^2 + \xi B_L^2}{B^2 + B_L^2}. \tag{5.28}$$

In the magnetic field $\boldsymbol{B}$, the spin transfer is accompanied by the energy transfer and the corresponding energy flux can be written simply as

$$q_\varepsilon = -\frac{\mu_I}{I}(\boldsymbol{B} \cdot \boldsymbol{q}_s) = -\frac{4}{3}(I+1)\mu_I \frac{(\boldsymbol{B} \cdot \boldsymbol{S})}{T_{1s}}. \tag{5.29}$$

On the other hand, the relaxation of the nuclear spin temperature is described by Eq. (5.21) with $\Theta_0 \to \infty$. As a result, for the reciprocal spin temperature relaxation we have

$$\frac{d\beta}{dt} = -\frac{\beta}{T_1}. \tag{5.30}$$

Let us evaluate the energy accumulated in the nuclear spin system making use of the nuclear spin temperature concept. Making use of Eq. (5.20) one has $\langle \mathcal{H}_N \rangle = -Z\beta\, \mathrm{Sp}\{\mathcal{H}_N^2\}$. Similar to derivation of Eqs. (3.27) and (5.24), the trace of the squared nuclear Hamiltonian is easily evaluated $\propto \beta(B^2 + B_L^2)$. Hence, the energy stored in the nuclear spin system per one nucleus reads

$$E_N = -\frac{(I+1)\mu_I^2}{3I}\beta(B^2 + B_L^2), \tag{5.31}$$

and the energy dissipation rate

$$r_\varepsilon = \frac{dE_N}{d\beta}\frac{d\beta}{dt} = \frac{(I+1)\mu_I^2}{3I}(B^2 + B_L^2)\frac{\beta}{T_1}. \tag{5.32}$$

The balance of the incoming energy flux and its dissipation, $q_\varepsilon + r_\varepsilon = 0$, gives the steady-state reciprocal nuclear spin temperature

$$\frac{1}{\Theta_0} = \frac{4k_B I}{\mu_I}\frac{(\boldsymbol{B} \cdot \boldsymbol{S})}{B^2 + \xi B_L^2}. \tag{5.33}$$

Equation (5.33) is the generalization of Eq. (5.22) for the case of arbitrary orientation of the nonequilibrium electron spin and the external magnetic field. It is derived under assumption that the Knight field is negligible compared with the external and local fields. Its role is discussed in more detail in Section 5.4. In accordance with the nuclear spin temperature concept and in agreement with

Eq. (5.25) the resulting nuclear spin polarization, $\langle \boldsymbol{I} \rangle$, can be evaluated as the equilibrium one in the external magnetic field $\boldsymbol{B}$ at the temperature Θ_0:

$$\langle \boldsymbol{I} \rangle = \frac{I+1}{3}\frac{\mu_I \boldsymbol{B}}{k_B \Theta_0} \propto \boldsymbol{B}(\boldsymbol{B}\cdot\boldsymbol{S}). \tag{5.34}$$

Equation (5.34) is particularly convenient for the analysis of the dependence of the nuclear polarization on the experimental geometry.

The expressions here were derived assuming that the only source of the nuclear spin relaxation is the hyperfine interaction with optically oriented electrons. As a result, the relation (5.28) between T_{1s} and T_1 holds. In order to account for other processes which contribute to the total longitudinal nuclear spin relaxation rate $1/T_{1N} = 1/T_1 + 1/T_d$ we add the terms

$$-\frac{\langle I_z \rangle}{T_d}, \quad -\frac{\Theta_N^{-1}}{T_d},$$

to the right-hand sides of Eqs. (5.13) and (5.21), respectively, where T_d is the relaxation time unrelated to the hyperfine interaction. It allows us to introduce a leakage factor

$$f = \frac{T_{1N}}{T_1} = \frac{T_d}{T_1 + T_d} < 1, \tag{5.35}$$

which serves as a measure of the reduction of the nuclear spin polarization due to relaxation processes. Accounting for the leakage factor the expression (5.33) for the nuclear spin temperature takes the form

$$\frac{1}{\Theta_0} = f\frac{4k_B I}{\mu_I}\frac{(\boldsymbol{B}\cdot\boldsymbol{S})}{B^2 + \xi B_L^2}. \tag{5.36}$$

The inclusion of the leakage factor into Eq. (5.36) is equivalent to the replacement of T_1 by T_{1N} in Eq. (5.30) describing the relaxation of the nuclear spin temperature.

Let us now consider the situation where, for some reason, the nuclear spin temperature Θ_N is a function of coordinates. This can be the case where the dynamical nuclear polarization is formed as a result of the optical orientation of the localized electrons. In this situation, as already mentioned in Section 5.1, see Eqs. (5.14) and (5.15), the polarization rate is the larger the larger is the probability to find the electron at a given nucleus. The truncated or secular part of the dipole–dipole interaction Hamiltonian at $\boldsymbol{B} \parallel z$ [cf. Eqs. (3.24), (3.23a)]

$$\mathcal{H}'_{int} = -\frac{\mu_I^2}{I^2}\sqrt{\frac{\pi}{5}}\sum_{i\neq j}\frac{1}{r_{ij}^3}Y_{20}(\theta_{ij},\varphi_{ij})\left[3\hat{I}_{i,z}\hat{I}_{j,z} - (\hat{\boldsymbol{I}}_i\cdot\hat{\boldsymbol{I}}_j)\right], \tag{5.37}$$

conserves the total z-component of the nuclear magnetization. However, due to the term $(\hat{\boldsymbol{I}}_i \cdot \hat{\boldsymbol{I}}_j)$ it gives rise to the flip-flop processes of the interacting nucleus where the z-component of one nucleus decreases by 1, while the z-component of the other nucleus increases by 1. We recall that in Eq. (5.37) $\boldsymbol{r}_{ij}$ is the vector connecting nuclei i and j, θ_{ij} and φ_{ij} are its corresponding angles in the spherical coordinate system, with the z being the polar axis. As a result, the inhomogeneous in space nuclear spin polarization tends to become homogeneous. This process can be described as a diffusion of nuclear spin polarization (Bloembergen, 1949). Thus, the reciprocal nuclear spin temperature, $\beta \equiv \beta(\boldsymbol{r}, t)$, obeys the equation (Bloembergen, 1949; Khutsishvili, 1966; Paget, 1982; Abragam, 2002)

$$\frac{\partial \beta}{\partial t} = D\Delta\beta, \tag{5.38}$$

where we introduced the *nuclear spin diffusion coefficient* D. In Eq. (5.38) the diffusion was assumed to be isotropic, the condition can be violated in low-dimensional systems where, for example, inhomogeneous distribution of elastic strain is present. An estimate of the diffusion coefficient can be obtained from the following qualitative analysis: The rate of exchange of nuclear spins between the neighboring sites separated by the lattice constant a_0 is on the order of T_2^{-1}. Hence the diffusion coefficient

$$D = C\frac{a_0^2}{T_2}, \tag{5.39}$$

where C is the coefficient. Depending on the system, field orientation with respect to the crystalline axes, the coefficient C ranges from 10^{-2} to 1 (Abragam, 2002). The estimates on the basis of the experimental data performed in (Paget, 1982) demonstrate that in GaAs for ^{75}As nuclei the nuclear spin diffusion constant $D \approx 10^{-13}$ cm^2/s. It should be noted that Eq. (5.38) is just a model equation to describe the dynamics in the system of interacting nuclei and can be rigorously justified only in some specific situations (Khutsishvili, 1966; Abragam and Goldman, 1982). However, the model of nuclear spin diffusion is very efficient at describing the effects related to the inhomogeneous in space nuclear spin polarization.

For completeness, we also note that in sufficiently large magnetic fields, $B \gg B_L$, the dipole–dipole processes, which do not conserve the z-component of the nuclear spin, are strongly suppressed due to the energy mismatch. This suggests that there are two energy "reservoirs" in the nuclear spin system: the Zeeman one and the dipole–dipole one. The change of energy in the Zeeman reservoir occurs by $\pm\hbar\omega_{0N}$, where $\hbar\omega_{0N}$ is the nuclear Zeeman splitting. This energy exceeds by far the range of eigenenergies of the dipole–dipole interaction Hamiltonian (2.23). As a result, the nuclear spin system can be characterized by two temperatures and the spin density matrix can be recast in the form (Provotorov, 1962; Abragam and Goldman, 1982)

$$\rho \propto \hat{1} - \beta_Z \sum_i \left(-\frac{\mu_I}{I} I_{i,z}\right) - \beta_{dd}\mathcal{H}'_{int}.$$

Here, β_Z is the reciprocal spin temperature of the Zeeman reservoir and β_{dd} is that of the dipole–dipole reservoir. The high temperature approximation is used in the previous expression. In small magnetic fields, $B \sim B_L$ or in the presence of external perturbations that enable the nuclear spin-flip transitions, for example, the radio frequency field with the frequency $\omega \approx \omega_{0N}$, the temperatures of the Zeeman and dipole–dipole reservoirs equalize.

Previously, we disregarded the role of the quadrupole interaction in the effect of dynamical nuclear polarization. The quadrupole splittings as noted in Section 3.4 are practically inevitable in nanosystems. In many important cases, quadrupole effects stabilize nuclear spins in quantum dots and enhance dynamical nuclear polarization (Dzhioev and Korenev, 2007), though in certain situations quadrupole interaction gives rise to nuclear spin relaxation (Paget *et al.*, 2008; Kotur *et al.*, 2016; Mocek *et al.*, 2017). Typically in nanosystems, the quadrupole splittings exceed local field B_L due to dipole–dipole interaction, making it possible to describe the dynamical nuclear polarization in the kinetic equation approach presented in Section 5.1 taking the quadrupole Hamiltonian into consideration. This allows one to avoid invoking the nuclear spin temperature concept, whose applicability in the presence of strong quadrupole splittings has been questioned, for example (Maletinsky *et al.*, 2009). This issue is definitely worthy of further investigations, while here we just refer to two recent experimental works where spin temperature description is validated for quadrupole-split nuclei: (Chekhovich *et al.*, 2017; Vladimirova *et al.*, 2017).

5.3 Bistabilities and instabilities in the nuclear spin system

Since the end of 1990s with pioneering reports of dynamical nuclear polarization in quantum dots (Brown *et al.*, 1996; Dzhioev *et al.*, 1998, 1999) this effect has been actively studied in semiconductor nanosystems. The model of the dynamical nuclear polarization presented in Sections 5.1 and 5.2 does not only qualitatively and, in many cases quantitatively, describe the build up of nuclear polarization under optical orientation conditions in bulk semiconductors and semiconductor nanosystems, but it can also be applied to describe various nonlinear phenomena in the interacting system of the host lattice nuclei and electron (Meier and Zakharchenya, 1984; Bracker *et al.*, 2008; Krebs *et al.*, 2008; Urbaszek *et al.*, 2013; Dyakonov, 2017). To that end, the model should be somewhat extended in order to account for the feedback of the nuclear spin system on electron spin dynamics and vice versa.

First, let us take into account the effect of nuclear spin polarization on electron spin dynamics. We assume that the external magnetic field is applied in the Faraday geometry, $\boldsymbol{B} \parallel z$, and optically oriented electron spin $\boldsymbol{S} \parallel z$ as well. The external field magnitude B is assumed to exceed by far the local fields in the structure. Consequently, we disregard the dipole–dipole interaction between the nuclei and represent the kinetic equation for nuclear spin polarization, Eq. (5.13), in the form

$$\frac{d\langle I_z\rangle}{dt} = -\frac{1}{T_1}\left[\langle I_z\rangle - QS_z\right] - \frac{\langle I_z\rangle}{T_d}, \tag{5.40}$$

with the nuclear spin polarization rate [cf. Eqs. (2.52) and (5.14)]

$$\frac{1}{T_1} = \left(\frac{A}{\hbar N_n}\right)^2 \frac{2F\tau_c}{1+(\Omega_L\tau_c)^2} \tag{5.41}$$

Here, as before, we have neglected thermal polarization of the electron, $S_T = 0$, introduced the factor

$$Q = \frac{4}{3}I(I+1),$$

and included all other mechanisms of the nuclear spin relaxation unrelated to the hyperfine interaction with the localized electron via the phenomenological relaxation time, T_d. We also applied the *box model* to describe the hyperfine interaction of the electron and nuclear spins assuming that the envelope function $\varphi_c(\boldsymbol{r}) = \sqrt{2/(N_n v_0)}$ is uniform within the quantum dot volume and is zero outside (Ryabchenko and Semenov, 1983), N_n is the number of nuclei within the quantum dot and we keep in mind a III-V or II-VI cubic semiconductor with two atoms within the primitive cell. For simplicity, we disregard here the difference of hyperfine coupling constants for different isotopes, it can be included into renormalization of the hyperfine interaction constant A and the factor Q (Krebs *et al.*, 2008; Urbaszek *et al.*, 2013). Although the box model imposes stringent limitations on the maximum value of the dynamical nuclear polarization, see, for example (Kozlov, 2007, 2008) it is very instructive for the semi-quantitative understanding of experiments.

In Eq. (5.41), the factor F ($F \ll 1$) represents the average fraction of the time where the dot or the donor is occupied, that is the hyperfine interaction is "on", and Ω_L denotes the Larmor precession frequency of the electron. In Section 5.1 the effect of nuclear spin polarization on the electron spin dynamics has been completely neglected and, correspondingly, the electron spin precession frequency has been determined solely by its Zeeman splitting in the external magnetic field. Here we take into account the fact that the nonzero average nuclear spin polarization $\langle I_z\rangle$ produces the average Overhauser field [cf. Eq. (2.51)]

$$B_N = \frac{2A}{g\mu_B}\langle I_z\rangle, \tag{5.42}$$

where g is the electron Landé factor, acting on the electron spin. The Overhauser field gives rise to the additional contribution to the spin precession frequency. Hence, in Eq. (5.41)

$$\Omega_L = \frac{g\mu_B}{\hbar}(B + B_N). \tag{5.43}$$

Even if the nuclear polarization is far from the maximum possible, the Overhauser field can be comparable with the external magnetic field and, therefore, can either increase the external field or, at least partially, compensate it. In the steady state, the time derivative of $\langle I_z \rangle$ vanishes and steady-state nuclear spin polarization can be found from the equation

$$\frac{1}{T_1}[\langle I_z \rangle - QS] + \frac{\langle I_z \rangle}{T_d} = 0. \tag{5.44}$$

Equation (5.44) is nonlinear due to the implicit dependence of T_1 on $\langle I_z \rangle$. Provided that the electron spin S is independent of the Overhauser field (this assumption is analyzed and lifted in what follows) Eq. (5.44) is the cubic equation for $\langle I_z \rangle$:

$$f(\langle I_z \rangle) \equiv \langle I_z \rangle \left[1 + \zeta + \zeta\alpha^2 \left(\frac{g\mu_B B_z}{A} + \langle I_z \rangle \right)^2 \right] = QS, \tag{5.45}$$

with the following dimensionless parameters

$$\zeta = \frac{N_n \hbar^2}{2FA^2 \tau_c T_d}, \qquad \alpha = \frac{A\tau_c}{\hbar}. \tag{5.46}$$

Physically, the ζ characterizes the ratio of the nuclear spin relaxation time due to the electrons, T_1, at zero magnetic and Overhauser fields to the spin relaxation time T_d unrelated to the hyperfine coupling. The parameter α is the ratio of the Overhauser field induced splitting of the electron spin levels (for fully polarized nuclei) to the electron level broadening related to the finite correlation time $\hbar/\tau_c$. The product $\zeta\alpha^2$ can be rewritten in the form $N_n \tau_c/(2FT_d)$.

Depending on the parameter values, Eq. (5.45) admits one or three solutions. In the limit of short correlation times where $\zeta(\alpha QS)^2, \zeta(\alpha g\mu_B B/A)^2 \ll 1$, or, in another words

$$\Omega_L \tau_c \ll 1,$$

where Ω_L is the electron Larmor frequency in the external field, the only physical solution of Eq. (5.45) is, in agreement with Eq. (5.36),

$$\langle I_z \rangle = QS/(1 + \zeta).$$

The leakage factor is $f = 1/(1+\zeta)$ in accordance with the definition (5.35). In this regime the feedback of dynamical nuclear polarization $\langle I_z \rangle$ on the spin flip rates is negligible.

In the limit of relatively long electron correlation time the situation becomes more involved. Mathematically, this problem resembles the situation of a driven anharmonic oscillator (Landau and Lifshitz, 1976). Here, an interplay of the Overhauser and external magnetic fields contributions to the electron Larmor frequency can give rise to the situation where three solutions of Eq. (5.45), one of which is unstable, are possible. Such a situation occurs where the signs of QS and $g\mu_B B/A$ are opposite, that is, where the Overhauser field and the external magnetic can compensate each other. Two stable solutions correspond to: (i) high nuclear spin polarization

$$\langle I_z \rangle_h \approx -\frac{g\mu_B B}{A}, \quad 0 < \frac{1}{\zeta\alpha^2}\left(\frac{QS}{\langle I_z \rangle} - 1 - \zeta\right) \ll 1, \tag{5.47a}$$

which describes the situation of almost compensated Zeeman and Overhauser splittings, and (ii) low nuclear spin polarization

$$\langle I_z \rangle_l \approx \frac{QS}{1+\zeta+\alpha\zeta^2(g\mu_B B/A)^2}, \quad |A\langle I_z \rangle_l| \ll |g\mu_B B|, |AQS|, \tag{5.47b}$$

which corresponds to the significant difference of the Zeeman and Overhauser splittings and, in fact, to significant leakage of the nuclear spin polarization. Hence, one can expect the bistable behavior and hysteresis of the nuclear spin polarization as a function of external parameters, such as the magnetic field, pumping power, and so on. Such a bistable behavior has indeed been observed in various experiments performed on quantum dot structures (Eble *et al.*, 2006; Braun *et al.*, 2006*b*; Maletinsky *et al.*, 2007; Tartakovskii *et al.*, 2007; Kaji *et al.*, 2008; Skiba-Szymanska *et al.*, 2008; Belhadj *et al.*, 2008; Vidal *et al.*, 2016).

Let us discuss in more detail the bistable behavior of the dynamical nuclear spin polarization as a function of the external magnetic field. Figure 5.2 demonstrates the set of experimental results obtained on a single quantum dot grown from InGaAs on the (001) crystalline plane in (Krebs *et al.*, 2008), see also (Urbaszek *et al.*, 2013). In this experiment the X^+ trion consisting of the electron and two holes in spin singlet state was excited and its emission detected as a function of the magnetic field and the pump power. In the particular quantum dot under study, the electron g-factor was negative, $g = -0.6$. As follows from Table 4.1, the hyperfine coupling constants for the electron with nuclei in InGaAs systems are positive. Moreover, in accordance with the selection rules the excitation with the σ^+ radiation creates the electron spin-z component $S < 0$, while the excitation with the σ^- radiation creates the electrons with $S > 0$, see Section 2.4. Therefore, under σ^+ excitation and positive magnetic field, where the z component of magnetic field $B_z > 0$ the nuclear spin polarization tends to be antiparallel to

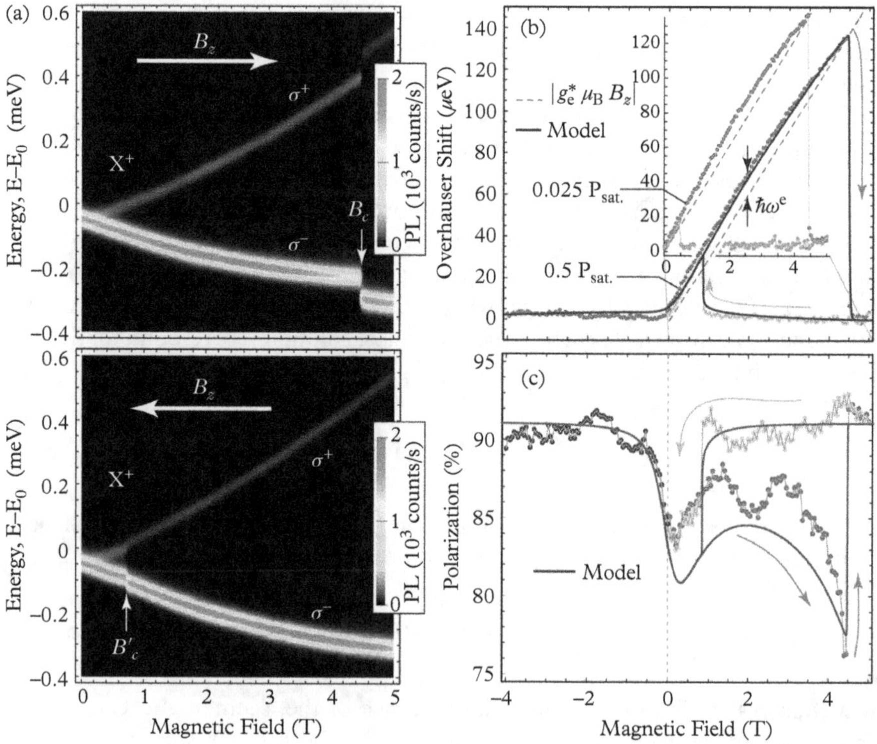

Figure 5.2 *(a) Density plots of X^+ trion (two holes and one electron) photoluminescence intensity from a single InGaAs QD grown along z ∥ [001]. Data are taken at T = 1.7 K as a function of an increasing or decreasing magnetic field **B** ∥ z, E_0 = 1.3143 eV. The quasi resonant σ^- polarized excitation is at ~ 60 meV above the trion line in zero field. (b) Absolute Overhauser shift of the Zeeman splitting for two excitation powers, and (c) circular polarization as a function of magnetic field. The electron Zeeman splitting is denoted $\hbar\omega^e = |g\mu_B B_z|$ [$g_e^* = g$ in (b)]. The saturation power P_{sat} roughly corresponds to the electron filling factor F = 1 in the quantum dot. From (Urbaszek et al., 2013).*

z-axis, $\langle I_z \rangle < 0$ and, hence, the Overhauser fields adds to the external magnetic field. Similarly, for σ^- excitation and $B_z < 0$ the B_N is negative as well. In the described configurations the build up of the Overhauser field increases the T_1 and, correspondingly, bistability is not observed in agreement with the analysis here. The situation changes drastically if σ^- excitation takes place at $B_z > 0$ or σ^+ excitation takes place at $B_z < 0$. In such a case, the Overhauser and external fields are antiparallel and the conditions are favorable for bistability observation.

Figure 5.2(a) demonstrates the emission spectra of the X^+ trion as a function of magnetic field at the σ^- pumping with an increase in the B_z (top panel) and decrease in the B_z (bottom panel). Both σ^+ and σ^- components of the trion emission are plotted, with the σ^- component being the strongest and lower in energy due to particular relation between the electron and hole g-factors (Urbaszek *et al.*, 2013). One can see that in the process of the increase in B_z at the certain critical field $B_c \approx 4.5$ T the splitting of the trion emission lines abruptly increases. During the sweep of magnetic field in the opposite direction the similar abrupt change in the splitting is observed at lower critical field $B_c' \approx 1$ T. Such abrupt changes in the splitting are related to the switching of the nuclear spin polarization between the stable values $\langle I_z \rangle_h$ and $\langle I_z \rangle_l$, see Eqs. (5.47). The bistability effect is most pronounced in Figure 5.2(b) where the Overhauser shift is plotted by subtracting the trion Zeeman splitting from the data in panel (a). One can see that the increase in magnetic field up to B_c is accompanied by the increase in the Overhauser field. The Overhauser splitting of the electron spin states is very close to the absolute value of the electron Zeeman splitting, but, as noted previously, has an opposite sign. It can even slightly overcompensate for the Zeeman splitting. This branch of solution corresponds to the higher absolute value of the nuclear spin polarization, $\langle I_z \rangle_h$, see Eq. (5.47a). The compensation of the external field and the Overhauser field increases the rate of the electron-nuclei flip-flop processes and results in the efficient dynamical nuclear polarization. At $B_z = B_c$, the nuclear spin polarization rate can no longer overcome the nuclear polarization losses, and a further increase in B_z results in rapid reduction of Overhauser splitting and switching to the lower value of $\langle I_z \rangle_l \approx 0$; see Eq. (5.47b). If the external magnetic field is decreased then, just like in typical bistable systems, the lower value of $\langle I_z \rangle_l \approx 0$, Eq. (5.47b), is realized until smaller critical field B_c' is reached.

Interestingly, the circular polarization of the X^+ trion emission, P_c, also demonstrates the bistable behavior shown in Figure 5.2(c). Taking into account that, in accordance with the selection rules, $P_c = -2S$, where S is the electron spin z component, the electron spin polarization S also demonstrates the bistability. Such a bistability may arise if the feedback of nuclear spin polarization on the electron spin dynamics, in particular on the electron spin relaxation, is taken into account. This effect has not been included in the model developed previously. However, the variation of the electron polarization is minor $\sim$10%...15% compared with the variation of the Overhauser field $\sim$ 100% in the vicinity of B_c, B_c'. Hence, to model simultaneously the bistability of the nuclear spin polarization and the electron spin polarization one can introduce the effect of nuclear spin polarization on electron spin dynamics in the X^+ trion using the framework of the kinetic equation for the electron spin polarization [cf. Eq. (2.33)]

$$\frac{dS}{dt} + \frac{S}{\tau_r} + \frac{S}{T_{1s}} = G_s. \tag{5.48}$$

Here, G_s is the spin generation rate, $G_s = \mp 1/2$ for the $\sigma^\pm$ polarization of the incident radiation with account for the strict selection rules described in Section 2.4, τ_r is the trion recombination time and the term with S/T_{1s} accounts for the rate of the spin polarization transfer to the nuclear spin system with $T_{1s} \sim T_1$, see Eq. (5.27). Under the experimental conditions of (Krebs *et al.*, 2008;

Urbaszek *et al.*, 2013) $\tau_r \sim 1$ ns and $T_{1s} \sim 10$ ns, hence, an effect of the nuclear spin dynamics on the electron spin polarization is indeed expected in the order of 10%. The complete model description of the experimental data in Figure 5.2(b), (c) represented by the solid lines also requires one to account for the magnetic field dependence of the nuclear spin relaxation time T_d unrelated to the hyperfine interaction:

$$\frac{1}{T_d} = \frac{1}{T_{d\infty}} + \frac{1}{T_{d0}} \frac{1}{1 + (B_z/B_0)^2}.$$

Here, $T_{d\infty}$ corresponds to the (additional) nuclear spin relaxation time at $B_z \gg B_0$ and $T_{d\infty}T_{d0}/(T_{d\infty} + T_{d0})$ is the nuclear spin relaxation time at $B_z \to 0$. The magnetic field B_0 in (Krebs *et al.*, 2008; Urbaszek *et al.*, 2013) was related to the typical strength of the quadrupole fields in strained InGaAs quantum dots grown on the (001) crystalline plane. The fit of the experimental data presented in Figure 5.2 allowed the authors of (Urbaszek *et al.*, 2013) to obtain a reasonable set of parameters $F_e T_{d0} = 65$ μs, $F_e T_{d\infty} = 5$ ms, $\tau_c = 75$ ps. The parameters of the hyperfine interaction were fixed to correspond to typical quantum dots, $N_n = 5 \times 10^4$, $Q = 13$ and maximum Overhauser splitting at the complete nuclear spin polarization of 230 μeV. The hole-nuclear coupling was disregarded in (Krebs *et al.*, 2008; Urbaszek *et al.*, 2013) based on the estimate that it contributes by $\lesssim 10\%$ to observed Overhauser shifts and plays practically no role in the process of the dynamical nuclear spin polarization.

The basic physics of the bistability of nuclear spin polarization as a function of the external magnetic field remains the same in strain-free GaAs/AlGaAs quantum dots grown on (111) plane (Belhadj *et al.*, 2008; Vidal *et al.*, 2016). As an example, we demonstrate in Figure 5.3 the dependence of the total Overhauser shift of the X^+ trion [panel (a)] and the photoluminescence polarization degree P_c of the X^+ emission [panel (b)] as a function of magnetic field measured on a single GaAs (111) quantum dot in (Vidal *et al.*, 2016). Both nuclear spin polarization and the electron spin polarization dependence on the magnetic field magnitude and sweep direction are similar to those in the strained (001) quantum dots shown in Figure 5.2.

The bistable behavior of the nuclear spin system under the conditions of the electron optical orientation can be controlled not only by the magnetic field but also by other parameters, such as, for example, incident radiation power. Figure 5.4 demonstrates the experimental results obtained in (Tartakovskii *et al.*, 2007) on the GaAs/AlGaAs quantum dot embedded in the Schottky diode. The studied quantum dot was neutral and the zero-dimensional exciton, that is the electron hole pair, was optically excited. Figure 5.4(a,b) presents the dependence of the bright exciton level splitting, being the combination of the spin splittings of the electron and the hole states (see Section 4.4 for details on the bright and dark excitonic states), as a function of the radiation power. The hysteresis loops of spin splittings are clearly observed. Similar to the data in Figures 5.2 and 5.3, the

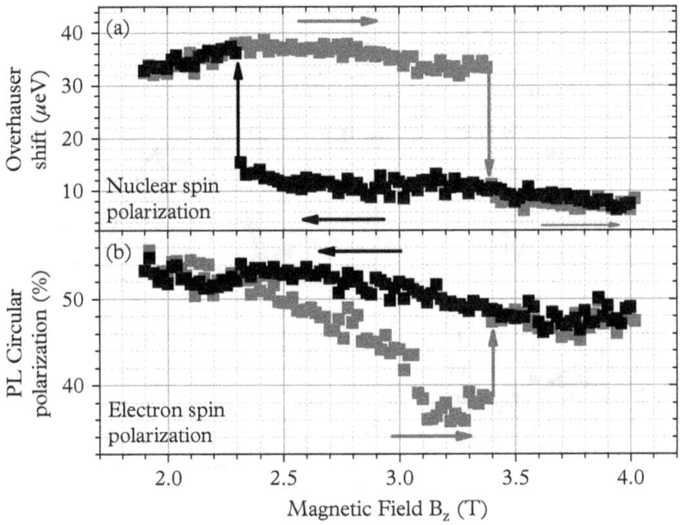

Figure 5.3 *Bistability of nuclear polarization in a (111) grown GaAs/AlGaAs quantum dot. (a) The total Overhauser shift (electron and hole contributions) is plotted as a function of magnetic field* $\mathbf{B} \parallel [111]$ *for different sweep directions. (b) The polarization degree of the* X^+ *emission is plotted, showing a bistability of the electron spin polarization as* $P_c = -2S$*. From (Vidal et al., 2016).*

bistable behavior of the exciton splitting is related to the bistability of the dynamical nuclear polarization. The model description of dynamical nuclear polarization via excitons has been put forward in (Korenev, 1999; Merkulov, 2002). The model developed in (Korenev, 1999) and elaborated on in (Tartakovskii *et al.*, 2007) accounts for the interplay of bright and dark excitons in the process of the nuclear spin polarization. The authors of (Tartakovskii *et al.*, 2007) assumed that the recombination rate of the bright exciton is large enough, so the flip-flop processes due to the hyperfine interaction are inefficient, and no considerable spin transfer from the electron towards the nuclear spin system is provided by the bright excitons. Also, for bright excitons, to realize the electron-nucleus flip-flop process one has to overcome the exchange splitting between the bright and dark states. This would require unrealistically short correlation times so that the mechanism is essentially blocked (Eble *et al.*, 2006). By contrast, the nominally dark excitons with the $(+1/2_e, +3/2_h)$ or $(-1/2_e, -3/2_h)$ spin configurations of the electron and the hole can recombine in the following two-step process: (i) the electron spin is flipped due to the hyperfine interaction with the nuclei, giving rise to the "virtual" bright exciton, and (ii) the bright exciton recombines. In the described process, the dynamical nuclear polarization builds up (Erlingsson *et al.*, 2001; Gammon *et al.*, 2001), see also (Chekhovich *et al.*, 2010). The resulting equation for the nuclear spin polarization $\langle I_z \rangle$ is somewhat more complex compared with Eq. (5.45):

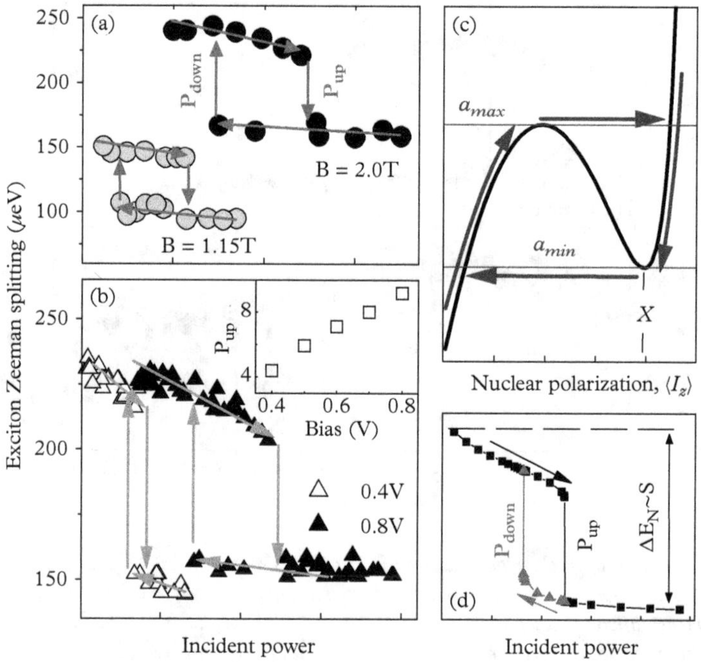

Figure 5.4 *(a) Power dependence of the electron-hole pair Zeeman splitting under σ^- circularly polarized excitation measured at the magnetic field $B = 2$ T and 1.5 T in GaAs/AlGaAs quantum dot. The arrows show the direction in which the hysteresis loop is measured with two thresholds P_{up} and P_{down}. (b) The same as in (a) but for two different bias voltages of 0.4 and 0.8 V. The inset shows the P_{up} dependence on the reverse bias applied to the diode. (c) The full line shows the function $f(\langle I_z\rangle)$ in Eq. (5.49). Arrows show how the hysteresis loop is formed when the parameter a is varied for fixed b, see text for details. (d) Hysteresis loop of the electron-hole pair Zeeman splitting as a function of incident power calculated using Eq. (5.49). From (Tartakovskii et al., 2007).*

$$f(\langle I_z\rangle) \equiv \langle I_z\rangle \left[1 + b\frac{(g\mu_B B_z/A + \langle I_z\rangle)^2}{I - \langle I_z\rangle}\right] = a. \tag{5.49}$$

Here, I is the maximal nuclear spin polarization, a and b are the parameters that depend on the pumping and relaxation rates. Equation (5.49) is also the cubic equation for $\langle I_z\rangle$ and, just like Eq. (5.45), can give rise to the bistable behavior of the nuclear spin polarization as is illustrated in Figure 5.4(c) where the function $f(\langle I_z\rangle)$ is plotted for different sweep directions of the parameter a, which is proportional to the pumping rate. Correspondingly, the calculated behavior of the exciton spin splitting shown in Figure 5.4(d) agrees with the experimental findings. It is remarkable that the hysteresis loop in these experiments can be controlled also by the magnetic field and by the bias voltage. The latter is due to the fact that in the studied Schottky diode structure the bias voltage affects the charge carrier population and its correlation time via charging and recharging processes.

An interesting instability effect, the *dynamic nuclear self-polarization*, has been predicted in (D'yakonov and Perel', 1972) for nuclei interaction with electrons that are maintained in the unpolarized state, for example, by the pumping with linearly polarized light. Let us assume for simplicity that $I = 1/2$ and assume that the only magnetic field acting on the electron spins is the Overhauser field caused by the nuclei. It follows from Eq. (5.12), (5.9), and (5.42) that the nuclear spin polarization obeys the following nonlinear equation

$$\langle I \rangle = \frac{1}{2} \tanh\left(\frac{2T_c\langle I \rangle}{T}\right), \tag{5.50}$$

with the parameter $T_c = A/4k_B$. It can be derived by assuming that the argument of the Brillouin function in Eq. (5.12) simply equals $x = -4IS_T = A\langle I \rangle / k_B T$ (note that condition $S = 0$ should be fulfilled in order to observe the self-polarization effect).

For $T > T_c$, Eq. (5.50) has just one solution with $\langle I \rangle = 0$ corresponding to the unpolarized nuclear system. For $T < T_c$, another stable solution of Eq. (5.50) with $\langle I \rangle \neq 0$ is possible. Hence, T_c is a critical temperature of the dynamical nuclear self-polarization. For several nuclei in the primitive cell one has $T_c = \sum_i I_i(I_i + 1)/(3k_B)$, where i enumerates the nuclei. For typical semiconductors T_c is in the order of 1 K. Precise conditions where the instability described by Eq. (5.50) could be observed experimentally are discussed in (D'yakonov and Perel', 1972; Dyakonov, 1975; Meier and Zakharchenya, 1984). The self-polarization of nuclei spins in quantum dots via excitons have been studied theoretically in (Korenev, 1999), where it has been shown that the self-polarization arises as a result of the conversion of the alignment of excitons (i.e., correlations of electron and hole spins) into nuclear orientation in the effective magnetic field of the polarized nuclei. A similar effect for near-resonant excitation of trions in quantum dots has been put forward in (Korenev, 2007). Detailed discussion of the possibilities of the self-polarization of nuclei in quantum dots is also presented in (Abalmassov, 2013).

5.4 Anomalous Hanle effect in an interacting electron-nuclear system

The important ingredient of the model presented in Section 5.3 and describing the bistability of the nuclear spin polarization was the account for the nuclear polarization feedback on the electron spin dynamics, particularly on the electron spin precession frequency via the Overhauser field, see Eq. (5.43). It is natural to address the effects of the Knight field induced by the spin polarized electron on the dynamics of nuclei. In order to analyze the effects we disregard the possible bistabilities related to the Overhauser field and represent the nuclear spin temperature in the form

$$\frac{1}{\Theta_0} = \frac{4k_B I}{\mu_I} \frac{(\boldsymbol{B} \cdot \boldsymbol{S}) + b_e S^2}{(\boldsymbol{B} + b_e \boldsymbol{S})^2 + \xi B_L^2}, \tag{5.51}$$

where we omitted the leakage factor for simplicity and introduced the Knight field $\boldsymbol{B}_e = b_e \boldsymbol{S}$ with

$$b_e = -\frac{A v_0 |\varphi_c(\boldsymbol{R})|^2}{g_n \mu_N}. \tag{5.52}$$

We recall that the g_n and μ_N are nuclear g-factor and magneton, respectively. Equation (5.51) follows from Eq. (5.33) with the replacement $\boldsymbol{B} \to \boldsymbol{B} + \boldsymbol{B}_e$, that is, accounting for both the external magnetic field and the Knight field action on the nuclei.

Let us focus on the situation where $\boldsymbol{B} \perp \boldsymbol{S}$, that is, where the Voigt geometry is realized. This situation is most interesting because, in absence of the Knight field, $b_e \to 0$, the reciprocal nuclear spin temperature is effectively zero and the dynamical nuclear polarization is absent. Accounting for the Knight field the nuclear polarization becomes possible with

$$\frac{1}{\Theta_0} = \frac{4k_B I}{\mu_I} \frac{b_e S^2}{b_e^2 S^2 + \xi B_L^2}. \tag{5.53}$$

The sign of the Θ_0 in Eq. (5.53) is determined by the sign of the parameter b_e. The nuclear polarization is given by the standard expression (5.34) with $\boldsymbol{B} \to \boldsymbol{B} + b_e \boldsymbol{S}$. Particularly, for a not too high value of B_L, an observable nuclear spin polarization at $\boldsymbol{B} = 0$ may arise solely due to the Knight field acting on the nuclear spins.

The nonlinear behavior in the electron-nuclei spin system can be realized if one takes into account the feedback of the nuclear spin polarization on the electron dynamics. To that end we present the kinetic equation (2.33) for the electron spin $\boldsymbol{S}$ in the form

$$\frac{d\boldsymbol{S}}{dt} + \frac{g\mu_B}{\hbar} \boldsymbol{S} \times (\boldsymbol{B} + \boldsymbol{B}_N) + \frac{\boldsymbol{S}}{T_s} = \boldsymbol{G}, \tag{5.54}$$

where $\boldsymbol{G}$ is the spin generation rate, T_s is the electron spin lifetime, $\boldsymbol{B}_N = A\langle \boldsymbol{I} \rangle / (g\mu_B)$ is the Overhauser field and g is electron g-factor. Under conditions $\boldsymbol{G} \parallel z$, $\boldsymbol{B} \perp z$ one reduce Eq. (5.54) to the cubic equation for the dimensionless z-component of the electron spin polarization $s_z = S_z/(GT_s)$ (Meier and Zakharchenya, 1984):

$$1 - s_z = s_z(\Omega_B T_s)^2 \left(1 + \frac{\hbar^2}{g^2\mu_B^2 T_s^2} \frac{\kappa s_z}{B^2 + \xi B_L^2}\right)^2, \tag{5.55}$$

where $\Omega_B = g\mu_B B/\hbar$, and

$$\kappa = -\frac{4I^2(I+1)S_0^2}{3} \frac{A^2 v_0 |\varphi_c(\boldsymbol{R})|^2}{g\mu_B\mu_I} \frac{\hbar^2}{g^2\mu_B^2 T_s^2}. \tag{5.56}$$

The parameter κ describes the efficiency of the hyperfine coupling and the "depth" of nuclear spin cooling. In the limit of negligible hyperfine interaction, $\kappa \to 0$, Eq. (5.55) yields

$$S_z = \frac{GT_s}{1 + (g\mu_B BT_s/\hbar)^2}, \tag{5.57}$$

in agreement with the model of the Hanle effect in absence of the nuclear polarization, see Eq. (2.40). Figure 5.5 demonstrates the dependence of the reduced electron spin polarization, s_z, on the external magnetic field calculated for different values of the parameter κ.

Depending on the relative signs of the electron transverse g-factor and nuclear magnetic moment the parameter κ can be either negative or positive. For instance, in InGaAs system where $g < 0$ and $\mu_I > 0$ for all relevant isotopes the $\kappa > 0$. In this case, both external field and Overhauser fields add up in the Voigt geometry and

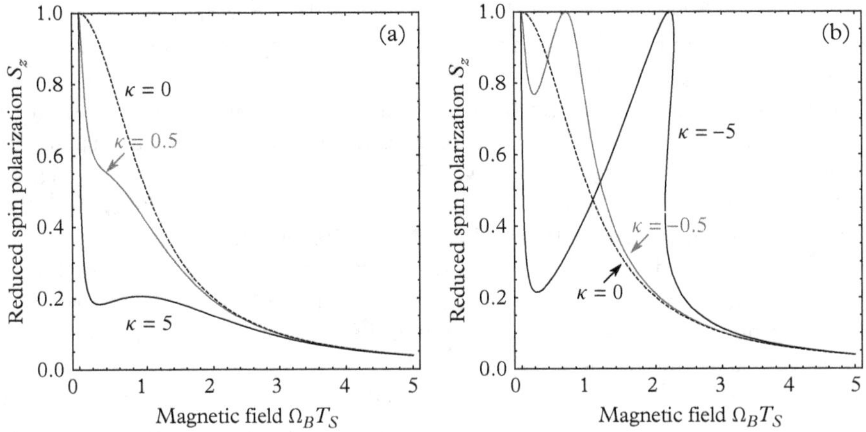

Figure 5.5 *Reduced electron spin polarization $s_z = S_z/(GT_s)$ as a function of transversal magnetic field in the units of $\Omega_B T_s$ calculated after Eq. (5.55) for the positive κ [panel (a)] and negative κ [panel (b)]. For comparison, dashed lines show the standard Hanle curves, Eq. (5.57), which correspond to $\kappa = 0$. The dimensionless strength of the local field is $\sqrt{\xi} g\mu_B B_L T_s/\hbar = 0.25$.*

Eq. (5.55) has one physical solution. The form of $s_z(B)$ curve, however, markedly differs from the standard Hanle curve, Eq. (5.57), see Figure 5.5(a). In particular, for a sufficiently large κ the dependence of $s_z(B)$ becomes nonmonotonic. In $Al_xGa_{1-x}As$ crystals the electron g-factor becomes positive for $x \gtrsim 0.12$. A similar situation can be realized in GaAs/AlGaAs trigonal quantum dots grown by the droplet epitaxy (Sallen *et al.*, 2014). This results in the $\kappa < 0$ giving rise to the partial compensation of the external and Overhauser fields. In this case the bistability may take place, see Figure 5.5(b), as shown theoretically and experimentally in (Novikov and Fleisher, 1976). The situation becomes more involved in an oblique magnetic field, see (Meier and Zakharchenya, 1984) for further details. The electron g-factor anisotropy also leads to a bistable behavior of the dynamical nuclear polarization as observed in (Kalevich and Korenev, 1993). In fact, for systems with an anisotropic electron g-factor, a variety of nonlinear effects, also termed catastrophes, is predicted (Kiselev, 1993).

It is also worth noting that in solid solutions like $Al_xGa_{1-x}As$ and in quantum dots significant quadrupole effects occur. The nonlinearities and instabilities of interacting electron-nuclear spin system in the bulk semiconductors with strong quadrupole interactions are discussed in (D'yakonov *et al.*, 1980; Meier and Zakharchenya, 1984).

It has been suggested in (Dzhioev and Korenev, 2007) that the quadrupole splittings of the nuclear spin states may stabilize the nuclear spin polarization in InGaAs quantum dots in the zero external field. This is because in quantum dots grown on (001), the growth axis $z \parallel [001]$ is expected to be the axis where the electric field gradients $\mathcal{V}_{zz}$ are the largest, see Section 3.4. To the first order in the transverse magnetic field only the doublet with $|I_z| = 1/2$ is affected by the field, see Figure 3.5. The estimates in (Urbaszek *et al.*, 2013) show that the quadrupole effects indeed stabilize the nuclear spin orientation along the z-axis in transverse magnetic fields up to ~ 100 mT.

The interplay of various mechanisms related to nuclear spin polarization in weak magnetic fields or in absence of external field is a very active field of research (Lai *et al.*, 2006; Krebs *et al.*, 2010; Urbaszek *et al.*, 2013; Sallen *et al.*, 2014). Here, we discuss the results of (Krebs *et al.*, 2010) where the interplay of the dynamical nuclear polarization and the Zeeman effect in the transversal magnetic field, $\boldsymbol{B}$, has been studied experimentally on self assembled InGaAs/GaAs quantum dots. Under circularly polarized excitation of the X^+ trion the experiments reveal the nuclear polarization at $\boldsymbol{B} = 0$ with the nuclear field $\boldsymbol{B}_N$ directed along the growth axis and $B_N \approx 0.35$ T. In a transverse magnetic field the circular polarization degree of the trion luminescence, $P_c = -2S$, differs strongly from the Hanle curve, Eq. (5.57), demonstrating hysteresis loops and bistable behavior as can be seen from Figure 5.6(a, b). The width of this "anomalous" feature in the $P_c(B)$ behavior is about 0.5 T and exceeds by far the width expected from Eq. (5.57) with the spin lifetime being equal to the trion lifetime. The modulation of the circular polarization of radiation with a frequency of a few KHz allows one, at least to some extent, to recover the conventional behavior of the luminescence

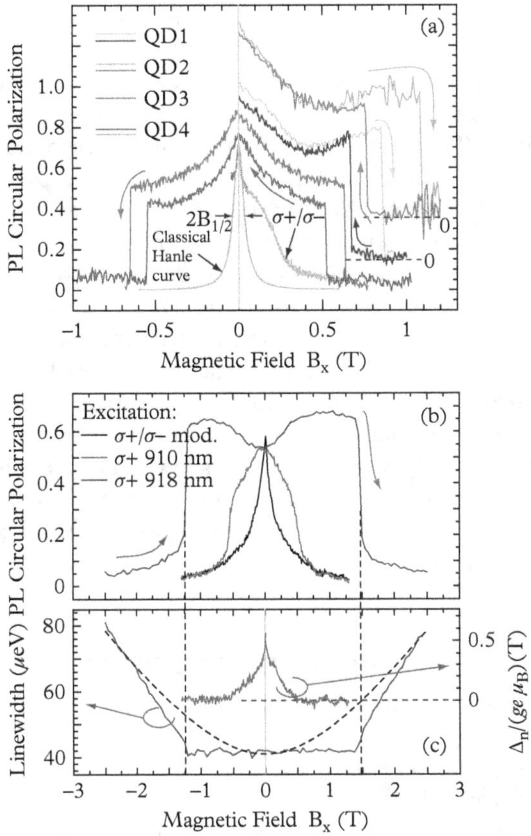

Figure 5.6 *(a) Hanle depolarization curves of X^+ trions in four different InAs/GaAs quantum dots (QD1, …, QD4), under a constant σ^+ polarized excitation. QD4 was also measured under the modulated σ^+/σ^0 excitation. Arrows indicate the magnetic field sweep directions. Curves for QD1 and QD2 are vertically shifted for clarity. (b) Hanle depolarization curves of an X^+ in another sample under different excitation conditions denoted in the legend. (c) X^+ apparent linewidth and the Overhauser shift estimated from the the trion lines splitting. The dashed line shows the calculated evolution of effective linewidth due to the electron Zeeman splitting. From (Krebs et al., 2010).*

depolarization and leads to the narrowing down of the $P_c(B)$ profile. These experimental findings demonstrate that the observed anomalous Hanle effect is governed by nuclear polarization. In contrast to the dynamical nuclear polarization observed in the longitudinal field, see Section 5.3, the observed depolarization curves are independent of the sign of the transversal field component, B_x, being in full agreement with the symmetry of the problem.

The significant polarization degree of the luminescence P_c observed in the transverse magnetic field can be explained in the framework of two possible scenarios: (a) The total magnetic field, $\boldsymbol{B}_{tot} = \boldsymbol{B} + \boldsymbol{B}_N$, acting on the electron spin has a significant z-component, $|B_{tot,z}| \gg |B_{tot,x}|, |B_{tot,y}|$, or (b) the field is in the quantum dot plane, but its in-plane components are small as compared with $\hbar/(g\mu_B T_s)$. Although option (a) definitely holds at $B = 0$, further experiments performed in (Krebs *et al.*, 2010) rule out this scenario for substantial magnetic fields $B \gtrsim 0.5$ T. Indeed, Figure 5.6(c) shows the Overhauser splitting of the X^+ trion as a function of external field. The nuclear field z component evaluated from these data rapidly decreases with an increase in the external field. Moreover, the linewidth of the X^+ resonance deduced from the widths of σ^+ and σ^- emitting lines remains practically constant until the collapse in the nuclear polarization occurs, see Figure 5.6(c), and then abruptly increases, which indicates the appearance of trion Zeeman splitting. The presence of the in-plane nuclear field is also confirmed by additional experiments carried out in (Krebs *et al.*, 2010) where trion emission has been measured in linear polarizations while the excitation remains circularly polarized, which allows an increase in the sensitivity of the measurements. These observations corroborate the scenario (b) that the significant nuclear field builds up in the quantum dot plane and compensates for the external field.

5.5 Locking of optical quantum dot resonances via hyperfine coupling

In the last section of this chapter, we briefly discuss another effect related to the dynamical nuclear polarization manifested in single InGaAs quantum dots embedded in a charge tunable device (Warburton *et al.*, 2000). The experiments reported in (Latta *et al.*, 2009; Xu *et al.*, 2009; Högele *et al.*, 2012) demonstrate that the absorption line of the quantum dot may, under certain conditions, follow the energy of the laser. Typically, in magnetic fields exceeding 1 T the quantum dot absorption abruptly turns on if the detuning $\delta = \omega_X - \omega_L$ between the quantum dot resonance frequency, ω_X, and the laser frequency, ω_L, becomes smaller than $\sim 2\gamma$, where γ is the linewidth. Moreover, the absorption remains practically constant if the detuning is scanned across the resonance within the range of about $\pm 10\gamma$. The behavior is, however, different for two Zeeman-split transitions: The higher energy (denoted "blue") transition follows the laser energy, exhibiting the so-called *locking effect*. For this transition the resonance energy and the laser energy coincide. By contrast, the lower energy ("red") transition energy remains either higher or lower than the laser energy, so that the exact resonance between the laser energy and the transition energy is avoided. The effect is qualitatively the same for any orientation of the magnetic field and occurs for any laser polarization (linear or circular). Moreover, the behavior is practically the same for neutral and charged excitons. Typical experimental results illustrating the locking effect are presented in Figure 5.7. The analysis performed in (Latta *et al.*, 2009; Högele *et al.*, 2012;

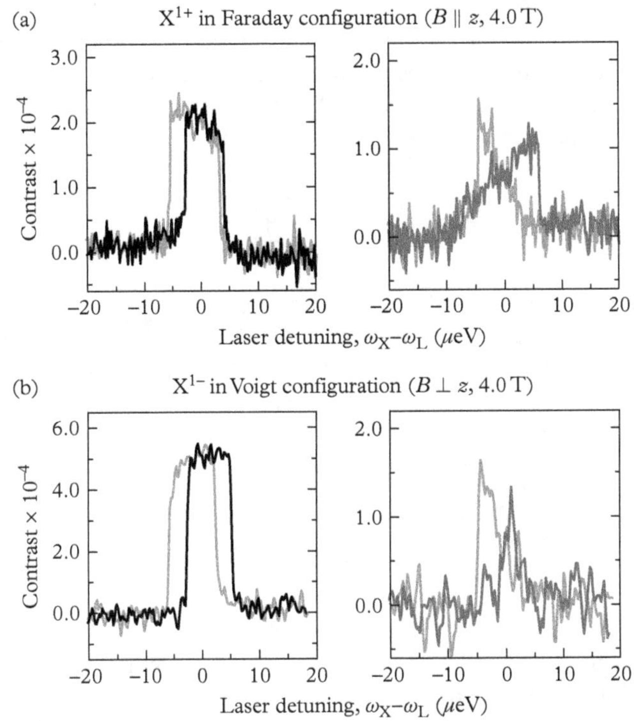

Figure 5.7 *Optical transitions in quantum dots at $B = 4$ T as a function of quantum dot charge state [X^{1-} (X^{1+}) denotes the excitation of the negatively (positively) charged trion in the electron (hole) charged dot] and magnetic field orientation. Both X^+ and X^- trion exhibit characteristic flat-top resonances on the higher energy ("blue", left panel) Zeeman transition and triangular line shapes on the lower energy ("red", right panel) Zeeman transition in a magnetic field oriented (a) parallel (Faraday) and (b) perpendicular (Voigt) to the sample growth axis z. The curves demonstrate the resonance behavior at the increase and decrease of the detuning, respectively. Here, ω_X is the trion resonance frequency and ω_L is the laser frequency. From (Högele et al., 2012).*

Urbaszek *et al.*, 2013) shows that the origin of such a behavior of quantum dot absorption resonances is related to nuclear polarization.

The hyperfine interaction in the form of Eq. (5.1) alone does not allow one to explain the experimental findings. In particular, as discussed in Sections 5.1 and 5.2, the dynamical nuclear polarization depends on the sign of the radiation helicity, that is, the orientation of the electron spin polarization $\boldsymbol{S}$ and the magnetic field $\boldsymbol{B}$. Moreover, in sufficiently strong magnetic fields the efficiency of the dynamical nuclear polarization decreases, see Section 5.3, while in experiments the efficiency of the locking of the resonances increases with an increase in the field (Högele *et al.*, 2012). In order to explain the locking effect, the concept of the *noncollinear hyperfine interaction* has been invoked (Huang and Hu, 2010; Yang and Sham, 2012). Without going into the details of the modeling of the experimental data, let us discuss the noncollinear interaction of the electron and nuclear spins and briefly address its consequences on the nuclear spin polarization.

The noncollinear interaction Hamiltonian can be presented in the form

$$\mathcal{H}_{nc} = v_0\delta(\boldsymbol{r}-\boldsymbol{R})\sum_{\alpha\neq\beta} A_{\alpha\beta}^{(nc)}\hat{s}_\alpha\hat{I}_\beta, \tag{5.58}$$

where $\alpha,\beta = x,y,z$, $A_{\alpha\beta}^{(nc)}$ are the coefficients, $\boldsymbol{r}$ and $\boldsymbol{R}$ are the position vectors of the electron and nucleus, respectively, v_0 is the primitive cell volume. Note that in summation $\alpha\neq\beta$, that is, the interaction takes place between different Cartesian components of the electron, $\hat{\boldsymbol{s}}$, and nuclear, $\hat{\boldsymbol{I}}$, spins. Such noncollinear coupling is also typical for the dipole–dipole interaction of magnetic moments, see Eqs. (2.23) and (4.5). Usually, a simplified form of Eq. (5.58) is used [cf. (Urbaszek *et al.*, 2013)]

$$\mathcal{H}_{nc} = v_0\delta(\boldsymbol{r}-\boldsymbol{R})A^{(nc)}\hat{s}_z\hat{I}_x, \tag{5.59}$$

with $A^{(nc)} = A_{zx}^{(nc)}$. Equations (5.58) and (5.59) imply that the symmetry of the nucleus is reduced as compared with the C_{2v} point group typical for quantum dots grown along [001]. Such a symmetry reduction can be expected, for example, due to inhomogeneous strain profiles or due to overall structure imperfections. Microscopically, the noncollinear interaction can be expected for holes in trigonal quantum dots grown on (111) oriented substrates due to specific form of the Bloch functions, see Eq. (4.31) and (Vidal *et al.*, 2016) for details. For the conduction band electrons the Bloch functions are mainly of the *s*-type and the hyperfine interaction is practically isotropic, as discussed in Chapter 4. However, the effective noncollinear interaction of the conduction band electron spins and the nuclear spins arises due to the combined action of the hyperfine interaction, Eq. (5.1), and the quadrupole splittings, see Eq. (2.24) (Huang and Hu, 2010).

As an illustrative example of the effect of noncollinear interaction, we consider a $I = 3/2$ nucleus in the situation where the electric field gradients have a nonzero xz component, $V_{xz}\neq 0$. Hence, the quadrupole interaction Hamiltonian can be written as

$$\mathcal{H}_Q = E_Q(\hat{I}_x\hat{I}_z + \hat{I}_z\hat{I}_x), \tag{5.60}$$

where the parameter E_Q determines the quadrupole splitting strength. We also assume that the external magnetic field $\boldsymbol{B}$ is applied along the z-axis giving rise to the nuclear Zeeman splitting

$$\mathcal{H}_{B,N} = -\frac{\mu_I}{I}B_z\hat{I}_z, \tag{5.61}$$

and take into account the collinear term

$$\mathcal{H}_{hf} = Av_0|\varphi_c(\boldsymbol{R})|^2 \hat{s}_z \hat{I}_z, \tag{5.62}$$

in the hyperfine interaction. The possibility to disregard remaining terms is justified by the significant electron Zeeman splitting induced by magnetic field $\boldsymbol{B}$. In sufficiently strong magnetic field where $|\mu_I B_z/I| \gg E_Q$ the quadrupole interaction admixes nuclear $\pm 1/2$ state to nuclear $\pm 3/2$ as vice versa, for example,

$$\Psi_{\pm 3/2} = |\pm 3/2\rangle + \gamma|\pm 1/2\rangle, \quad \Psi_{\pm 1/2} = |\pm 1/2\rangle - \gamma|\pm 3/2\rangle,$$

where $|I_z\rangle$ denotes the nuclear states with the spin component I_z, $\gamma \sim E_Q/(\mu_I B_z)$ is the admixture coefficient and $|\gamma| \ll 1$. One can readily check that the hyperfine coupling Hamiltonian (5.62) has nonzero matrix elements between the pairs

$$\langle \Psi_{\pm 3/2}|H_{hf}|\Psi_{\pm 1/2}\rangle = \mp \gamma A v_0 |\varphi_c(\boldsymbol{R})|^2 \hat{s}_z. \tag{5.63}$$

Hence, in the presence of the quadrupole splittings in the form of Eq. (5.60) the hyperfine interaction, $\propto \hat{I}_z \hat{s}_z$, produces the nuclear spin flips. As a result, the effective Hamiltonian of such a noncollinear interaction can be written as

$$\mathcal{H}'_{nc} = \frac{Av_0|\varphi_c(\boldsymbol{R})|^2 E_Q I}{\mu_I B_z} \hat{s}_z (\hat{I}_x \hat{I}_z + \hat{I}_z \hat{I}_x). \tag{5.64}$$

It differs from Eq. (5.59) by an extra factor, $\sim \hat{I}_z/B_z$. The presence of the nuclear Zeeman splitting in the denominator ensures the time reversal symmetry of the effective Hamiltonian (5.64). The noncollinear interaction Eq. (5.64) has been derived in (Högele *et al.*, 2012; Urbaszek *et al.*, 2013) by means of the Schrieffer–Wolf transformation (Schrieffer and Wolff, 1966), which, in this situation, allows to eliminate the quadrupole Hamiltonian. Interestingly, a similar procedure applied to the full hyperfine coupling Hamiltonian (5.1) allows one to obtain additional terms responsible for the electron-mediated nuclear spin diffusion (Latta *et al.*, 2011; Urbaszek *et al.*, 2013).

Under experimental conditions the magnitude of the noncollinear coupling is about 1–2 orders of magnitude smaller compared with the collinear hyperfine interaction in Eq. (5.62). Despite this smallness, the effect of the noncollinear interaction can be rather pronounced because, in the magnetic field, the rate of transitions between the nuclear spin levels described by Eq. (5.64) is not suppressed due to the Zeeman splitting of the electron states in $\boldsymbol{B} \parallel z$. This is because the fluctuations of the electron spin z-component do not precess around the magnetic field in the Faraday geometry. The analysis performed in (Högele *et al.*, 2012) demonstrates that the coupling in the form of Eq. (5.64) produces (i) the nuclear spin relaxation and (ii) the nuclear spin polarization responsible for the locking and dragging of quantum dot resonances. Numerical simulations in (Högele *et al.*, 2012; Yang and Sham, 2012) provide a reasonably good description of the experimental data: For the higher energy Zeeman sublevel ("blue" sublevel) the stable solution of the kinetic equation for the nuclear spin distribution function corresponds to the zero detuning between the laser and the quantum dot resonance frequencies. For the lower energy Zeeman sublevel ("red" sublevel) the solution with zero detuning turns out to be unstable and the resonance is avoided. It is shown that the main prerequisite of the dragging effect is the presence of the electron (either in the initial or in the final state of the optical transition) with a relatively long spin lifetime (Urbaszek *et al.*, 2013). This condition is fulfilled in the experiments.

To conclude the chapter, we stress the main points related to the dynamical nuclear polarization:

- Thermal polarization of the nuclear spins is negligible if the nuclear spin system is in the equilibrium with the crystalline lattice.
- The hyperfine interaction of the nuclei with polarized or depolarized (with respect to thermal polarization in magnetic field) electrons produces the efficient transfer of the electron spin to the nuclear spin system.
- The dipole–dipole interaction between the nuclei establishes quasi-equilibrium within the nuclear spin system and makes it possible to describe the nuclear spins by effective spin temperature Θ_N, which can be either positive or negative.
- The feedback of the nuclear spin polarization on the electron spin dynamics via the Overhauser field or of the electron spin dynamics on the nuclei via the Knight field (treated here within the mean field approach) can provide bistable behavior of nuclear and electron spin polarization.
- The flip-flop hyperfine interaction induced dynamical spin polarization is, as a rule, suppressed in strong magnetic fields due to a mismatch of the electron and nuclear Zeeman energies, in contrast to the polarization induced by the noncollinear interaction in strained quantum dots.

6

Interaction of Spins with Light

6.1 Spin dynamics probed by the spin-Faraday and spin-Kerr effects in pump-probe experiments

Due to the rather strong spin-orbit interaction in semiconductors, the spin dynamics of electrons is very pronounced in the optical response. The optical orientation technique and detection of polarized luminescence makes it possible to study spin states of the charge carriers and nuclei as we already discussed in previous chapters. A special role in optical studies of the spin effects in bulk semiconductors and semiconductor nanosystems is played by the pump-probe technique (Awschalom *et al.*, 1985; Zheludev *et al.*, 1994), which has made it possible to study electron and nuclear spin dynamics with high temporal resolution. This method based on the spin-induced Kerr or Faraday effect takes a very important place in the arsenal of experimental tools allowing one to visualize the spin dynamics, in particular, to perform the measurements of ultra long electron spin relaxation times in bulk materials (Kikkawa and Awschalom, 1998), quantum wells (Kikkawa *et al.*, 1997) and quantum dots (Greilich *et al.*, 2006*b*), directly observe spin-orbit coupling-induced electron spin precession and relaxation (Brand *et al.*, 2002; Leyland *et al.*, 2007; Griesbeck *et al.*, 2009), study nuclear spin dynamics (Giri *et al.*, 2013) as well as the spin transport (Kikkawa and Awschalom, 1999; Kato *et al.*, 2004; Crooker and Smith, 2005). The pump-probe technique provides an insight into the intertwined dynamics of electron and nuclear spins. In contrast with the luminescence polarization dynamics, in the pump-probe technique the charge carrier spin dynamics can be studied on time scales that exceed by far the luminescence decay time giving direct access resident (major) charge carrier spin physics. Moreover, application of an additional circularly or linearly polarized "control" pulse (O'Leary and Wang, 2008; Greilich *et al.*, 2009*a*; Phelps *et al.*, 2009; Zhukov *et al.*, 2010) opens up possibilities to manipulate spins by nonmagnetic means. This motivates us to address the pump-probe technique applied to semiconductor nanostructures in detail. In this chapter, we focus on the basic theoretical principles of ultrafast spin manipulation by optical pulses and review some experimental data, further details can be found in (Gywat *et al.*, 2010; Glazov, 2012; Yakovlev and Bayer, 2017).

Electron and Nuclear Spin Dynamics in Semiconductor Nanostructures. M. M. Glazov, Oxford University Press (2018). © M. M. Glazov. DOI: 10.1093/oso/9780198807308.001.0001

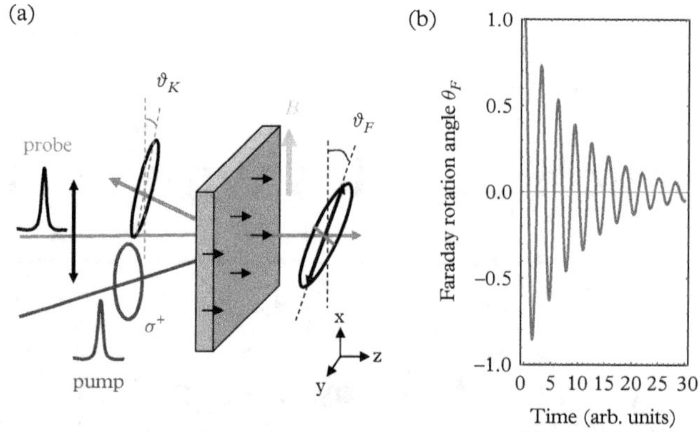

Figure 6.1 *(a) Schematic illustration of the pump-probe Faraday and Kerr rotation measurement techniques in the transverse magnetic field B. Angles ϑ_F and ϑ_K are the spin-Faraday and spin-Kerr rotation angles, respectively. (b) Spin beats of the electrons visualized by the dependence of the Faraday rotation angle on the pump-probe delay.*

The basic principles of the pump-probe technique are schematically shown in Figure 6.1 (a): A sample is subject to a sufficiently strong circularly polarized pump pulse. It causes, in agreement with the selection rules described in Section 2.4, the optical orientation of charge carriers and their Coulomb-correlated complexes such as neutral excitons and charged excitons (trions). After a variable delay, a much weaker linearly polarized probe pulse arrives at the sample. The presence of the nonequilibrium spin polarization in the sample makes the system optically active: The polarization plane of the probe pulse rotates in the transmission geometry (magneto-optical or spin-Faraday effect) and in the reflection geometry (spin-Kerr effect) (Aronov and Ivchenko, 1973). Due to possible absorption processes, the transmitted or reflected probe also acquires a certain degree of a circular polarization; that is, the spin-induced ellipticity. The polarization plane rotation angle as well as the induced ellipticity are proportional to the spin polarization in the system. If the sample is subject to a magnetic field in the Voigt geometry [the field is applied in the plane perpendicular to the pump and probe pulses propagation direction, $\boldsymbol{B} \parallel x$ in the coordinate frame shown in Figure 6.1 (a)], the spins of electrons, holes, and Coulomb-correlated complexes – excitons and trions – precess around the external field. Therefore, the spin ellipticity, Faraday, and Kerr rotation angles oscillate and decay as functions of the delay between the pump and probe pulses, reflecting spin precession and relaxation, see Figure 6.1 (b). Typically, the time resolution of the method is better than 1 ps and the values of detected Faraday or Kerr rotation angles range from microradians in case of single quantum dots to radians for bulk semiconductors placed in microcavities.

To be specific, we focus here, as before, on systems based on GaAs-like semiconductors. In bulk n-type semiconductors with low density of donors N_d: $N_d a_B^3 \ll 1$, Eq. (2.53), quantum wells with a low density electron gas, where $N_e a_B^2 \ll 1$, where N_e is the electron density and a_B is the exciton Bohr radius, Eq. (2.54), and in singly charged (with electrons) quantum dots the optical absorption results in the formation of the singlet X^- trions (also known as D^0X neutral donor-bound excitons in bulk semiconductors): the three particle complexes, consisting of an electron pair with opposite spins and a hole with unpaired spin. The trion formation process is spin-dependent. In accordance with the selection rules outlined in Section 2.4, the σ^+ polarized photon creates an electron with the spin component $-1/2$. Hence, only electrons with spin components equal to $s_z = +1/2$ participate in trion formation for the σ^+ polarized pump pulse, as is schematically shown in Figure 6.2. This is because the spin of the resident electron should be opposite to that of the photocreated one to form a singlet state, see Eq. (2.16). To illustrate resident electron spin coherence generation, we consider the situation where the hole-in-trion spin relaxation time τ_s^T is small compared with the trion radiative lifetime τ_r^T. Hence, after the trion recombination the depolarized electrons return to the system. Therefore, due to the trion formation, a σ^+ pump pulse depolarizes electrons with the spin component $+1/2$ and does not affect electrons with the spin component $-1/2$. Since at low temperatures typical resident electron spin relaxation time $\tau_s \gtrsim 1$ ns and exceeds by far the trion lifetime $\tau_r^T \sim 10 \div 100$ ps, see (Dzhioev *et al.*, 1997; Kikkawa and Awschalom, 1998; Greilich *et al.*, 2006*b*; Fokina *et al.*, 2010) and following Chapters 7 and 8, after the trion recombination the imbalance of electrons with opposite spin projection arises; that is spin polarization (Shabaev *et al.*, 2003; Tribollet *et al.*, 2003). Note that the spin accumulated in the system is directed in the same way

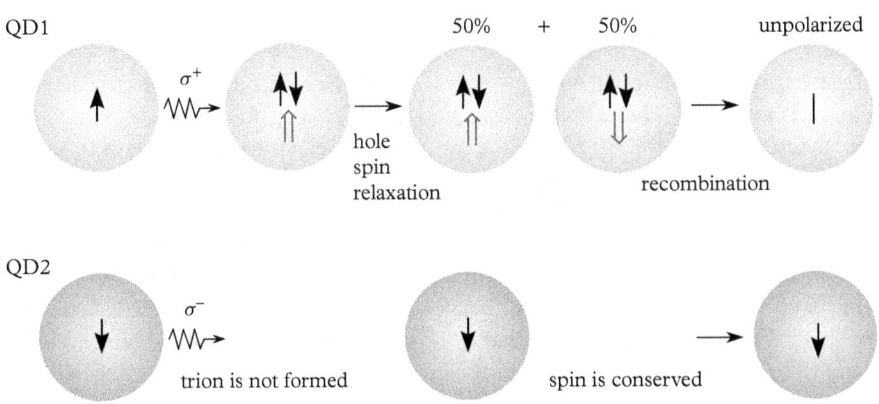

Figure 6.2 *Scheme of resident electron spin orientation at the resonant trion excitation. QD1 and QD2 are two quantum dots where the electron spins before the pump pulse arrival are opposite. As a result of the σ^+ photon absorption the trion is formed in the dot QD1 only.*

as the spin of photocreated electrons: opposite to the z-axis direction at the σ^+ pumping and along the z-axis direction in the case of σ^- pumping. The model of the long living electron spin polarization generation at a trion resonant excitation presented here is quite analogous to the classical spin pumping of majority carriers in semiconductors (Dyakonov and Perel, 1984). If hole spin relaxation is suppressed, $\tau_s^T \gg \tau_r^T$, then in the absence of a magnetic field the electron spin orientation is ineffective: Spin of electrons returning after trions recombination exactly compensates the polarization of carriers that did not participate in the trion formation. The magnetic field presence results in the electron and hole spin precession and, due to the significant difference of the in-plane g-factors for the electrons and holes, in the imbalance of the spins of the carriers that participated and did not take part in the trion formation. Hence, after trion recombination the spin compensation is broken in the magnetic field that leads to the appearance of resident electron spin polarization (Dutt *et al.*, 2005; Kennedy *et al.*, 2006).

To quantify this scenario, we introduce the electron spin pseudovector $\boldsymbol{S} = (S_x, S_y, S_z)$, describing the mean values of spin components of the charge carriers in a quantum well or in quantum dot ensemble. Trion spin states can also be characterized by an effective spin $\mathcal{J}_z = (T_+ - T_-)/2$, where $T_\pm$ is the number of trions with the heavy-hole spin projections onto the z-axis equal to $\pm 3/2$, respectively. Kinetic equations describing spin dynamics of excitons and trions after the system photoexcitation have the form (Chen *et al.*, 2007*b*; Zhukov *et al.*, 2007).

$$\frac{\mathrm{d}S_z}{\mathrm{d}t} = S_y\Omega_B - \frac{S_z}{\tau_s} + \frac{\mathcal{J}_z}{\tau_r^T}, \tag{6.1a}$$

$$\frac{\mathrm{d}S_y}{\mathrm{d}t} = -S_z\Omega_B - \frac{S_y}{\tau_s}, \tag{6.1b}$$

$$\frac{\mathrm{d}\mathcal{J}_z}{\mathrm{d}t} = -\frac{\mathcal{J}_z}{\tau_s^T} - \frac{\mathcal{J}_z}{\tau_r^T}. \tag{6.1c}$$

These expressions follow from the general kinetic Eqs. (2.32) for the spin density matrix. It is assumed here that the magnetic field $\boldsymbol{B}$ is applied along the x-axis in the structure plane, therefore, the x component of electron spin $S_x(t) \equiv 0$, $\Omega_B = g\mu_B B/\hbar$ is the electron spin precession frequency in the external field, and we recall that g is the electron g-factor, μ_B is the Bohr magneton, see Section 2.3, and τ_s is the resident electron spin relaxation time. A simple form of Eq. (6.1c) for the trion spin follows from the fact that the hole spin rotation angle during its lifetime is small compared with that of an electron, because, as a rule, the heavy-hole g-factor in the quantum well or dot plane is small compared with the electron g-factor (Marie *et al.*, 1999). The analysis of the general case is presented in (Zhukov *et al.*, 2007; Yugova *et al.*, 2009*b*). The initial conditions for the system (6.1) are

$$S_y(0) = 0, \quad S_z(0) = -\mathcal{J}_z(0) = -\frac{N_0^T}{2}, \tag{6.2}$$

where N_0^T is the number of photoexcited trions. The system of kinetic equations (6.1) describes well the spin beats in the quantum well structures containing low density electron gas (Zhukov *et al.*, 2007; Astakhov *et al.*, 2008). In quantum dot structures, hyperfine interaction effects in the spin dynamics are more pronounced, see Chapter 9 for further details. Note that in *p*-type quantum wells and quantum dots the physical principles of hole spin coherence excitation are analogous to those considered previously. The specifics of these structures are related to the previously mentioned fact that the hole *g*-factor in the structure plane is much smaller than the *g*-factor of the electron, see (Korn, 2010; Korn *et al.*, 2010; Machnikowski and Kuhn, 2010; Eble *et al.*, 2010) for details. We also note that in the case of resonant generation of excitons in quantum well structures the mechanisms of the spin coherence generation are different and related to (i) capture of electrons by excitons accompanied with the trion formation and (ii) spin-dependent electron-exciton scattering (Zhukov *et al.*, 2007; Fokina *et al.*, 2010; Glazov, 2012).

The spin polarization of charge carriers and electron-hole complexes leads to an optical activity of a medium: Interaction efficiencies for right- and left- circularly polarized electromagnetic waves with such a system turn out to be different. The response of the quantum well structures and planar quantum dot ensembles to the electromagnetic radiation can be conveniently characterized by frequency and polarization dependent light reflection coefficients, $r_\pm(\omega)$, which in the vicinity of the exciton or trion resonance have the form

$$r_\pm(\omega) = \frac{\mathrm{i}\Gamma_{0,\pm}}{\omega_{0,\pm} - \omega - \mathrm{i}(\Gamma_{0,\pm} + \Gamma_\pm)}. \tag{6.3}$$

Here, ω is the probe optical (carrier) frequency, subscripts "+" and "−" refer to the σ^+ and σ^- components of the pulse, respectively, ω_0 is the exciton or trion resonant frequency, Γ_0 is its radiative, and Γ is its nonradiative damping. The difference of the resonance parameters for right- and left- circularly polarized radiation is related to the spin polarization of charge carriers:

$$(\Gamma_{0,+} - \Gamma_{0,-}),\ (\Gamma_+ - \Gamma_-),\ (\omega_{0,+} - \omega_{0,-}) \propto S_z.$$

The spin-Faraday and spin-Kerr rotation, as well as induced ellipticity signals, are formed because of this difference. The signal sensitivity to the electron spin z-component is determined by the selection rules related to the heavy hole excitation under normal incidence of light. Making use of the light hole related resonances one can investigate the dynamics of all spin pseudovector components (Kosaka *et al.*, 2009).

Before discussing the physical mechanisms of the reflection modulation by the electron spin polarization, let us establish the link between the light reflection coefficients and the spin-Faraday, spin-Kerr, and induced ellipticity signals. Assume that the probe pulse propagates along the structure normal, that is z-axis, and let its electric field be polarized along the x-axis. In Faraday effect studies, the probe pulse is split into two components being linearly polarized at $\pm 45^{\circ}$ angles with respect to the initial polarization. The time integrated difference of intensities of these pulses as a function of the pump-probe delay is measured. Hence, the Faraday rotation signal equals (Yugova *et al.*, 2009*a*)

$$\mathcal{F} = \lim_{z\to+\infty} \int_0^{T_{\rm exp}} \left[\left| E_{x'}^{(t)}(z,t) \right|^2 - \left| E_{y'}^{(t)}(z,t) \right|^2 \right] \mathrm{d}t$$
$$= \lim_{z\to+\infty} \int_0^{T_{\rm exp}} 2\,\mathrm{Re}\{E_x E_y^*\} \mathrm{d}t. \tag{6.4a}$$

Here, the set of x', y' axes is rotated by 45° with respect to the initial frame x, y, the

$$E_{x'}^{(t)} = \frac{E_x^{(t)} + E_y^{(t)}}{\sqrt{2}}, \quad \text{and} \quad E_{y'}^{(t)} = \frac{E_x^{(t)} - E_y^{(t)}}{\sqrt{2}}$$

are the components of the field transmitted through the sample. The integration in Eq. (6.4a) is carried out over the measurement time, $T_{\rm exp}$, which exceeds by far all other time constants in the system. The Kerr effect is studied in the reflection geometry and its magnitude is defined as

$$\mathcal{K} = \lim_{z\to-\infty} \int_0^{T_{\rm exp}} \left[\left| E_{x'}^{(r)}(z,t) \right|^2 - \left| E_{y'}^{(r)}(z,t) \right|^2 \right] \mathrm{d}t, \tag{6.4b}$$

where the superscript r indicates that the fields of reflected wave enter Eq. (6.4b). In pump-probe experiments the induced ellipticity effect is also measured, which in transmission geometry is described by the following expression

$$\mathcal{E} = \lim_{z\to+\infty} \int_0^{T_{\rm exp}} \left[\left| E_{\sigma+}^{(t)}(z,t) \right|^2 - \left| E_{\sigma-}^{(t)}(z,t) \right|^2 \right] \mathrm{d}t. \tag{6.4c}$$

In this case, the difference of transmitted wave circularly polarized components, $E_{\sigma\pm}^{(t)} = (E_x^{(t)} \mp \mathrm{i} E_y^{(t)})/\sqrt{2}$, is analyzed. The ellipticity of the reflected light is given by the same expression (6.4c) where the reflected field components $E_{\sigma\pm}^{(r)}$ are used instead of the transmitted ones. We also introduce the Faraday and Kerr rotation angles, ϑ_F and ϑ_K, respectively, as

$$\vartheta_F = \frac{\mathcal{F}}{2\mathcal{I}^{(t)}}, \quad \vartheta_K = \frac{\mathcal{K}}{2\mathcal{I}^{(r)}}, \tag{6.5}$$

where

$$\mathcal{I}^{(t,r)} = \lim_{z\to\pm\infty} \int_0^{T_{\rm exp}} \left[\left| E_{x'}^{(t)}(z,t) \right|^2 + \left| E_{y'}^{(t)}(z,t) \right|^2 \right].$$

In some situations it is also instructive to introduce, by analogy with Eq. (6.5), the ellipticity angle, $\vartheta_E = \mathcal{E}/(2\mathcal{I}^{(t)})$. Equation (6.5) holds for not too large rotation angles, $\vartheta_{F,K,E} \ll 1$, the general expressions are given, for example, in (Smirnov *et al.*, 2015).

In the single quantum well structure and single layers of quantum dots, the transmission coefficients through the layer, $t_\pm(\omega)$, are related to the reflection coefficients, $r_\pm(\omega)$, by the simple expression

$$t_\pm(\omega) = 1 + r_\pm(\omega),$$

which means that the planar layer emits light symmetrically to the both sides of the layer. In realistic systems one has, as a rule, $|r_\pm(\omega)| \ll 1$ and $|r_+(\omega) - r_-(\omega)| \ll |r_\pm(\omega)|$. As a result, the spin Faraday and ellipticity signals are described by a simplified formula (Zhukov *et al.*, 2007; Fokina *et al.*, 2010):

$$\mathcal{E} + \mathrm{i}\mathcal{F} \propto r_+(\omega) - r_-(\omega). \tag{6.6}$$

The Kerr effect is associated with the light reflection from the sample, therefore, it is determined by the interference of the beams reflected from the sample surface (vacuum-sample boundary) and from the well or dot layer. The interference brings about an additional phase equal to $2qL$, where L is the cap layer thickness (distance from the boundary with vacuum and the well or dot array) and q is the light wavevector in the cap layer (Ivchenko *et al.*, 1997; Ivchenko, 2005), see Section 6.4 and Figure 6.8 for details. As a result, the Kerr signal is related to the reflection coefficients from the system as (Zhukov *et al.*, 2007)

$$\mathcal{K} \propto \mathrm{Im}\{e^{2iqL}[r_+(\omega) - r_-(\omega)]\}. \tag{6.7}$$

Having introduced the spin-Faraday, -Kerr, and ellipticity effects, let us now turn to the mechanisms of the spin-induced signal formation. The trion oscillator strength in the σ^+ and σ^- circular polarizations is directly proportional to the number of electrons with corresponding spin projection onto the z-axis, $N_{\pm 1/2}$ (Astakhov *et al.*, 2000, 2002). Indeed, as discussed before, only the resident electrons with $s_z = \pm 1/2$ take part in the singlet trion formation by $\sigma^\pm$ polarized light. This is illustrated in Figure 6.3(a). Therefore, the trion radiative damping being proportional to its oscillator strength can be represented in the form:

$$\Gamma^{\mathrm{T}}_{0,\pm} = \alpha_{\mathrm{T}} \Gamma_0^{\mathrm{X}} N_{\pm 1/2}.$$

Here, α_{T} is a constant and Γ_0^{X} is the exciton radiative damping. Trion resonance frequency renormalization ω_0^{T} due to electron spin polarization and the modification of its nonradiative damping Γ^{T} are negligibly small because they are determined mainly by the exchange interaction of an electron and hole. As a result,

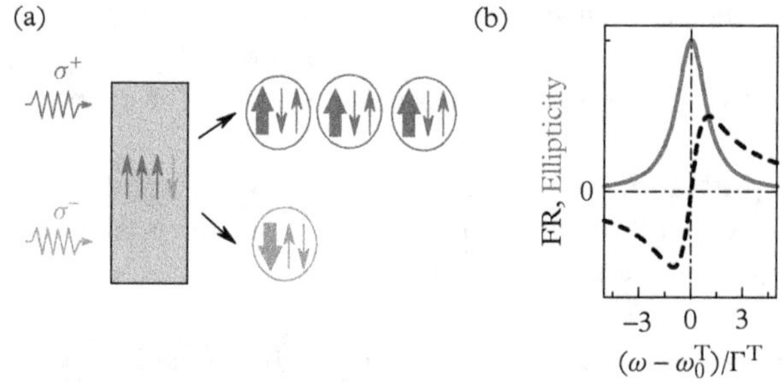

Figure 6.3 *(a) Schematic illustration of the trion oscillator strength polarization dependence in the presence of spin-polarized electrons. The right circularly polarized component of the probe pulse (σ^+) is absorbed more strongly compared with the left circularly polarized one (σ^-), since the number of resident electrons with the spin projection +1/2 onto the z-axis is larger than that of electrons with spin projection −1/2. (b) Spectral dependence of the Faraday rotation (dashed line) and ellipticity (solid line) at trion resonance detection.*

the trion resonance contribution to the spin Faraday and ellipticity signals can be written as (Fokina *et al.*, 2010)

$$\mathcal{E} + \mathrm{i}\mathcal{F} \propto \frac{\mathrm{i}\alpha_{\mathrm{T}}\Gamma_0^{\mathrm{X}}(N_{+1/2} - N_{-1/2})}{\omega_0^{\mathrm{T}} - \omega - \mathrm{i}\Gamma^{\mathrm{T}}} \propto S_z. \tag{6.8}$$

We took into account that $\Gamma_0^{\mathrm{T}} \ll \Gamma^{\mathrm{T}}$ in the derivation of Eq. (6.8). Hence, the spin ellipticity and Faraday rotation signals detected at the trion resonance are indeed proportional to the total electron spin z-component. Frequency dependence of these signals calculated according to Eq. (6.8) is shown in Figure 6.3(b). It can be seen that the Faraday rotation signal is an odd function of the detuning between the probe pulse and the trion resonance frequencies, while the ellipticity signal is an even function. Physically, it is related to the fact that the spin ellipticity is related to the absorption dichroism, which takes its maximum value at the resonance, while Faraday rotation is related to the refraction of electromagnetic waves in the medium. In a similar way, one can describe the spin-Faraday, -Kerr, and ellipticity signal formation for the detection at the exciton resonance or at the band-to-band transitions related to the free electron states (Fokina *et al.*, 2010; Giri *et al.*, 2012; Glazov, 2012).

6.2 Microscopic model of the optical orientation of localized electron spins

Previously, we have formulated the phenomenological description of the electron spin orientation and detection in pump-probe experiments. Now let us turn to the microscopic theory of the spin orientation effect by short optical pulses. We also present the mechanisms of the nonmagnetic control of the electron spin.

For illustration, we consider the electron localized in a single quantum dot. A similar description is also valid for other systems with localized charge carriers, for example, bulk semiconductors with donor-bound electrons and quantum wells where electrons are localized; for example, at interface fluctuations.

The optical excitation of a quantum dot involves the ground two-fold degenerate electron state corresponding to the electron spin components $\pm 1/2$, and the excited state, the singlet trion, whose state is also two-fold degenerate with respect to the unpaired hole spin. For simplicity, we disregard all remaining states assuming that the spectral width of the pump pulse is small as compared to the distance to these "remote" states. In such a four-level model the quantum dot states can be conveniently described by a four component wave function (Yugova *et al.*, 2009*a*)

$$\Psi = \left[\psi_{1/2}, \psi_{-1/2}, \psi_{3/2}, \psi_{-3/2}\right], \tag{6.9}$$

where the subscripts $\pm 1/2$ refer to the resident electron spin states and subscripts $\pm 3/2$ to the photocreated hole in the trion trion spin states. Electron spin components are expressed as quantum mechanical averages of the spin operator $\hat{\boldsymbol{s}} = (\sigma_x, \sigma_y, \sigma_z)/2$ in the form:

$$\begin{gathered} S_z = \left(|\psi_{1/2}|^2 - |\psi_{-1/2}|^2\right)/2, \\ S_x = \mathrm{Re}(\psi_{1/2}\psi_{-1/2}^*), \quad S_y = -\mathrm{Im}(\psi_{1/2}\psi_{-1/2}^*)\,. \end{gathered} \tag{6.10}$$

We assume that the pump pulse duration, τ_p, is long enough compared to the field oscillations period at a carrier (optical) frequency of electromagnetic wave, which we denote as ω_{P}, that is $\tau_p \gg 2\pi/\omega_{\mathrm{P}}$, but it is short enough compared with the electron spin precession period in the quantum dot in an external magnetic field, $\tau_p \ll 2\pi/\Omega_B$, field of nuclear spin fluctuations, or Overhauser fields of polarized nuclei, if present, and compared to the trion lifetime in the quantum dot, $\tau_p \ll \tau_r^T$. These relations between the time scales are typical for pump-probe experiments (Dyakonov, 2017). Making use of the Schrödinger equation for the quantum dot wave function Ψ we obtain the set of equations describing the dynamics of the components $\psi_{\pm 1/2}, \psi_{\pm 3/2}$ under the optical pulse action:

$$\begin{aligned} \mathrm{i}\hbar\dot{\psi}_{3/2} &= \hbar\omega_0^{\mathrm{T}}\psi_{3/2} + V_+(t)\psi_{1/2}\,, \\ \mathrm{i}\hbar\dot{\psi}_{1/2} &= V_+^*(t)\psi_{3/2}\,, \end{aligned} \tag{6.11a}$$

$$\begin{aligned} \mathrm{i}\hbar\dot{\psi}_{-3/2} &= \hbar\omega_0^{\mathrm{T}}\psi_{-3/2} + V_-(t)\psi_{-1/2}\,, \\ \mathrm{i}\hbar\dot{\psi}_{-1/2} &= V_-^*(t)\psi_{-3/2}\,. \end{aligned} \tag{6.11b}$$

Here, $\dot{\psi} \equiv \mathrm{d}\psi/\mathrm{d}t$, and time dependent matrix elements

$$V_\pm(t) = -\int \mathsf{d}(\boldsymbol{r})E_{\sigma^\pm}(\boldsymbol{r},t)\mathrm{d}\boldsymbol{r} \tag{6.12}$$

describe the interaction of the circularly polarized components of the incident field $E_{\sigma^\pm} = (E_x \mp \mathrm{i}E_y)/\sqrt{2} \propto e^{-\mathrm{i}\omega_\mathrm{P} t}$ with the quantum dot. Function $\mathsf{d}(\boldsymbol{r})$ in Eq. (6.12) is an effective dipole moment (Yugova *et al.*, 2009*a*). We stress that the absence of contributions responsible for the electron spin precession in Eqs. (6.11) is related to the small duration of the pump pulse. Under the condition $\tau_p \ll \tau_r^T$ where the quantum mechanical description is still valid, and for $\tau_p \gtrsim 1/\Omega_B$ one has to allow for the influence of an external field in Eqs. (6.11), see (Spatzek *et al.*, 2011). Moreover, if $\tau_p \gtrsim \tau_r^T$, then the description based on the spin density matrix is required and such a pump pulse can be considered to be a quasi-stationary one; see the end of this section and (Loo *et al.*, 2011).

Let us consider in more detail the interaction of the quantum dot with the σ^+ polarized pulse. Under the experimental conditions, see (Greilich *et al.* 2006*a*; Greilich *et al.* 2006*b*; Carter *et al.* 2009), the pumping is carried out by a train of the pulses following with a repetition period $T_R \sim 10$ ns. Usually, this period exceeds by far the trion lifetime, $T_R \gg \tau_r^T$, hence the quantum dot is in the ground state at the moment of the next pulse arrival: $\psi_{3/2} = \psi_{-3/2} = 0$. However, the time T_R is, as a rule, smaller than the relaxation time of the localized electron spin (Dzhioev *et al.*, 1997; Kikkawa and Awschalom, 1998; Greilich *et al.*, 2006*b*), hence by pulse arrival the resident electron can be spin polarized. It follows from Eqs. (6.11b), that under the action of σ^+ polarized pulse the $\psi_{-1/2}(t) = \text{const}$, while system (6.11a) can be rewritten as a single second-order differential equation:

$$\ddot{\psi}_{1/2} - \left(\mathrm{i}\omega' + \frac{\dot{f}(t)}{f(t)}\right)\dot{\psi}_{1/2} + f^2(t)\psi_{1/2} = 0\,. \tag{6.13}$$

Here, $\omega' = \omega_\mathrm{P} - \omega_0^\mathrm{T}$ is the detuning between the pump carrier (photon) frequency and the trion resonance frequency and $f(t)$ is the pump pulse smooth envelope defined as

$$f(t) = -\frac{e^{\mathrm{i}\omega_\mathrm{P} t}}{\hbar}\int \mathsf{d}(\boldsymbol{r})E_{\sigma_+}(\boldsymbol{r},t)\mathrm{d}^3 r\,. \tag{6.14}$$

Equation (6.13) is valid on the time scale $\sim \tau_p$ where other processes like spin precession, spin relaxation and recombination are irrelevant. It follows from Eq. (6.13), that the values of $\psi_{1/2}(-\infty)$, that is before the pump pulse arrival, and the values $\psi_{1/2}(\infty)$ (after the pump pulse arrival) are connected linearly. This linear relation can be written as

$$\psi_{1/2}(\infty) = Qe^{\mathrm{i}\Phi}\psi_{1/2}(-\infty)\,, \tag{6.15}$$

where the real coefficient Q satisfies the condition $0 \leqslant Q \leqslant 1$, and the phase Φ can be chosen in the interval from $-\pi$ to π. These parameters are determined by the pulse shape, its duration, and the detuning from the resonance frequency. Making use of Eqs. (6.10) and (6.15) and the relation $\psi_{-1/2}(t) = \mathit{const}$, we relate the electron spin values before the pump pulse arrival, $\boldsymbol{S}^- = (S_x^-, S_y^-, S_z^-)$, and right after the pump pulse arrival, $\boldsymbol{S}^+ = (S_x^+, S_y^+, S_z^+)$ as:

$$S_z^+ = \frac{Q^2 - 1}{4} + \frac{Q^2 + 1}{2} S_z^- , \tag{6.16a}$$

$$S_x^+ = Q \cos \Phi S_x^- + Q \sin \Phi S_y^- , \tag{6.16b}$$

$$S_y^+ = Q \cos \Phi S_y^- - Q \sin \Phi S_x^- . \tag{6.16c}$$

The system of equations (6.16) describes the electron spin orientation and transformation by a short σ^+ polarized pulse in the quantum dot. One can check that the trion spin polarization defined as $\mathcal{J}_z = (|\psi_{3/2}(\infty)|^2 - |\psi_{-3/2}(\infty)|^2)/2$ right after the pulse arrival equals to

$$\mathcal{J}_z = S_z^- - S_z^+ . \tag{6.17}$$

The transformation of electron spin under the action of left circularly polarized, σ^-, pump pulse is described by the analogous set of equations. In such a case, one has to change the sign in the first term of Eq. (6.16a), and the Φ should be replaced by $-\Phi$ in Eqs. (6.16b) and (6.16c).

Expressions (6.16) and (6.17) provide a quantum mechanical basis for the initial conditions, Eq. (6.2), in the set of equations (6.1) describing coupled electron and trion spin dynamics. It follows from Eqs. (6.16) that σ^+ pulse changes electron spin z component by $S_z^+ - S_z^- = (Q^2 - 1)(1 + 2S_z^-)/4$. The pulse also leads to the rotation of electron spin in the structure plane at $\Phi \neq 0$ [see Eqs. (6.16b) and (6.16c)]. We briefly address the spin rotations induced by circularly polarized pulses in Section 6.3.

Let us now discuss the dependence of the spin coherence generation efficiency on the pump pulse power. To that end it is instructive to start with the resonant pulse where $\omega' = 0$. In this situation the solution of Eq. (6.13) is particularly simple:

$$\psi_{1/2}(t) = \psi_{1/2}(-\infty) \cos \left[\int_{-\infty}^{t} f(t') \mathrm{d}t' \right]. \tag{6.18}$$

Hence the quantities Q and Φ in Eq. (6.15) equal to

$$Q = \cos(\Theta/2), \quad \Phi \equiv 0, \tag{6.19}$$

where

$$\Theta = 2 \int_{-\infty}^{\infty} f(t') \mathrm{d}t' \tag{6.20}$$

is an effective *pulse area*. It follows from Eqs. (6.19) and (6.16) that electron spin excited by a single pulse depends periodically on the pulse area Θ, that is on the field amplitude in the pulse. The power dependence of the electron spin should

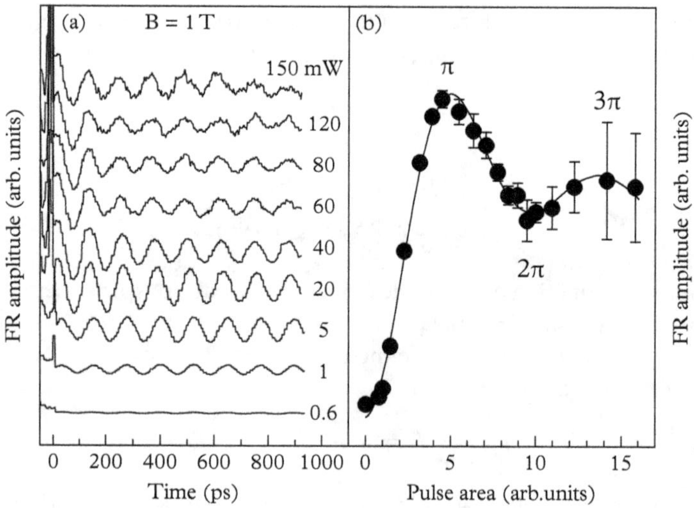

Figure 6.4 *(a) Faraday signal dependence on the delay between the pump and probe pulses measured in the InGaAs/GaAs n-type quantum dot array (20 layers with dot density per layer of 10^{10} cm^{-2}, structure is doped in a such a way that there is one electron per dot on average) at $B = 1$ T. Different curves correspond to the different pump pulse powers. (b) Spin beats amplitude dependence on the pump pulse area obtained by the scaling of the amplitudes power dependence in panel (a). From (Greilich et al., 2006a).*

also have an oscillatory character typical for two-level systems, known as the *Rabi effect* (Landau and Lifshitz, 1977): A strong enough pulse does not merely transfer the system from the ground to the excited state, but can also transfer it back from the excited state to the ground one. It is demonstrated in Figure 6.4, where the spin beat signals and their amplitudes measured in the InGaAs/GaAs quantum dot ensemble in (Greilich *et al.*, 2006*a*) are shown as a function of the pump pulse power.

The solution of Eq. (6.13) at arbitrary detuning ω' can be found analytically for the important case of the so-called Rosen and Zener pulse (Rosen and Zener, 1932) illustrated in the inset of Figure 6.5:

$$f(t) = \frac{\mu}{\cosh(\pi t/\tau_p)}, \tag{6.21}$$

where the coefficient μ characterizes the amplitude of the field in the maximum of the pulse. For the pump pulse in the form of Eq. (6.21) the effective area $\Theta = 2\mu\tau_p$ and one can express the solution Eq. (6.13) at via the hypergeometric function as

$$\frac{\psi_{1/2}(t)}{\psi_{1/2}(-\infty)} = {}_2\mathrm{F}_1\left[\frac{\Theta}{2\pi}, -\frac{\Theta}{2\pi}; \frac{1}{2} - \frac{\mathrm{i}\omega'\tau_p}{2\pi}; \frac{1}{2}\tanh\left(\frac{\pi t}{\tau_p}\right) + \frac{1}{2}\right]. \tag{6.22}$$

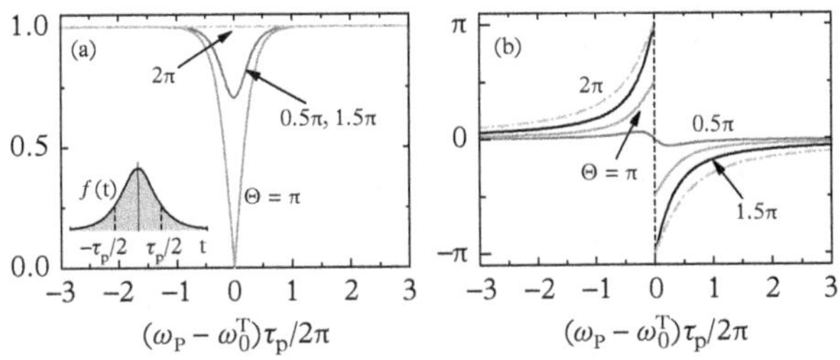

Figure 6.5 *Dependence of the parameters Q (a) and Φ (b), see Eqs. (6.15) and (6.16), on the dimensionless detuning $(\omega_P - \omega_0^T)\tau_p/2\pi$ for the Rosen–Zener pulse (6.21), calculated for different pulse areas $\Theta = \pi/2, \pi, 3\pi/2, 2\pi$. Inset illustrates the pulse shape. From (Yugova et al., 2009a).*

The explicit expressions for the parameters Q and Φ in Eqs. (6.15) are given in (Yugova *et al.*, 2009*a*), their dependence on the detuning for different areas Θ is shown in Figure 6.5. It can be seen from Figure 6.5 that, for relatively large detunings, $|\omega'\tau_p/(2\pi)| \gg 1$, the value of Q is close to 1, while the phase Φ tends to 0. Therefore, as expected from the qualitative arguments, the detuned pulses do not affect the quantum dot state. At $\Theta = \pi$, function $Q(y)$ has a sharp dip at $\omega_P = \omega_0^T$. Such a pulse makes in-plane spin components S_x^+ and S_y^+ zero, and results in the most efficient spin z-component generation.

The four-level model, Eq. (6.9), described previously relies on the strict selection rules valid in quantum wells and most of quantum dot systems where the heavy- and light-hole states are split with the splitting exceeding by far the spectral width of the pump pulse. An interplay of the heavy- and light-hole states results in a more complex relation between the electron spin components $\boldsymbol{S}^-$ and $\boldsymbol{S}^+$ before and after the pump pulse, as demonstrated theoretically in (Smirnov and Glazov, 2012).

At the end of this section let us briefly address the spin initialization by sufficiently long pulses where the spectral width of the pulse $\sim \hbar/\tau_p$ (valid for the bandwidth- or Fourier-transform-limited pulses) can be comparable or even smaller than the Zeeman splitting between the spin sublevels, $\hbar\Omega_B$. Provided that the pulse duration is short as compared with the lifetime of the trion, $\tau_p \ll \tau_r^T$ the action of the pulse can be described within the Schrödinger equation approach, including the Zeeman splitting in the set (6.11), see (Spatzek *et al.*, 2011) for details. The efficiency of the spin initialization along the z-axis decreases with an increase of the dimensionless parameter $\Omega_B\tau_p$, which describes the angle through which the electron spin rotates during the pulse action. Moreover, after the pump pulse the initially unpolarized electron also acquires nonzero S_x and S_y components: The x spin component appears as a result of nonequal depletion of eigenstates of the electron spin in $\boldsymbol{B} \parallel x$ by the pulse, see Figure 6.6, while the y spin component results from the electron spin precession around the x-axis during the pulse action. Hence, phenomenologically, the spin generated by the circularly polarized pulse at $\boldsymbol{S}^- = 0$ can be written as

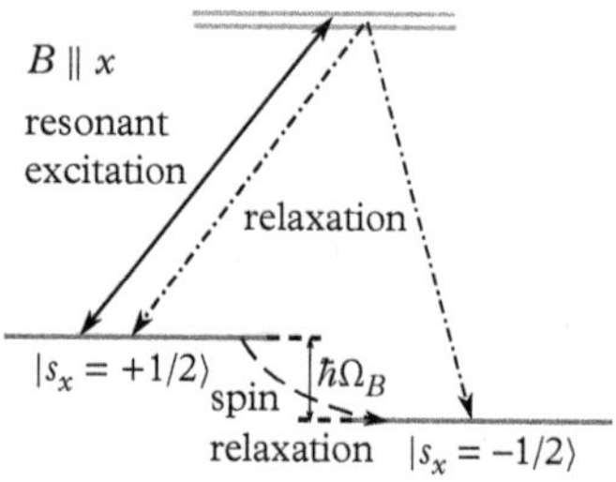

Figure 6.6 *Illustration of the spin initialization by a long pulse in the presence of the magnetic field. The radiation is in resonance with the transition from the $|s_x = +1/2\rangle$ upper Zeeman sublevel to the trion state. The spin relaxation process toward the lower Zeeman sublevel and the relaxation of the trion are shown by the dashed and dash-dotted arrows, respectively.*

$$\boldsymbol{S}^{+} = g_1 [\boldsymbol{e} \times \boldsymbol{e}^*] + g_2 \boldsymbol{B} + g_3 \left[[\boldsymbol{e} \times \boldsymbol{e}^*] \times \boldsymbol{B} \right], \tag{6.23}$$

where g_1, g_2, and g_3 are the parameters that depend on the pump pulse shape, duration, area, and are even functions of $\Omega_B \tau_p$, $\boldsymbol{e}$ is the complex polarization vector of the incident pump pulse, and we remind the reader that the circular polarization degree is $P_c = [\boldsymbol{e} \times \boldsymbol{e}^*]_z$ for the light propagating in the positive direction of the z-axis, see Eq. (2.26a). Here, as before, we disregard for the illustration purposes the actual symmetry of the studied system and consider it as spherically symmetric.

It follows from Eq. (6.23) that in the presence of magnetic field the nonequilibrium electron spin polarization can be achieved by the linearly polarized or unpolarized radiation yielding $\boldsymbol{S}^{+} \parallel \boldsymbol{B}$. The qualitative principle of this process is illustrated in Figure 6.6. It is similar to the laser cooling or coherent population trapping and is based on the realization of the Λ-scheme: The optical excitation depletes the upper Zeeman sublevel by resonantly coupling it with the trion, while the relaxation processes tend to populate the lower Zeeman sublevel. Such a process has been introduced in atomic spin physics (Brossel, Jean *et al.*, 1952) and is also popular for initializing the localized electron spins in semiconductors, see for example (Fu *et al.*, 2005; Atature *et al.*, 2006; De Greve *et al.*, 2011).

6.3 Spin control by short optical pulses

It follows from Eqs. (6.16) that the circularly polarized light transforms the spin, which is already present in the system due to, for example, previous pump pulses. For example, pulses with $Q = 0$ erase completely the electron spin components in the structure plane. Such pulses result in the electron spin alignment along the z-axis. On the contrary, detuned pulses perform the spin rotation through the angle Φ around the z-axis as is schematically illustrated in Figure 6.7. The most favorable conditions for the spin rotation correspond to the so-called 2π-pulses, where the pulse area $\Theta = 2\pi$. As is seen from Figure 6.5 (dash-dot lines), for such pulses the

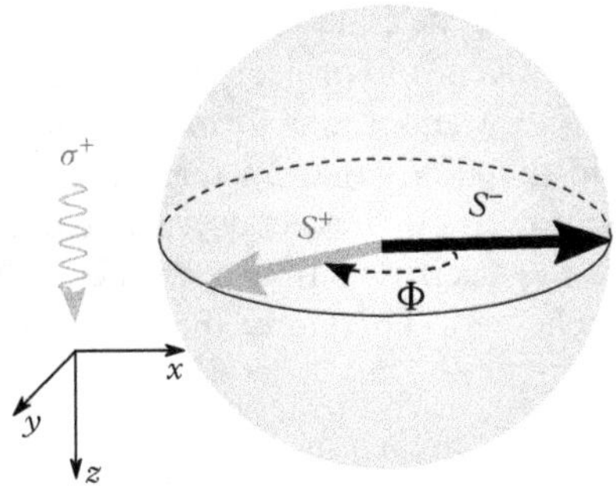

Figure 6.7 *Sketch of the electron spin rotation by the circularly polarized light pulse with $Q = 1$. Thick arrows demonstrate the electron spin orientation before, $\mathbf{S}^-$, and after, $\mathbf{S}^+$, the control pulse arrival, Φ denotes the spin rotation angle in Eqs. (6.16).*

parameter $Q = 1$, that is, there is no suppression of the in-plane spin components, while the spin rotation angle, Φ, is maximal. Naturally, the same condition, $Q = 1$, holds for $\Theta = 2N\pi$ with N being integer.

Qualitatively, the spin rotation by the circularly polarized pulse can be interpreted as the *inverse Faraday effect* (Pitaevskii, 1961; Pershan *et al.*, 1966): The circularly polarized pulse induces the splitting of electron spin sublevels with spin projections $\pm 1/2$ onto the z-axis. This splitting is equivalent to an effective magnetic field directed along the z-axis and results in the spin rotation in the (xy) plane. This phenomenon can be also considered as the dynamic Zeeman effect.

The spin rotation by light pulses opens up the possibility to control electron spins by short optical pulses (Economou *et al.*, 2006). Corresponding experiments are carried out in the three pulse "pump-control-probe" method. The first circularly polarized pulse orients spins of resident electrons, the second one serves to control the spins, and the third pulse is used for the spin polarization detection via the spin-Faraday or spin-Kerr effects. Experimentally, spin rotation by polarized pulses has been demonstrated in (Phelps *et al.*, 2009) and (Carter *et al.*, 2007) for CdTe and GaAs quantum well structures, respectively, and for GaAs quantum dot structures in (Bonadeo *et al.*, 1998; Berezovsky *et al.*, 2008; Greilich *et al.*, 2009*a*; Kim *et al.*, 2011). The detailed analysis of the spin polarization control mechanisms is presented in (Economou *et al.*, 2006; Takagahara, 2010), while the experimental achievements are reviewed in (Ramsay, 2010).

Interestingly, despite the fact that a linearly polarized pump pulse does not lead to spin coherence generation (at least, under the model assumptions formulated in Section 6.2, where, e.g., symmetry reduction effects are disregarded), such a pulse can significantly suppress the spin beats as shown experimentally and theoretically in (Zhukov *et al.*, 2010) in the three-pulse pump-control-probe experiments performed on CdTe/CdMgTe quantum well structures. Qualitative explanation of this effect is as follows: Let $N_{+1/2}$ be the number of the "spin-up" electrons and $N_{-1/2}$ be the number of the "spin-down" electrons at the moment of the control pulse arrival. We represent the linearly polarized pulse as a superposition of two circularly polarized ones. Since the trion oscillator strength is proportional to the number of electrons with a given spin projection, then, due to the σ^+ component of the control pulse, $WN_{+1/2}$ trions are formed, while the absorption of the σ^- component results in the $WN_{-1/2}$ trion formation, where W is a constant proportional to the pulse power. Further, let us assume that the hole-in-trion spin relaxation is faster than the trion recombination, hence, all electrons returned from the trions are depolarized. Therefore, the z component of electron spin changes by

$$\Delta S_z = -(WN_{+1/2} - WN_{-1/2})/2 = -WS_z^-, \tag{6.24}$$

where S_z^- denotes the spin z component before the control pulse arrival. Equation (6.24) describes the suppression of the total spin in the system. Quantitative description of this effect can be carried out within the framework of the equation system (6.11). Calculation shows that the resident electron spin after the control pulse arrival, $\boldsymbol{S}^+$, and spin before the pulse arrival, $\boldsymbol{S}^-$, are linked by a simple relation (Zhukov *et al.*, 2010)

$$\boldsymbol{S}^+ = Q_l^2 \boldsymbol{S}^-. \tag{6.25}$$

Here, Q_l is a constant defined according to Eq. (6.15) for individual circular components of the control pulse. Equation (6.25) clearly demonstrates that the linearly polarized pulse results in the suppression of all spin components.

6.4 Spin coherence detection via spin-Faraday and spin-Kerr effects

The detection of electron spin by the linearly polarized probe pulse via the spin-Faraday, -Kerr, and ellipticity effects can be described using the four level model of the electron and trion states introduced in Section 6.2. We focus here on the quantum dot structure schematically illustrated in Figure 6.8. This structure comprises a planar array of quantum dots covered by a cap layer with the thickness L.

We decompose a probe pulse whose electric field $\boldsymbol{E}^{\mathrm{pr}}(z,t)$ oscillates along the x-axis into a superposition of two circularly polarized ones. Making use of the system (6.11) one can readily calculate the first-order response of the quantum dot on the incident field and evaluate the field-induced dielectric polarization in a dot as (Yugova *et al.*, 2009*a*)

$$\delta P_x^{QD}(\boldsymbol{r},t) = \frac{n_{tr} - n_e}{2i\hbar} \mathrm{d}^*(\boldsymbol{r}) \int d\boldsymbol{r}' \int_{-\infty}^{t} dt' e^{i\omega_0^{\mathrm{T}}(t'-t)} \mathrm{d}(\boldsymbol{r}') E_x^{\mathrm{pr}}(\boldsymbol{r}',t') + \mathrm{c.c}, \tag{6.26a}$$

$$\delta P_y^{QD}(\boldsymbol{r},t) = -\frac{S_z - \mathcal{J}_z}{\hbar} \mathrm{d}^*(\boldsymbol{r}) \int d\boldsymbol{r}' \int_{-\infty}^{t} dt' e^{i\omega_0^{\mathrm{T}}(t'-t)} \mathrm{d}(\boldsymbol{r}') E_x^{\mathrm{pr}}(\boldsymbol{r}',t') + \mathrm{c.c.}. \tag{6.26b}$$

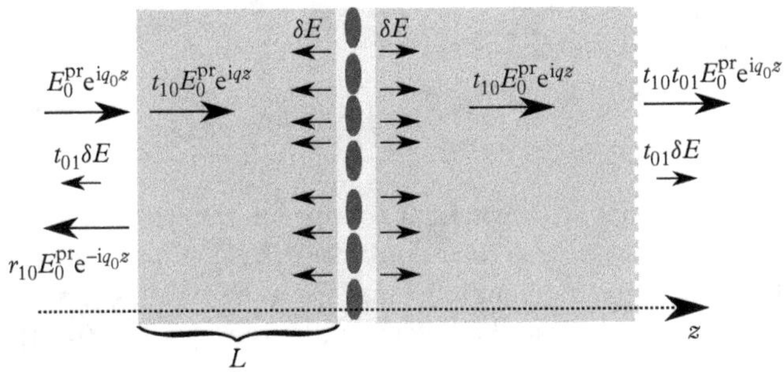

Figure 6.8 *Scheme of the structure, which contains a quantum dot layer covered by a cap layer of the thickness L. Long arrows show the transmission and reflection of the incident probe light with the electric-field amplitude* $\mathbf{E}_0^{pr}$ *on the external surface while short arrows illustrate the formation of the secondary field,* $\delta\mathbf{E}$*, due to scattering of the transmitted light by dots. The* r_{10}*,* t_{10}*, and* t_{01} *are the reflection and transmission coefficients at the corresponding boundaries. Light reflection and the rightmost boundary is neglected.*

Here we assumed that, at the moment of the probe pulse arrival, both the electron and trion states have nonzero occupations: $n_e = |\psi_{1/2}|^2 + |\psi_{-1/2}|^2$ and $n_{tr} = |\psi_{3/2}|^2 + |\psi_{-3/2}|^2$, respectively, and that the spin polarization of the electron and trion can also be present in the system: $S_z = \left(|\psi_{1/2}|^2 - |\psi_{-1/2}|^2\right)/2 \neq 0$ and $\mathcal{J}_z = \left(|\psi_{3/2}|^2 - |\psi_{-3/2}|^2\right)/2 \neq 0$. It follows from Eqs. (6.26) that the induced polarization in the quantum dot has two components. One of those, δP_x^{QD}, is parallel to the polarization plane of the probe pulse, and its value is proportional to the difference of the electron and trion states populations, $n_e - n_{tr}$. Another component, δP_y^{QD}, is orthogonal to the probe pulse polarization plane, and its magnitude is determined by the difference of the electron and trion spin projections onto the z-axis, $S_z - \mathcal{J}_z$. It is the component responsible for the probe pulse polarization plane rotation as well as the appearance of its ellipticity.

Once the quantum dot polarization is known, one can readily calculate the electric field produced by the quantum dots solving the Maxwell equations and determine the spin-Faraday, -Kerr, and ellipticity signals by Eqs. (6.4). For a planar quantum dot ensemble where the distances between the dots are small compared with the wavelength, we have (Yugova *et al.*, 2009*a*)

$$
\mathcal{E} + \mathrm{i}\mathcal{F} = t_{01} t_{10} \frac{3\pi}{q^2 \tau_r^T} N_{QD}^{2D} (\mathcal{J}_z - S_z) \times \int_{-\infty}^{\infty} \mathrm{d}t \int_{-\infty}^{t} \mathrm{d}t' \mathrm{e}^{\mathrm{i}\omega_0^{\mathrm{T}}(t'-t)} E_0^{\mathrm{pr}*}(t) E_0^{\mathrm{pr}}(t'), \tag{6.27}
$$

where N_{QD}^{2D} is the two-dimensional density of the dots in the ensemble of quantum dots, $E_0^{\mathrm{pr}}(t)$ is the time-dependent amplitude of the electric field in the probe pulse, $q = \omega_{\mathrm{pr}}\sqrt{\varepsilon_b}/c$ is the wavevector of light in the system, ε_b is the background dielectric constant assumed to be the same both for dots and matrix, τ_r^T is the trion radiative lifetime:

$$\frac{1}{\tau_r^T} = \frac{4}{3}\frac{q^3}{\varepsilon_b \hbar}\left|\int d\boldsymbol{r} d(\boldsymbol{r})\right|^2. \tag{6.28}$$

The Kerr rotation signal is determined by the interference of the probe pulse reflected from the structure cap layer and from the quantum dot ensemble, see Figure 6.8. For simplicity, the light reflection from the rightmost boundary of the system is disregarded. It is described by the expression [cf. Eq. (6.7)]

$$\mathcal{K} = r_{01} t_{01} t_{10} [\cos(2qL)\mathcal{F} + \sin(2qL)\mathcal{E}]. \tag{6.29}$$

In Eqs. (6.27) and (6.29) r_{01} is the reflection coefficient at the air/cap layer boundary, t_{01} and t_{10} are the transmission coefficients of this boundary inside and outside, respectively, found from the Fresnel formulae:

$$r_{01} = \frac{1-\sqrt{\varepsilon_b}}{1+\sqrt{\varepsilon_b}}, \quad t_{01}t_{10} = \frac{4\sqrt{\varepsilon_b}}{(1+\sqrt{\varepsilon_b})^2}. \tag{6.30}$$

Here, as before, it is assumed that the light reflection from the quantum dot ensemble is weak enough and multiple round trips of light within the structure are disregarded.

It follows from Eqs. (6.27) and (6.29) that the spin signals amplitudes are determined by the difference of the spin z-components of trion and electron in a quantum dot. To analyze the probe carrier frequency dependence of the Faraday rotation and ellipticity signals, we represent the probe pulse field as $E_0^{\rm pr}(t) = E^{(0)} s(t) e^{-i\omega_{\rm pr} t}$, where $s(t)$ is the probe pulse envelope function introduced by analogy with $f(t)$ in Eq. (6.14). One can check that

$$\mathcal{F} \propto \mathrm{Im}\, G(\omega_{\rm pr} - \omega_0^{\rm T}), \quad \mathcal{E} \propto \mathrm{Re}\, G(\omega_{\rm pr} - \omega_0^{\rm T}), \tag{6.31}$$

where

$$G(\Lambda) = \int_{-\infty}^{\infty} dt \int_{-\infty}^{t} dt' s(t) s(t') e^{i\Lambda(t-t')}, \tag{6.32}$$

and $\Lambda = \omega_{\rm pr} - \omega_0^{\rm T}$. The function $G(\Lambda)$ can be calculated for the rectangular, Rosen–Zener, and exponential pulse shapes as follows

$$G(\Lambda) = \begin{cases} \dfrac{1}{\Lambda^2}(i\Lambda\tau_p + 1 - e^{i\Lambda\tau_p}) & \text{for } s(t) = 1 \text{ if } -\tau_p/2 \le t \le \tau_p/2 \\ \dfrac{\tau_p^2}{\pi^2}\zeta\left(2, \frac{1}{2} - \frac{i\Lambda\tau_p}{2\pi}\right) & \text{for } s(t) = \cosh^{-1}(\pi t/\tau_p), \\ \tau_p^2 \dfrac{2 + i\Lambda\tau_p(3+\Lambda^2\tau_p^2)}{(1+\Lambda^2\tau_p^2)^2} & \text{for } s(t) = e^{-|t|/\tau_p}, \end{cases} \tag{6.33}$$

where $\zeta(a,b) = \sum_{k=0}^{\infty}(k+b)^{-a}$ is the generalized Riemann ζ-function. Qualitatively, the behavior of Faraday rotation and ellipticity as functions of detuning between the trion resonance and the probe pulse carrier frequency are similar to those obtained in Section 6.1 and shown in Figure 6.3. It is worth noting that the maximum sensitivity of Faraday rotation and ellipticity signals corresponds to different detunings: Faraday rotation signal takes its maximum value for detuned pulses $|\Lambda|\tau_p \approx 1$, while ellipticity takes its maximum for the resonant ones, in agreement with qualitative arguments in Section 6.1.

In this analysis, we have neglected the temporal dynamics of the electron and trion spins during the probe pulse action τ_p. In agreement with applicability criteria of Eqs. (6.11) this is possible for short enough pulses where $\Omega_B\tau_p \ll 1$. The effect of spin precession during the probe pulse action is analyzed in (Spatzek *et al.*, 2011); basically, the spin precession results in the suppression of the magnitude of the spin-Faraday, -Kerr, and ellipticity signals. The spin coherence detection by the *cw* probe beams is briefly analyzed next.

6.5 Addressing single electron spin with light

Typical values of the spin-Faraday and spin-Kerr rotation angles, Eq. (6.5), in quantum dot structures with $N_{2D}^{QD} \sim 10^{10}$ cm^{-2} probed by relatively short pulses $\tau_p \sim 1$ ps amount to milliradians. Similar magnitudes of signals are expected from quantum well structures with the same density of electrons. In bulk semiconductors the Faraday rotation accumulates when the light propagates through the sample, which gives rise to an increase in the Faraday rotation signal proportional to the sample thickness. In fact, the sensitivity of the detection of the polarization plane rotation by the balance detectors, where the difference in the intensities of the field components polarized at $\pm 45°$ with respect to the initial orientation of polarization plane is measured, see Eqs. (6.4a) and (6.4b), is much higher. This has made it possible to measure the Faraday rotation induced by a single electron spin in a quantum dot (Atature *et al.*, 2007; Mikkelsen *et al.*, 2007) and realize the coherent spin control in a single dot (Berezovsky *et al.*, 2008). In these experiments typical Faraday and Kerr rotation angles are about 10 μrad. Further experiments demonstrated the coherent control of one or two hole spins in single quantum dots (De Greve *et al.*, 2011; Greilich *et al.*, 2011; Godden *et al.*, 2012).

The optical response of a single spin in a Faraday or Kerr signal can be enhanced by placing the quantum dot in a microcavity where the rotation of the polarization plane is accumulated due to multiple round trips of light between the mirrors (De Greve *et al.*, 2011). An especially interesting situation is realized for single quantum dots embedded in micropillar cavities like those shown in Figure 6.9(a). Such micropillar devices allow for deterministic coupling of the photon mode supported by the cavity with the quantum dot embedded between the mirrors (Dousse *et al.*, 2010; Arnold *et al.*, 2015): They support two orthogonally linearly polarized and almost degenerate cavity modes with frequencies close to that of the resonant transition in a quantum dot. In such a quantum dot in

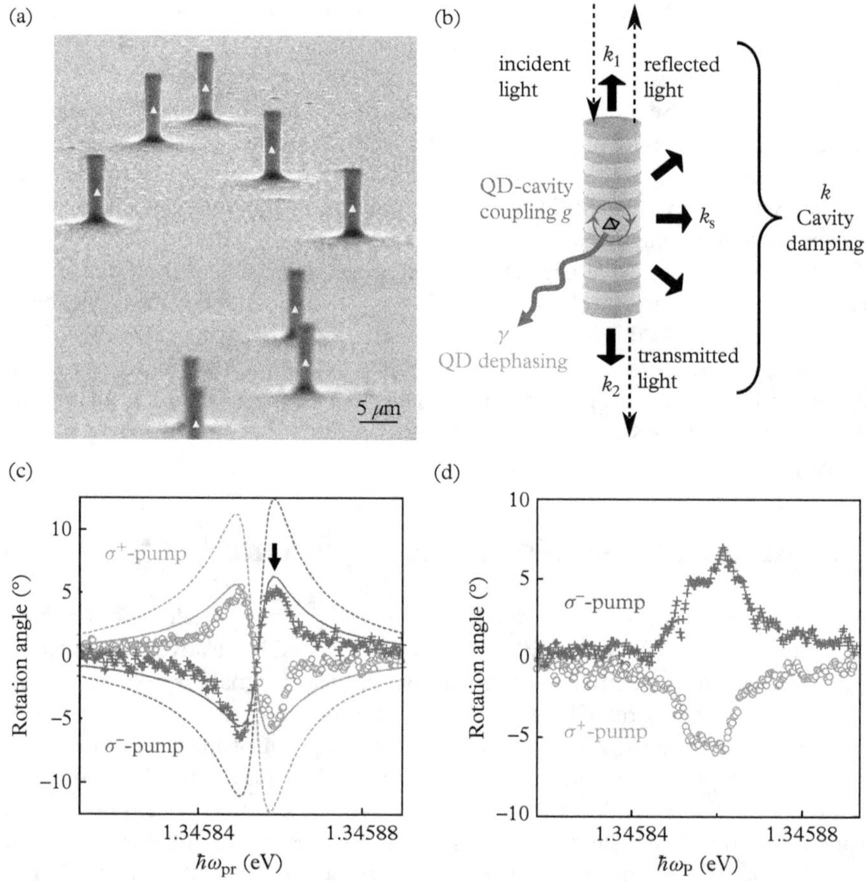

Figure 6.9 *(a) Scanning electron microscope view of various micropillar cavity devices. Triangles denote the positions of the deterministically coupled quantum dots. (b) Sketch of the quantum dot in micropillar device and corresponding interaction processes (circle denotes the light-matter coupling, g; straight arrows show the cavity damping $\kappa_1, \kappa_2, \kappa_s$; wavy arrow shows the dot dephasing, γ). Dashed arrows demonstrate incident, reflected, and transmitted radiation, respectively. (c) Spin-Kerr rotation angle as a function of the probe energy $\hbar\omega_{pr}$, with the pump energy $\hbar\omega_P$ fixed at 1.345857 eV. Symbols: experimental data (σ^--pump in crosses, σ^+-pump in circles): macroscopic Kerr rotation angles up to $+6°$ or $-6°$ are obtained, depending on the pump helicity. Solid line: theoretical fit with partial spin initialization. Dashed line: theoretical prediction with perfect spin initialization. (d) Kerr rotation angle as a function of pump energy $\hbar\omega_P$, with the probe energy $\hbar\omega_{pr}$ fixed at a maximum of Kerr rotation (thick vertical arrow in c). Adapted from (Arnold et al., 2015).*

a microcavity structure, one can expect a substantial modulation of reflectivity and, correspondingly, the large Faraday rotation angles induced by a single charge carrier spin confined in a quantum dot (Hu *et al.*, 2008*b*). Indeed, the reflection coefficients, $r_\pm$, of the monochromatic $\sigma^\pm$ polarized light from such a micropillar device with resident electron, see Figure 6.9(b), can be written as (Walls and Milburn, 2008; Hu *et al.*, 2008*b*; Arnold *et al.*, 2015; Smirnov *et al.*, 2015)

$$r_\pm = \frac{r_0 + r_1}{2} \mp S_z(r_0 - r_1), \tag{6.34}$$

where

$$r_0(\omega) = -1 + \frac{2i\kappa_1}{\omega - \omega_c + i\kappa}, \tag{6.35a}$$

$$r_1(\omega) = -1 + \frac{2i\kappa_1}{\omega - \omega_c + i\kappa - \frac{g^2}{\omega - \omega_0^T + i\gamma}}. \tag{6.35b}$$

Here, $S_z = \pm 1/2$ is the electron spin projection onto the pillar axis z, ω_c and ω_0^T are the cavity and quantum dot resonance frequencies, respectively, κ_1 and κ_2 are the photon escape rates through the mirrors, $\kappa = \kappa_1 + \kappa_2 + \kappa_s$, where κ_s describes the photon escape through the cavity side walls, γ is the trion nonradiative damping rate in a quantum dot, and g is the coupling constant between the quantum dot and the cavity mode. In derivation of Eqs. (6.34) the magnetic field effect is neglected and it is assumed that the light is incident on the mirror characterized by the escape rate κ_1 (see Figure 6.9). Note that the particular choice of phases in Eqs. (6.35) depends on the choice of the origin of coordinates and presence or absence of the cap layer. Depending on mutual orientation of the electron spin and light helicity the reflection coefficient of the structure is either r_0, which corresponds to situation where the trion in quantum dot cannot be formed due to selection rules ($S_z = +1/2$ and light is σ^- polarized or $S_z = -1/2$ and σ^+ polarized light), or r_1, which corresponds to the situation where the trion resonance is excited. In the latter case, the photon enters the cavity and can get absorbed by the trion, when it is emitted back to the cavity and leaves the cavity or gets reabsorbed again, and so on. We stress that Eqs. (6.35) are valid both in the *weak* and *strong* coupling regimes, where $|\gamma - \kappa| < 2g$ or $|\gamma - \kappa| > 2g$, respectively (Ivchenko, 2005; Kavokin *et al.*, 2011). The analysis in (Hu *et al.*, 2008*b*; Arnold *et al.*, 2015) demonstrates that, for example, spin-Kerr rotation angle detected by the *cw* linearly polarized probe [cf. Eq. (6.5)]

$$\vartheta_K = \frac{1}{2} \mathrm{Arg}\{r_+(\omega) r_-^*(\omega)\}, \tag{6.36}$$

can be as high as $\pi/2$, that is, 90°, for the strong coupling regime.

Qualitatively, the substantial optical response of a single spin placed in the microcavity can be readily understood considering the "doubly-resonant" situation where $\omega = \omega_c = \omega_0^{\mathrm{T}}$, that is, where probe cavity and quantum dot resonance frequencies coincide. Let us assume that the cavity is symmetric, $\kappa_1 = \kappa_2$, and the losses through the side mirrors are not substantial, $\kappa_s \ll \kappa_1$. Furthermore, we assume that the cooperativity parameter $g^2/(\gamma\kappa)$, exceeds unity by far. In accordance with Eqs. (6.35) one circular polarization is almost fully transmitted, $|r_0| = \kappa_s/\kappa \ll 1$, while another one is almost fully reflected, $|r_1| = 1 - \gamma\kappa/g^2 \approx 1$. This yields almost fully circularly polarized reflected light, that is, the giant ellipticity of the reflected light and, hence, significant Faraday and Kerr rotation for the probe frequency detuned from the resonance. This has led to proposals for entangling the photon polarization and the single charge carrier spins (Hu *et al.*, 2008*a*).

Figure 6.9(c,d) demonstrates the spin-Kerr rotation angle measured on a single micropillar device with embedded quantum dot with a resident hole (Arnold *et al.*, 2015). The hole spin polarization has been achieved by the pumping with sufficiently long circularly polarized pulse. Figure 6.9(c, d) illustrates "macroscopic", up to $\pm 6°$, Kerr rotation angle by a single hole spin. The dependence of the Kerr rotation angle on the detuning between the probe frequency and micropillar device resonance [panel (c)] and on the detuning between the pump frequency and the device resonance [panel (d)] are in good agreement with model expectations: The pumping is most efficient at the resonance, while the detection is most efficient for slightly detuned pulse in agreement with Eqs. (6.34) and (6.36). Similar results have been demonstrated in (Androvitsaneas *et al.*, 2016) for a quantum dot with a resident electron. We note that in both cases the micropillar devices operated in the weak coupling regime, $2g < |\kappa - \gamma|$.

While the description of the spin coherence detection via Faraday or Kerr rotation in microcavities can be performed considering the electromagnetic radiation as a classical field both in the weak and strong coupling regimes, the electromagnetic field quantization inside the cavity is needed in order to study the spin orientation and control for the microcavity structures operating in the strong light-matter coupling regime. This is because the coupling of a two-level system with a single mode of electromagnetic field results in the formation of the Jaynes–Cummings ladder of coupled states with the splitting between the states scaling as $g\sqrt{m}$, where m is the number of excitations in the system (Jaynes and Cummings, 1963; Kasprzak *et al.*, 2010; Kavokin *et al.*, 2011). The presence of the resident charge carrier in the quantum dot brings about the spin degree of freedom in the Jaynes–Cummings model. As a result, the four-state approach presented in Section 6.2 is not applicable for the strongly coupled quantum dot in microcavity systems. The theory of the optical spin control in such systems is developed in (Smirnov *et al.*, 2015).

6.6 Nuclei-induced Faraday rotation

A spin polarized nuclear system with $\langle \boldsymbol{I} \rangle \neq 0$, just like any other spin-polarized system in a semiconductor, should give rise to Faraday rotation. The transitions between the nuclear spin sublevels typically correspond to the radio frequency range, see Chapter 3, while the internal nuclear transitions correspond typically to the gamma radiation range (Kagan, 1999). The hyperfine interaction of the electron and nuclear spin provides the feedback from the spin-polarized nuclei on the electron spin system. Hence, one may expect that the nuclear spin polarization should manifest itself in the Faraday or Kerr rotation effects at the optical frequencies as well. This has been explored in some detail in atomic physics, where the Faraday rotation induced by polarized nuclei of ytterbium has been observed for atomic beams (Takeuchi *et al.*, 2006) and even for single atoms (Takei *et al.*, 2010).

Artemova and Merkulov have demonstrated theoretically that nuclei-induced Faraday rotation is possible in semiconductors as well and proposed a protocol to probe the Faraday rotation induced by nuclear spin (Artemova and Merkulov, 1985). The protocol consists of two steps: (i) preparation of the spin polarized nuclear system via pumping in the longitudinal magnetic field (Faraday geometry) and (ii) detection of nuclei-induced Faraday rotation by the linearly polarized probe beam. This protocol has been realized in (Giri *et al.*, 2013) on *n*-type GaAs samples. The samples were placed in microcavities to enhance the Faraday rotation (Berger *et al.*, 1996; Cherbunin *et al.*, 2015).

Figure 6.10(a–c) demonstrates microscopic mechanisms behind the Faraday rotation in the bulk semiconductor. Panel (a) shows the situation realized at the nuclear spin pumping stage, where the dynamical nuclear polarization is achieved due to the spin transfer from the electrons to nuclei as discussed in detail in Chapter 5. In this situation, the Faraday signal is induced by optically polarized electrons mainly due to Pauli blocking in the conduction band (Aronov and Ivchenko, 1973; Giri *et al.*, 2012; Glazov, 2012). The temporal dynamics of the Faraday rotation signal in this regime is associated with the electron spin dynamics that takes place on a time scale of $\lesssim 200$ ns (Giri *et al.*, 2012). After the pumping stops the signal drops on a time scale related to the electron spin relaxation, and the remaining Faraday rotation is much weaker and is only due to spin polarized nuclei. The Faraday rotation angle slowly decays to zero on a scale of several minutes 6.10(f). Note that the static rotation due to the external magnetic field is subtracted from the data. The authors of (Giri *et al.*, 2013) checked that this slowly decaying signal is indeed associated with the host lattice nuclei by observing nuclear magnetic resonance induced variation of the amplitude of the decaying contribution.

Microscopically, two dominant microscopic mechanisms of nuclei-induced Faraday rotation are identified, as sketched in Figure 6.10 (b) and (c). Both are related to the effective Overhauser field, see Eq. (5.42),

$$\boldsymbol{B}_N = \frac{A}{g\mu_B}\langle \boldsymbol{I} \rangle,$$

produced by the nuclei and exerted on the electron spins, g is the electron g-factor. The field $\boldsymbol{B}_N$ produces the splitting of the electron spin states and the electron spin polarization. The electron spin polarization induced by the nuclear field $\boldsymbol{B}_N$ is illustrated in Figure 6.10(b) and it favors absorption at the highest spin

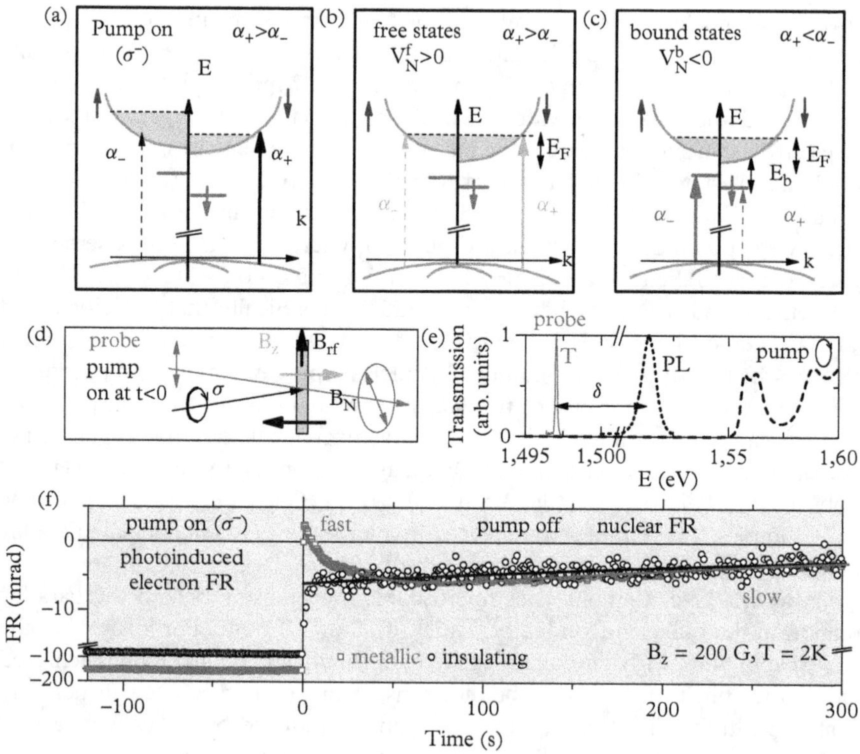

Figure 6.10 *(a)–(c) Band diagrams with relative absorption probabilities α_+, α_- for the two opposite light helicities: (a) under σ^- optical pumping, (b, c) in the presence of nuclear field only, for the transitions involving free (b) and bound (c) electron states. For clarity, only transitions from the heavy hole band are shown. (d) Sketch of the experimental setup. (e) Calculated transmission of the microcavity sample above the stop-band (dashed), measured transmission of the probe beam (solid), and photoluminescence spectrum (dotted). (f) Faraday rotation scans for two samples, including the preparation ($t < 0$) and measurement ($t > 0$) stage. Total pumping time is 10 minutes. From (Giri et al., 2013).*

state due to the Pauli exclusion principle. By contrast, the Overhauser splitting favors absorption at the lowest spin band, see Figure 6.10(c). Thus, the circular dichroism denoted as $\alpha_+ - \alpha_-$ and, correspondingly, the Faraday rotation has opposite sign in these two cases. Quantitatively, in bulk systems it is convenient to introduce the *nuclear Verdet constant* V_N. In analogy with the traditional Verdet constant, this quantity is defined from the relation

$$\vartheta_F = V_N L B_{N,z}, \tag{6.37}$$

where L is the sample thickness. By convention, the sign of the Verdet constant is positive, if the rotation is clockwise for the observer looking in the direction of the magnetic field. The analysis in (Giri *et al.*, 2013) demonstrates that under the experimental conditions for free electrons the nuclei-induced Faraday rotation is dominated by the Pauli-blocking effect, while for the donor-bound electrons the dominant contribution results from the nuclei-induced splitting. For corresponding Verdet constants one has

$$V_N^f = -\frac{e^2\hbar}{48mcn}\frac{g\mu_B}{\sqrt{\delta}}\left[\left(\frac{2\mu_{hh}}{\hbar^2}\right)^{3/2} + \left(\frac{2\mu_{lh}}{\hbar^2}\right)^{3/2}\right], \quad (6.38a)$$

$$V_N^b = \frac{\pi\hbar^3c^2}{(\hbar\omega)^2\tau_r n}\frac{N_b g\mu_B}{4k_BT\delta}. \quad (6.38b)$$

Here, the subscripts f and b correspond to the free and donor-bound electrons, c is the speed of light, m is the electron effective mass, $\mu_{hh} = m_e m_{hh}/(m_e + m_{hh})$ $[\mu_{lh} = m_e m_{lh}/(m_e + m_{lh})]$ are the reduced masses of electron-heavy (light) hole pair, with m_{hh} and m_{lh} being the heavy- and light-hole masses, respectively, n is the background refractive index, τ_r is the radiative lifetime, N_b is the density of bound electrons, δ is the detuning between the fundamental absorption edge energy and the probe energy (the latter is equal to $\hbar\omega$). In derivation of Eqs. (6.38) we assumed that the detuning exceeds by far the electron Fermi energy and the binding energy of a localized state. For a GaAs sample with the density of localization centers $N_b = 2\times10^{15}$ both contributions due to free and localized electrons are comparable in magnitude but have opposite signs, $V_N^f = 0.2$ mrad/(cm·G) and $V_N^b = -0.05$ mrad/(cm·G). These values are in good agreement with experimental results obtained in (Giri *et al.*, 2013).

Moreover, the presence of two contributions to the nuclei-induced Faraday rotation explains the intriguing observation presented in Figure 6.10(f): While in the insulating sample (electron density $N_e = 2\times10^{15}$ cm^{-2}) a monotonous decay of the signal associated with nuclear spin relaxation is observed, a nonmonotonous behavior is clearly seen in the metallic sample (with $N_e = 2\times10^{16}$ cm^{-2}). The analysis shows that in the metallic sample there are two contributions to the Faraday rotation, with strongly different decay times, comparable amplitudes, and opposite signs. The characteristic decay times of the "fast" and "slow" components are of the order of several seconds and several hundreds of seconds, respectively. The interpretation of "fast" and "slow" components in the metallic sample is based on the fact that the complexes of two nearby donors (with the distance on the order of Bohr radius) can bind the electron. As a result, in the metallic sample one observes two contributions to the Faraday effect: (i) the contribution due to nuclei interacting mainly with localized electrons, and (ii) the contribution of other nuclei interacting with free carriers. In accordance with Eqs. (6.38) these contributions have the opposite signs. Moreover, the nuclear spin relaxation rate is expected to be strongly different for these two contributions. The nuclei experiencing the enhanced hyperfine interaction with the localized electron lose their spin faster as compared with nuclei interacting with free electrons.

Hence, the "fast" component is associated with the nuclei interacting with bound carriers and the "slow" component results from the hyperfine interaction with free electrons. Accordingly, in the insulating sample only "fast" component of the nuclei-induced Faraday rotation is observed.

To conclude this section, we note that the observation of the nuclei-induced Faraday rotation in semiconductors provides an additional flexible tool to study experimentally the nuclear spin dynamics without destroying the nuclear spin polarization as in NMR experiments.

7

Electron Spin Decoherence by Nuclei

7.1 Central spin model and its extensions

In quantum dot structures and other systems with localized electrons or holes, the hyperfine interaction of the charge carrier and nuclear spins is enhanced compared with free electrons in semiconductors and, as a first approximation, one can consider the system "carrier spin+nuclear spins" as isolated. This is because, due to the electron localization in a quantum dot with $N_n \sim 10^4 \ldots 10^5$ nuclei, the effective Knight field acting on the nuclear spins from the electron is quite significant and amounts to about ~ 100 mT, see Section 2.6 and (Kalevich *et al.*, 2017). This field exceeds by far the local fields induced by the dipole–dipole interaction with neighboring nuclei and can also exceed the quadrupole splittings of nuclear spin states discussed in Chapter 3. As a result, the electron mainly exchanges its spin with nuclei, while other effects, such as interactions between the nuclei, play a relatively minor role. The necessity to describe the physical effects in the system of interacting electron and nuclear spins leads to formulation of the *central spin model*. In its simplest form the central spin model aims to study the dynamics of the electron spin interacting with a mesoscopic number, $\sim 10^3$–10^6, of nuclei via the hyperfine coupling:

$$\mathcal{H}_{CSM} = \sum_n A_n(\hat{\boldsymbol{s}} \cdot \hat{\boldsymbol{I}}_n). \tag{7.1}$$

Here, $\hat{\boldsymbol{s}}$ is the electron spin operator, $\hat{\boldsymbol{I}}_n$ is the operator of nth nucleus interacting with electron, A_n is the effective hyperfine interaction constant of the electron with the nth nucleus,

$$A_n = Av_0|\varphi_c(\boldsymbol{R}_n)|^2, \tag{7.2}$$

Electron and Nuclear Spin Dynamics in Semiconductor Nanostructures. M. M. Glazov, Oxford University Press (2018). © M. M. Glazov. DOI: 10.1093/oso/9780198807308.001.0001

with $\boldsymbol{R}_n$ being the nucleus position, φ_c the conduction band envelope function, v_0 the primitive cell volume, and A the bulk parameter introduced in Eq. (4.8). In Eqs. (7.1) and (7.2) it is assumed that all isotopes are identical. These expressions can be naturally extended to account for different nuclei species interacting with the charge carrier.

The Hamiltonian (7.1) is directly applicable for the conduction band electron confined in III–V or II–VI quantum dot or bound to a donor in a bulk cubic semiconductor where the hyperfine interaction is isotropic or almost isotropic, see Section 4.3 for details. The most straightforward generalization of the Hamiltonian (7.1) is the account of the anisotropy of the hyperfine interaction by the replacement

$$A_n(\hat{\boldsymbol{s}} \cdot \hat{\boldsymbol{I}}_n) \to \sum_{\alpha,\beta} A_{n,\alpha\beta} \hat{s}_\alpha \hat{I}_{n,\beta},$$

where $\alpha, \beta = x, y$, and z are the Cartesian components and $A_{n,\alpha\beta}$ are the components of the hyperfine interaction tensor constants. In particular, the uniaxial model of $A_{n,zz} \equiv C_\parallel \neq A_{n,xx} = A_{n,yy} \equiv C_\perp$ is usually applied to account for the hyperfine coupling of the nuclear and heavy-hole spins in quantum dots grown along $z \parallel [001]$ crystalline axis, see Eqs. (4.24) and (4.29). The parameters $A_{n,\alpha\beta}$ depend, in general, on the particular system, for example a quantum dot with resident electron or hole, and its point symmetry as discussed in more detail in Chapter 4.

In Eq. (7.1) the hyperfine coupling parameters depend, in general, on the nucleus position via the envelope function $\varphi_c(\boldsymbol{R}_n)$, Eq. (7.2). As a first approximation one may assume that $|\varphi_c(\boldsymbol{r})|$ is constant for coordinates $\boldsymbol{r}$ within the quantum dot and $|\varphi_c(\boldsymbol{r})| = 0$ for $\boldsymbol{r}$ outside the dot. This corresponds to the box model (Ryabchenko and Semenov, 1983) briefly introduced in Section 5.3. Assuming that all nuclei are identical, this corresponds to the situation where for all involved nuclei the hyperfine constants are the same, $A_n = A_0$. Similarly, one can describe within the box model the anisotropic hyperfine coupling. As we demonstrate next, such a model is instructive to elucidate the basic phenomena of an interacting electron-nuclear spin system, however, it has certain restrictions, see Section 7.2. The *graded* box model has been put forward in (Petrov *et al.*, 2009) to overcome some of these limitations.

The Hamiltonian (7.1) accounts for the hyperfine interaction only. One can supplement Eq. (7.1) with the additional terms describing the electron and nuclear Zeeman effects in the external magnetic field, see Eqs. (2.9) and (2.21). Usually, one neglects the nuclear Zeeman effect because for not too large magnetic fields the Knight field exceeds the external magnetic field. In some situations, where the anisotropy of the electron *g*-factor is negligible and the difference of nuclear magnetic moments, if any, is disregarded, one can eliminate the nuclear Zeeman effect by a unitary transformation that corresponds to the transition to the reference frame rotating around the external field with the nuclear Larmor frequency $\boldsymbol{\omega}_0$, see Section 3.1. In a similar way, one may extend the central spin model by taking into account the quadrupole splittings of the nuclear spin states, see Eq. (2.24), and dipole–dipole interaction of nuclei, see Eq. (2.23). The account for the quadrupole effects is particularly important in quantum dot structures where elastic strain is inevitable and even more so for nuclei interacting with resident holes. In the latter case, the hyperfine coupling is about an order of magnitude weaker than for electrons, and, correspondingly, the relative role of the quadrupole interactions strongly increases.

Moreover, the charge carrier can sometimes leave the dot or another electron or hole can be captured due to the tunneling or recharging processes and the photoexcitation, if any. This results in the extra ingredient of the electron dynamics usually accounted for by the effective correlation time of the charge carrier τ_c, see Eq. (5.15), as we already discussed in Chapter 5 while presenting the theory of the dynamical nuclear polarization. Typically, within the central spin model the correlation time τ_c is assumed to be long enough, such that $\tau_c/\hbar \gg \sqrt{N_n}/A_n$, see Section 2.6, otherwise the effect of nuclei on the electron spin diminishes. Physically, this condition describes the situation where the electron spin precession in the field of nuclear fluctuation is faster than the correlation time. Effects related to the finite correlation time are briefly addressed in the end of Section 7.2 and in Chapter 10.

The fact that the charge carrier interacts with large number of nuclei, $N_n \gg 1$, suggests that the direct diagonalization of the Hamiltonian (7.1) and determination of the spin dynamics on the basis of the eigenstates of Eq. (7.1), although being possible in some particular cases, is not instructive. Rather, statistical methods can be applied to analyze the dynamics of electron and nuclear spins within the central spin model. In this regard, the semiclassical approach presented in Section 7.2 turns out to be very fruitful. Quantum approaches to the central spin model are outlined further in Section 7.3.

Before turning to the details of the spin dynamics within the central spin model, let us discuss a simplest possible limit that is exactly solvable and highlights basic physics behind the model in question. To that end we use the box model, assume that $I = 1/2$, and, moreover, consider the extreme uniaxial anisotropy of the hyperfine coupling parameters by putting $A_{n,zz} = A_0$, and $A_{n,\alpha\beta} = 0$ for $\alpha, \beta = x, y$. Such a simplified model with the Hamiltonian

$$\mathcal{H}'_{CSM} = A_0 \sum_n \hat{s}_z \hat{I}_{n,z}, \tag{7.3}$$

can be used as a first approximation to study the dynamics of the heavy hole spin interacting with nuclei in a quantum dot, as described in Section 4.3. The Hamiltonian (7.3) is already diagonal in the basis of states with given z-spin components of the charge carriers, s_z, and all nuclei, $I_{n,z}$ with the eigenenergies $E(s_z, \{I_z\}) = A_0 s_z M_z\{I_z\}$. The quantity [cf. Eq. (3.20)]

$$M_z\{I_z\} = \sum_n I_{n,z}, \tag{7.4}$$

is the z-component of the total nuclear spin calculated for a given configuration of nuclear spin system denoted as $\{I_z\}$. The set of eigenvalues is almost continuous for $N_n \gg 1$ and spans the range from $-|A_0|N_n/4$ to $|A_0|N_n/4$.

If the nuclei are unpolarized then all $I_{n,z}$ are independent and equal to either $+1/2$ or $-1/2$ with the same probability. Hence, in accordance with the central limit theorem in the probability theory (Feller, 1968) the quantity $M_z\{I_z\}$ is distributed at $N_n \to \infty$ in accordance with Gaussian (normal) distribution law. The probability density reads

$$\mathcal{F}(M_z) = \frac{1}{\sqrt{2\pi}\sigma} \exp\left(-\frac{M_z^2}{2\sigma^2}\right). \tag{7.5}$$

Its average value is 0, and the standard deviation is $\sigma = \sqrt{N_n}/2$ as illustrated in Figure 7.1.

As a result, the z-component of the charge carrier spin is conserved since $\hat{s}_z$ commutes with $\mathcal{H}'_{CSM}$. In contrast, in the eigenstates of $\mathcal{H}'_{CSM}$ the transverse spin components s_x and s_y of the charge carrier precess around the z-axis with the rate $A_0 M_z/\hbar$. After the averaging over the realizations of nuclear spins, $\{I_z\}$, one obtains the Gaussian decay of the transverse components with the rate $\sim A_0\sqrt{N_n}/\hbar \sim A/(\sqrt{N_n}\hbar)$, where we took into account that within the box model $|\varphi_c(\boldsymbol{R}_n)|^2 = N_n^{-1}$ and, hence, $A_0 = A/N_n$. Note, that physically the averaging over the realizations of the nuclear spins takes place if one measures the spin dynamics on the ensemble of similar quantum dots or if one measures the spin dynamics on a single quantum dot by performing multiple experiments over a sufficiently long period of time. During this time the nuclear spin configuration rearranges, for example, as a result of the dipole–dipole interaction between the nuclei. The decoherence rate of the in-plane spin components is the larger the smaller number of nuclei is in the localization volume. As we demonstrate next, (i) the relatively fast decoherence of the charge carrier spin with the rate $\sim A/(\sqrt{N_n}\hbar)$, and (ii) the

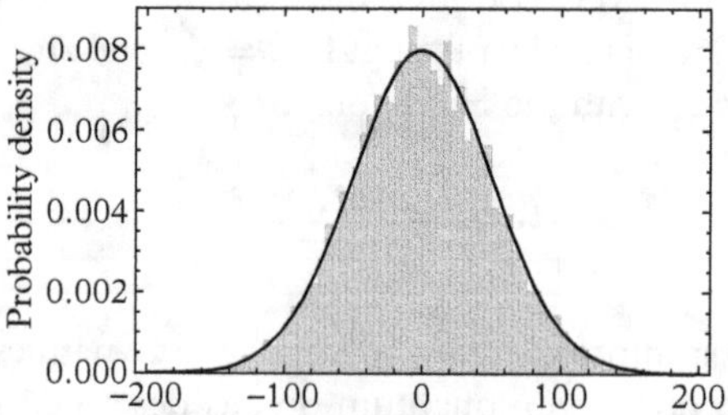

Figure 7.1 *Histogram of the distribution function $\mathcal{F}(M_z)$ calculated for $N_n = 10^4$ nuclear spins by averaging over 10^4 realizations of $\{I_z\}$ and probability density calculated after Eq. (7.5) (the curve).*

presence of nondecaying (or slowly decaying) fraction of the charge carrier spin polarization are the specific features of the central spin model.

7.2 Semiclassical approach

The semiclassical model of the electron and nuclear spin dynamics in quantum dots has been proposed in (Merkulov *et al.*, 2002). To begin with, let us consider the quantum dot with the single resident electron and isotropic hyperfine interaction in the form of Eq. (7.1). The semiclassical approach is based on the separation of time scales of the electron and nuclear spin dynamics. Indeed, as we already mentioned, for a typical quantum dot with $N_n \sim 10^5$ nuclei the time scale associated with the electron spin precession in nuclear field is $\sim \hbar\sqrt{N_n}/A \sim 1$ ns, while the time scale of nuclear spin precession in the electron-induced Knight field is about two or three orders of magnitude larger $\sim \hbar N_n/A \sim 1$ μs. As a result, on the time scales relevant for electron spin dynamics, the nuclear fields in quantum dot can be assumed to be frozen. Thus, in the Heisenberg equation for the electron spin operator

$$\frac{d\hat{\boldsymbol{s}}}{dt} + \frac{\mathrm{i}}{\hbar}[\hat{\boldsymbol{s}}, \mathcal{H}_{CSM}] \equiv \frac{d\hat{\boldsymbol{s}}}{dt} + \hat{\boldsymbol{s}} \times \hat{\boldsymbol{\Omega}}_N = 0, \tag{7.6}$$

with

$$\hat{\boldsymbol{\Omega}}_N = \sum_n \frac{A_n}{\hbar}\hat{\boldsymbol{I}}_n, \tag{7.7}$$

one can treat the operator of the nuclear field $\hat{\boldsymbol{\Omega}}_N$ as a time-independent classical vector $\boldsymbol{\Omega}_N$ whose components are numbers rather than operators. Physically, the vector $\boldsymbol{\Omega}_N$ plays a role of the electron spin precession frequency in the Overhauser field of nuclear spin fluctuation in the quantum dot as schematically illustrated in Figure 7.2. In the ensemble of identical quantum dots $\boldsymbol{\Omega}_N$ varies randomly from dot to dot.

Hence, instead of solving operator Eq. (7.6) one can solve its classical analogue [cf. Eq. (2.33)]:

$$\frac{d\boldsymbol{S}}{dt} + \boldsymbol{S} \times \boldsymbol{\Omega}_N = 0, \tag{7.8}$$

and average over the distribution of the spin precession frequency $\boldsymbol{\Omega}_N$. Its components are distributed according to the Gaussian law [cf. Eq. (7.5) and Fig. 7.1]

$$\mathcal{F}_e(\boldsymbol{\Omega}) = \frac{1}{\pi^{3/2}\delta_e^3} \exp\left(-\frac{\Omega^2}{\delta_e^2}\right). \tag{7.9}$$

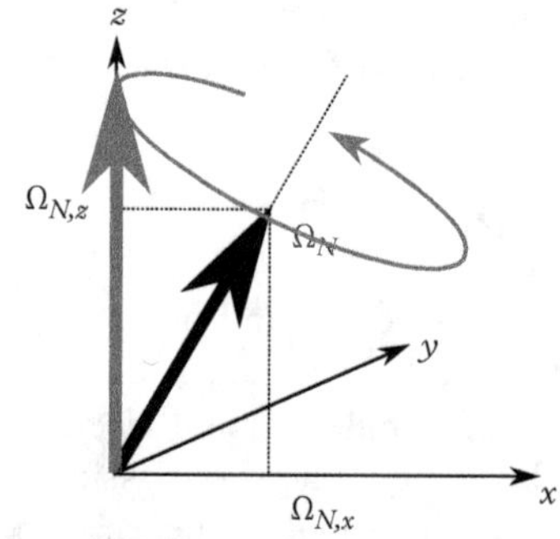

Figure 7.2 *Sketch of the electron spin precession in the field of nuclear fluctuation* $\mathbf{\Omega}_N$.

The parameter δ_e describes the dispersion of the field $\mathbf{\Omega}_N$, particularly,

$$\langle \Omega_N^2 \rangle = \int d\mathbf{\Omega}_N \Omega_N^2 \mathcal{F}_e(\mathbf{\Omega}_N) = \frac{3}{2}\delta_e^2,$$

and can be expressed via the constants A_n as

$$\delta_e^2 = \frac{2}{3\hbar^2} \sum_n I_n(I_n+1) A_n^2. \tag{7.10}$$

For GaAs-like semiconductors with two atoms in the primitive cell Eq. (7.10) can be recast in a somewhat different form (Merkulov *et al.*, 2002)

$$\delta_e^2 = \frac{2I(I+1)}{3\hbar^2} \frac{v_0}{V} \sum_j A_j^2, \quad V = \left(\int d\mathbf{r} |\varphi_c(\mathbf{r})|^4 \right)^{-1}, \tag{7.11}$$

with $I = 3/2$, j running through nuclei within the cell,

$$A_j = \frac{16\pi \mu_B \mu_j}{3 I v_0} |S(\mathbf{R}_j)|^2. \tag{7.12}$$

being the hyperfine coupling constants in the bulk material, see Eq. (4.8), where μ_B and μ_j are the Bohr magneton and nuclear magnetic moment, respectively, v_0 is the primitive cell volume, $S(\mathbf{R}_j)$ is the Bloch function at the nucleus position, and V is the electron localization volume. In GaAs, the sum $\sum_j (A_j)^2$ can be estimated as $\sim 10^{-3}$ meV^2. In the expressions (7.10)–(7.12) we disregarded the spread of $\varphi_c(\mathbf{r})$ in different quantum dots of the ensemble, the generalization is quite straightforward and does not lead to significant changes in the results (Hackmann *et al.*, 2014).

The classical Eq. (7.8) where the operators are replaced by the expectation values is readily solved as (Merkulov *et al.*, 2002)

$$\boldsymbol{s}(\boldsymbol{\Omega}_N;t) = (\boldsymbol{s}_0\cdot\boldsymbol{n})\boldsymbol{n} + [\boldsymbol{s}_0 - (\boldsymbol{s}_0\cdot\boldsymbol{n})\boldsymbol{n}]\cos\Omega_N t + [\boldsymbol{s}_0\times\boldsymbol{n}]\sin\Omega_N t, \tag{7.13}$$

where $\boldsymbol{s}_0 = \boldsymbol{s}(t=0)$ is the initial value of the electron spin, $\boldsymbol{n} = \boldsymbol{\Omega}_N/\Omega_N$ is a unit vector pointing along the nuclear spin fluctuation. In agreement with the schematics shown in Figure 7.2 the electron spin component along the nuclear field fluctuation is conserved, while the transversal to $\boldsymbol{\Omega}_N$ components precess with the frequency Ω_N. The solution (7.13) obtained for fixed $\boldsymbol{\Omega}_N$ should be averaged with the distribution function $\mathcal{F}_{\delta_e}(\boldsymbol{\Omega}_N)$ in Eq. (7.9) to obtain the electron spin evolution in the ensemble of the quantum dots in the form

$$\boldsymbol{S}(t) = \int d\boldsymbol{\Omega}_N \mathcal{F}_e(\boldsymbol{\Omega}_N)\boldsymbol{s}(\boldsymbol{\Omega}_N;t) = \frac{\boldsymbol{S}_0}{3} + \frac{2\boldsymbol{S}_0}{3}\left(1-\frac{t^2\delta_e^2}{2}\right)\exp\left(-\frac{t^2\delta_e^2}{4}\right). \tag{7.14}$$

The same equation holds for a single dot where the averaging is performed over many realizations of nuclear fluctuation.

This dependence is illustrated in Figure 7.3. It can be seen that the electron spin-z component, S_z, is nonmonotonous function of time: First, it drops to about one tenth of initial value and afterward increases up to one third of the initial value. This nondecaying fraction of one third of the initial spin remains constant in the model of frozen nuclear fluctuations. It is related to the fact the electron spin component along the nuclear field $\boldsymbol{\Omega}_N$ is conserved. The parameter δ_e describes the electron spin decoherence rate. A similar result was derived in (Khaetskii *et al.*, 2002) by a somewhat different method.

It is noteworthy that Eq. (7.14) appeared in a somewhat different context long before the papers (Merkulov *et al.*, 2002; Khaetskii *et al.*, 2002) were published. In fact, Kubo and Toyabe derived in 1967 the analogue of Eq. (7.14) for the magnetic resonance in the presence of local fields distributed with the Gaussian

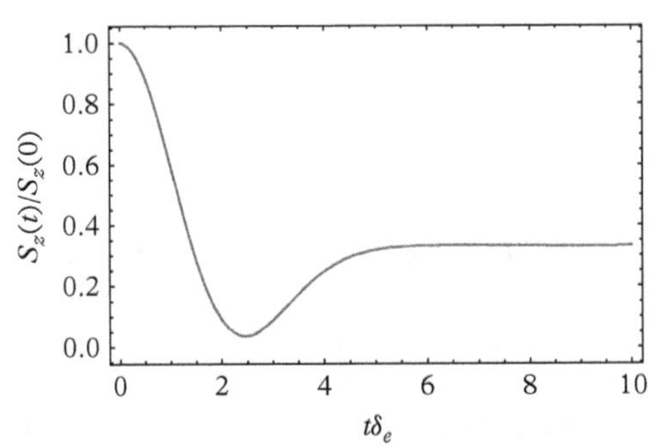

Figure 7.3 *Evolution of the electron spin initially polarized along the z-axis calculated after Eq. (7.14).*

function (Kubo and Toyabe, 1967). This result has been later exploited to study the depolarization of muons (Belousov *et al.*, 1979; Kehr *et al.*, 1978). Later, the same result Eq. (7.14) appeared in the field of the electron spin relaxation in free radicals (Schulten and Wolynes, 1978), see also (Werner *et al.*, 1977) where the same problem was treated numerically.

Let us now briefly discuss the effect of the anisotropic hyperfine interactions relevant; for example, for quantum dots with resident heavy holes. Assuming uniaxial symmetry of the structure we obtain for the distribution function for the hole spin precession frequencies in nuclear fields (Glazov and Ivchenko, 2012)

$$\mathcal{F}_h(\boldsymbol{\Omega}) = \frac{1}{\pi^{3/2}\delta_h^3\epsilon^2}\exp\left(-\frac{\Omega_z^2}{\delta_h^2} - \frac{\Omega_\perp^2}{(\epsilon\delta_h)^2}\right), \tag{7.15}$$

characterized by the two parameters, δ_h, the dispersion of the z-component of the hole spin precession frequency and ϵ being the anisotropy parameter. The case of $\epsilon = 1$ corresponds to the isotropic model, Eq. (7.9), while the $\epsilon = 0$ corresponds to the perfect anisotropy where the nuclear field is always directed along the z-axis, Eq. (7.5). It follows immediately from Eqs. (7.13) and (7.15) that the long-time asymptotic of hole spin at the initial condition $\boldsymbol{S}_0 \parallel z$, reads

$$\frac{S_z(t \gg \delta_h^{-1})}{S_0} = \frac{1-\epsilon^2+\epsilon\sqrt{1-\epsilon^2}\arccos\epsilon}{(1-\epsilon^2)^2}. \tag{7.16}$$

This function monotonously decreases with an increase in ϵ from 1 (at $\epsilon = 0$) to 1/3 (at $\epsilon = 1$) covering the range of perfectly anisotropic hyperfine interaction to the isotropic case. Equation (7.16) can be analytically continued to $\epsilon > 1$, which corresponds to the hyperfine fields predominantly lying in the (xy) plane relevant for the light holes, see Section 4.3 and (Testelin *et al.*, 2009) for details. The nondecaying fraction of z-spin component further decreases with an increase of ϵ and vanishes at $\epsilon \to \infty$. The decoherence rate for the nuclear distribution function in the form of Eq. (7.15) is in the order of δ_h.

In the presence of an external magnetic field the spin precession frequency $\boldsymbol{\Omega}_N$ in Eq. (7.13) should be replaced by the total spin precession frequency (Merkulov *et al.*, 2002; Semenov and Kim, 2003)

$$\boldsymbol{\Omega}_N \to \boldsymbol{\Omega}_N + \boldsymbol{\Omega}_B,$$

where $\boldsymbol{\Omega}_{B,\alpha} = \sum_\beta g_{\alpha\beta}\mu_B B_\beta/\hbar$ and $g_{\alpha\beta}$ are the Cartesian components of the charge carrier g-factor. Let us analyze the two most important limits: (i) $\boldsymbol{\Omega}_B \parallel z$, that is, the Faraday geometry where the magnetic field and initially created spin are parallel, and (ii) $\boldsymbol{\Omega}_B \perp z$, that is, the Voigt geometry with $\boldsymbol{\Omega}_B \perp \boldsymbol{S}_0$.

In the Faraday geometry, the magnetic field suppresses the effect of nuclear spin fluctuations and stabilizes the electron spin. For the isotropic hyperfine interaction, Eq. (7.9), the dynamics of z-spin component can be recast via the special functions as [Ref. (Bechtold *et al.*, 2015), supplemental information]:

$$\frac{S_z(t)}{S_0} = 1 - \frac{\delta_e^2}{\Omega_B^2}\left[1 - \cos(\Omega_B t)e^{-t^2\delta_e^2/4}\right] + \frac{\delta_e^3}{\Omega_B^3}D(\Omega_B/\delta_e)$$
$$- \frac{\sqrt{\pi}\delta_e}{2\Omega_B}e^{-\Omega_B^2/\delta_e^2}\,\mathrm{Re}\left\{\mathrm{erfi}\left[\delta_e\left(\Omega_B^{-1} - it/2\right)\right]\right\}, \tag{7.17}$$

where $D(x) = \exp(-x^2)\int_0^x \exp(t^2)dt$ is the Dawson integral and $\mathrm{erfi}(x) = (2/\sqrt{\pi})\exp(x^2)D(x)$ is the imaginary error function. As a result, the nondecaying fraction of the electron spin monotonously increases with an increase in magnetic field, see the solid curve in Figure 7.4. It can be recast as

$$\frac{S_z(t \gg \delta_e^{-1}, \Omega_B^{-1})}{S_0} = 1 - \frac{\delta_e^2}{\Omega_B^2} + \frac{\delta_e^3 D(\Omega_B/\delta_e)}{\Omega_B^3} \approx 1 - \frac{2}{3}\frac{1}{1 + \Omega_B^2/(2\delta_e^2)}, \tag{7.18}$$

where the approximate analytical formula presented in the last equality differs from the exact result shown in Figure 7.4 by less than 4% (Petrov *et al.*, 2008; Glasenapp *et al.*, 2016).

The situation is different in the magnetic field applied in the Voigt geometry where $\boldsymbol{\Omega}_B \perp \boldsymbol{S}_0$. Here the regular spin precession in the external magnetic field is manifested in the beats of the $S_z(t)$. For $\Omega_B \gg \delta_e$ the beats frequency equals to Ω_B and the oscillations of the spin-z component are described by (Merkulov *et al.*, 2002)

$$S_z(t) = S_0 \cos(\Omega_B t)\exp\left(-\frac{t^2\delta_e^2}{4}\right). \tag{7.19}$$

The spin beats decay according to the Gaussian law with the characteristic time scale; that is, transversal spin relaxation time, $T_2 = 2/\delta_e$. The nondecaying fraction of the electron spin decreases with an increase in magnetic field as shown in Figure 7.4 by the dashed curve, because the probability of finding the realization of $\boldsymbol{\Omega}_N$ such that $\boldsymbol{\Omega}_N + \boldsymbol{\Omega}_B \parallel \boldsymbol{S}_0$ is smaller, the larger Ω_B is.

The model presented here treats nuclear fluctuations as static and it provides an accurate description of the electron spin dynamics on the time scales $t \ll T_N = \sqrt{N_n}/\delta_e$. Let us now address the effects of the electron feedback on nuclear spin dynamics. These effects come into play on the time scale $\sim T_N$ related to the Knight field produced by the electron. Here, for simplicity, we resort to the case of isotropic hyperfine interaction and disregard an external magnetic field $\Omega_B = 0$. Due to the electron-induced nuclear spin dynamics, the effective magnetic field $\boldsymbol{\Omega}_N$ created by the nuclear fluctuation slowly varies with time. The electron spin quickly (during the time $\sim \delta_e^{-1}$) accommodates to the new direction of $\boldsymbol{\Omega}_N$. Averaging over the fast electron spin precession in the nuclear field, from the first term of Eq. (20) we obtain the following expression describing slow dynamics of the electron spin (Merkulov *et al.*, 2002)

$$\overline{\boldsymbol{S}(t)} = \overline{\frac{\boldsymbol{\Omega}_N(t)\left[\boldsymbol{\Omega}_N(t)\cdot\boldsymbol{S}(t)\right]}{\Omega_N^2(t)}}. \tag{7.20}$$

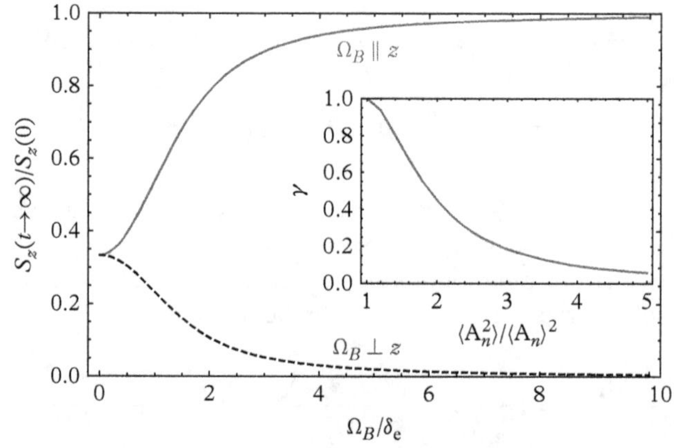

Figure 7.4 *Nondecaying fraction of the initial electron spin calculated in the model of frozen nuclear fluctuations for the longitudinal magnetic field, $\mathbf{\Omega}_B \parallel z$ (solid curve), and for the transverse magnetic field, $\mathbf{\Omega}_B \perp z$ (dashed curve). Inset shows the parameter γ in Eq. (7.22) calculated after Eq. (20) in (Merkulov et al., 2002).*

Here, the overline denotes the averaging over the ensemble of identical quantum dots and, as mentioned before, by the time interval $\gtrsim \delta_e^{-1}$. One can simplify Eq. (7.20) taking into account that the scalar product $[\mathbf{\Omega}_N(t)\cdot \boldsymbol{S}(t)]$ is, up to a constant factor, the energy of the electron-nuclear spin system, see Eqs. (7.1) and Section 5.2. Within the central spin model described by the Hamiltonian (7.1) the energy is conserved, hence, $[\mathbf{\Omega}_N(t)\cdot \boldsymbol{S}(t)] = [\mathbf{\Omega}_N(0)\cdot \boldsymbol{S}_0]$. Moreover, the absolute value of $\mathbf{\Omega}_N$ is also conserved because, as follows from the classical equation of motion of the nuclear spin

$$\frac{d\mathbf{\Omega}_N}{dt} \propto [\boldsymbol{S}(t)\times \mathbf{\Omega}_N] \perp \mathbf{\Omega}_N.$$

As a result, Eq. (7.20) can be conveniently recast as

$$\overline{\boldsymbol{S}(t)} = \overline{\frac{[\mathbf{\Omega}_N(t)\cdot \mathbf{\Omega}_N(0)]}{\Omega_N^2(0)}}\,\frac{\boldsymbol{S}_0}{3}. \tag{7.21}$$

Note that Eqs. (7.21) and Eq. (7.14) are consistent at the time scale $\delta_e^{-1} \ll t \ll T_N$ where the electron spin decoherence in random nuclear fields has already taken place, while the nuclear spin dynamics has not yet been activated.

Our next step is to determine the value of the nuclear field correlator $\overline{[\mathbf{\Omega}_N(t)\cdot \mathbf{\Omega}_N(0)]}$. The analysis performed in (Merkulov *et al.*, 2002) demonstrates that due to the total angular momentum conservation in the interacting "electron+nuclei" spin system the correlation function does not vanish in the limit of $t \gg T_N$:

$$\frac{\overline{[\boldsymbol{\Omega}_N(t)\cdot\boldsymbol{\Omega}_N(0)]}}{\Omega_N^2(0)} \to \gamma, \quad 0<\gamma\leqslant 1. \tag{7.22}$$

Indeed, the total spin of the system $\hat{\boldsymbol{F}} = \hat{\boldsymbol{s}} + \sum_n \hat{\boldsymbol{I}}_n$ remains constant. For $\sqrt{N_n} \gg 1$ the difference between expectation values of $\hat{\boldsymbol{F}}$ and total nuclear spin $\hat{\boldsymbol{M}} = \sum_n \hat{\boldsymbol{I}}$ is negligible; as a result, the total nuclear spin $\boldsymbol{M}$ is effectively conserved. In the framework of the box model the pseudovector $\boldsymbol{\Omega}_N$ and $\boldsymbol{M}$ differ by the constant factor only, as a result $\boldsymbol{\Omega}_N$ is time independent and, hence, $\gamma = 1$ in Eq. (7.22).

The spread of the hyperfine coupling constants A_n allows one to form various nuclear states with the same total spin $\boldsymbol{M}$ but different effective precession frequency $\boldsymbol{\Omega}_N$. In the simplest possible model the value of γ is determined by the probability to realize the state of nuclear spin system with given $\boldsymbol{M}$ and $\boldsymbol{\Omega}_N$, $\mathcal{P}(\boldsymbol{\Omega}_N,\boldsymbol{M})$. The quantity γ is the function of the parameter $\langle A_n^2\rangle/\langle A_n\rangle^2$ with angular brackets denoting the average over the distribution of A_n related to the spread of the electron envelope function $\varphi_c(\boldsymbol{R}_n)$. The ratio $\langle A_n^2\rangle/\langle A_n\rangle^2$ characterizes the spread of the hyperfine coupling constants: Within the box model $\langle A_n^2\rangle/\langle A_n\rangle^2 = 1$, while for smooth distributions of A_n the ratio is larger than 1. The quantity γ monotonously decreases with an increase of $\langle A_n^2\rangle/\langle A_n\rangle^2$ from unity, see the inset in Figure 7.4. Note, that the arguments presented in (Merkulov *et al.*, 2002) demonstrate that the averaging should be performed over the nuclei being in the statistical equilibrium with the electron, that is, over the nuclei whose spins have rotated through the significant angle during the time t. For times $t \gg T_N$, however, practically all nuclei enter the statistical equilibrium. To conclude the discussion of the nondecaying part of the electron spin within the semiclassical approach we stress that the expression (7.22) is valid at the time scales much shorter than those related to the dipole–dipole and quadrupole interactions. The rearrangement of the nuclear spin fluctuation $\boldsymbol{\Omega}_N$ due to these interactions results in the decay of the correlator $\overline{[\boldsymbol{\Omega}_N(t)\cdot\boldsymbol{\Omega}_N(0)]}$ and, hence, in the relaxation of the electron spin $\boldsymbol{S}(t)$ to zero.

At the end of this section, we briefly address the effects related to the finite correlation time τ_c of the electron in the quantum dot or of the donor-bound electron on the its spin decoherence induced by the nuclei. A more detailed discussion of the finite correlation time effects is presented in Chapter 10 (Section 10.3) devoted to the spin noise spectroscopy, while here we restrict ourselves with the analysis of the relevant limiting cases. To that end we consider an ensemble of identical quantum dots enumerated by the subscript $i = 1,\ldots,N_{QD}$, where N_{QD} is the number of quantum dots in the ensemble, and introduce the frozen nuclear fields $\boldsymbol{\Omega}_{N,i}$ in the dots. Here the dynamics of nuclear spins are completely neglected. We allow for the hopping transitions between the localization sites with the rates W_{ij}. Although the electron spin can rotate at the hopping transition between the sites due to the spin-orbit interaction (Kavokin, 2004, 2008; Intronati *et al.*, 2012; Smirnov and Golub, 2017), we disregard spin rotation at the hopping for simplicity. The set of Bloch equations describing the dynamics of electron spin $\boldsymbol{s}_i$ at the ith quantum dot reads (Glazov, 2015)

$$\frac{d\boldsymbol{s}_i}{dt} + \boldsymbol{s}_i \times \boldsymbol{\Omega}_{N,i} + \sum_{j=1}^{N_{QD}} [W_{ji}\boldsymbol{s}_i - W_{ij}\boldsymbol{s}_j] = 0. \tag{7.23}$$

The first two terms in Eq. (7.23) describe spin precession of the localized electrons and the last two terms describe the redistribution of the electrons between the localization sites. We assume that all transition probabilities between the dots are the same, $W_{ij} = W_0/N_{QD}$, and the number of quantum dots is macroscopically large, $N_{QD} \to \infty$. Hence, we can introduce the single correlation time of the electron in the dot as $\tau_c = W_0^{-1}$. Under these assumptions a closed-form solution of Eqs. (7.23) for the total spin in the ensemble $\boldsymbol{S} = \sum_i \boldsymbol{s}_i$ can be derived (Glazov, 2015). Let us now analyze the limiting cases of long and short correlation times.

For *long* correlation times, where $\delta_e \tau_c \gg 1$, the electron spin makes several revolutions before the electron hops to another quantum dot. Here the finite correlation time affects the spin dynamics of the electron only at $t \gg \delta_e^{-1}$ and the total electron spin dynamics is described by the following expression:

$$\boldsymbol{S}(t) = \frac{\boldsymbol{S}_0}{3} \exp\left(-\frac{2t}{3\tau_c}\right) + \frac{2\boldsymbol{S}_0}{3}\left(1 - \frac{t^2\delta_e^2}{2}\right)\exp\left(-\frac{t^2\delta_e^2}{4}\right). \tag{7.24}$$

Compared with Eq. (7.14) the decoherence contribution described by the second term in Eq. (7.24) is the same, while the long-living part of the spin polarization (related to $\boldsymbol{S}_0 \parallel \boldsymbol{\Omega}_{N,i}$) decays exponentially with the rate $2/(3\tau_c)$. Indeed, if the electron leaves the quantum dot i and arrives at another dot j its spin precession frequency changes randomly and spin relaxation occurs. The factor 2/3 in the decoherence rate results from the isotropic distribution of the spin precession frequencies where the ensemble average of $\boldsymbol{\Omega}_{N,i} \cdot \boldsymbol{\Omega}_{N,j}$ is zero, while the average of $(\boldsymbol{\Omega}_{N,i} \cdot \boldsymbol{\Omega}_{N,j})^2$ amounts to $(1/3)\langle \Omega_{N,i}^2 \rangle$.

In the case of a *short* correlation time, $\delta_e \tau_c \ll 1$, the electron spin rotation angle during the hop waiting time is small. Here the so-called *motional narrowing regime* is realized where the nuclear fields are effectively averaged due to the electron motion (Slichter, 1990; Abragam, 2002). Here the spin relaxation is exponential and described by [cf. Eq. (2.42)]

$$\boldsymbol{S}(t) = \boldsymbol{S}_0 \exp\left(-t\delta_e^2 \tau_c\right). \tag{7.25}$$

The rate of spin relaxation is smaller with the shorter correlation time.

In the analysis here we assumed that the electron hopping is at the origin of the finite correlation time τ_c. Generally, the distribution of hop waiting times in systems with localized electrons is quite broad due to the exponential dependence of the W_{ij} on the distance between the localization sites and on the energy (Shklovskii and Efros, 1984). This results in deviations of spin dynamics from simple exponential behavior at long times $t \gg \tau_c$ or $(\delta_e^2 \tau_c)^{-1}$ for long and short correlation times making spin dynamics quite involved and sensitive to system geometry (Harmon and Flatté, 2013; Roundy and Raikh, 2014; Shumilin and Kabanov, 2015; Mkhitaryan and Dobrovitski, 2015; Shumilin *et al.*, 2016). Furthermore, the exchange interaction between the localized electrons and the scattering between the localized and delocalized charge carriers, as well as the processes related to the photoexcitation of the localized electrons may provide sizable contribution to the inverse correlation time (Kavokin, 2008). In any case, Eqs. (7.24) and (7.25) provide a qualitative and, in some cases, quantitative descriptions of the spin dynamics of localized electrons interacting with the host lattice nuclei.

We also note that the quadrupole splitting induced dynamics of the nuclei can provide a contribution to an effective correlation time, τ_c, in Eq. (7.24), even in the absence of any real electron transitions. Indeed, if the dynamics of $\boldsymbol{\Omega}_N$ in each dot is controlled solely by the quadrupole interactions, see Eq. (2.24), then the electron spin experience the randomly varying effective magnetic field $\boldsymbol{\Omega}_N(t)$. The randomness is due to a wide distribution of the quadrupole splittings for nuclei in a quantum dot. The correlation time in this situation is just the correlation time associated with the random dynamics of the nuclei, it can be roughly estimated in the isotropic approximation as

$$\tau_c = \frac{2}{\delta_e^2} \int_0^\infty \langle \Omega_{N,\alpha}(t+\tau)\Omega_{N,\alpha}(t)\rangle d\tau, \quad \alpha = x,\ y,\ \text{or}\ z. \tag{7.26}$$

The angular brackets here denote averaging over t at a fixed τ.

7.3 Quantum mechanical treatment

Let us now proceed to an overview of quantum mechanical approaches to the central spin model. We have already briefly addressed the exact solution of the extreme anisotropy case, see Eq. (7.3). In such a situation, the spin dynamics is almost trivial. As a next step, it is instructive to consider the situation of the isotropic hyperfine interaction within the box model

$$\mathcal{H}_{CSM} = A_0 \sum_n (\hat{\boldsymbol{s}} \cdot \hat{\boldsymbol{I}}_n), \tag{7.27}$$

where A_0 is a constant. The energy spectrum of the Hamiltonian (7.27) can be also found analytically. Here we can make use of the total spin conservation and introduce the operators

$$\hat{\boldsymbol{M}} = \sum_n \hat{\boldsymbol{I}}_n, \quad \hat{\boldsymbol{F}} = \hat{\boldsymbol{s}} + \hat{\boldsymbol{M}}, \tag{7.28}$$

of the total spin of the nuclei, $\hat{\boldsymbol{M}}$, and of the total spin of the system $\hat{\boldsymbol{F}}$. Making use of the fact that $\hat{F}^2 = \hat{M}^2 + \hat{s}^2 + 2(\hat{\boldsymbol{s}} \cdot \hat{\boldsymbol{M}})$ the Hamiltonian (7.27) can be expressed as

$$\mathcal{H}_{CSM} = \frac{A_0}{2}\left(\hat{F}^2 - \hat{s}^2 - \hat{M}^2\right) = \frac{A_0}{2}\left[F(F+1) - M(M+1) - \frac{3}{4}\right], \tag{7.29}$$

This Hamiltonian (7.29) is readily diagonalized in the basis of states with the given total spin of the system F and total spin of the nuclei M (Bortz and Stolze, 2007*a*, *b*; Kozlov, 2007). For the fixed M, we have just two possible values of $F = M \pm 1/2$; as a result, the eigenenergies of Eq. (7.29) read

$$E_{M,\pm} = \frac{A_0}{2}\begin{cases} M, & F = M + 1/2, \\ -(M+1), & F = M - 1/2. \end{cases} \tag{7.30}$$

The eigenfunctions of Eq. (7.29) are spin functions with the given total spin F and its z-component F_z, which can be expressed as the linear combinations of the products of electron spinors $\uparrow, \downarrow$ and the eigenfunctions $|M, M_z\rangle$ of the operators $\hat{M}^2$, $\hat{M}_z$ of the nuclear spin system (Edmonds, 1974; Varshalovich *et al.*, 1988). The diagonalization of the box model opens a wide range of opportunities to study the dynamics of interacting electron and nuclear spin systems analytically for various excitation conditions and various initial states of the nuclear spin system (Bortz and Stolze, 2007*b*; Kozlov, 2007).

Here we just focus on the situation already considered previously in the semiclassical model, see Section 7.2: The nuclei are unpolarized and the electron at $t = 0$ is polarized along the z-axis. To model the electron decoherence quantum mechanically we consider the electron interacting with the ensemble of $I = 1/2$ nuclei and, first, construct the solution for $S_z(t; M, M_z) = \langle \hat{s}_z \rangle$ for the following initial conditions: At $t = 0$ the electron is in the spin-up state and the nuclear spin system is in the state with the total spin M and its z-component M_z. It follows from (Bortz and Stolze, 2007*b*; Kozlov, 2007), see also (Zhang *et al.*, 2006), that

$$S_z(t; M, M_z) = \frac{1}{2(2M+1)^2}\{(2M_z - 1)^2 + 4(M - M_z + 1)(M + M_z)\cos[A_0(M + 1/2)t]\}. \tag{7.31}$$

In a sense, Eq. (7.31) is a quantum-mechanical analogue of Eq. (7.13). Now, as a second step of evaluation, we need to average Eq. (7.31) over the realizations of nuclear spin system state making use of the fact that for given M there are $2M + 1$ values of $M_z = -M, -M + 1, \ldots, M$, and that the state with a given M can be combined from 2^{N_n} individual nuclear states in

$$\mathcal{N}(M) = \frac{(2M+1)N_n!}{\left(\frac{N_n}{2} + M + 1\right)!\left(\frac{N_n}{2} - M\right)!},$$

ways (Landau and Lifshitz, 1977). Evaluating the average in the limit $N_n \to \infty$ we obtain

$$S_z(t) = 2^{-N_n} \sum_{M,M_z} \mathcal{N}(M) S_z(t; M, M_z) = \frac{1}{6} + \frac{1}{3}\left(1 - \frac{t^2\delta_e^2}{2}\right) \exp\left(-\frac{t^2\delta_e^2}{4}\right), \tag{7.32}$$

with $\delta_e^2 = A_0 N_n/(2\hbar)$. It is in full agreement with semiclassical Eqs. (7.14), (7.10). We stress that Eq. (7.32) fully accounts for the nuclear spin dynamics within the box model. In agreement with the expectations based on the total spin angular momentum conservation the nondecaying fraction of the initial spin is one third. Hence, the exact quantum-mechanical solution of the central spin within the box model fully corroborates with the semiclassical results.

As a next step, we need to analyze how the spread of the hyperfine coupling constants affects the electron spin decoherence. Formally, the central spin model defined by the Hamiltonian (7.1) is exactly solvable by means of the Bethe ansatz (Gaudin, 1976). It means that one reduce the eigenvalue problem $\mathcal{H}_{CSM}\Psi = E\Psi$ to a set of algebraic equations (Gaudin, 1976; Sklyanin, 1989; Bortz and Stolze, 2007*a*). However, this set cannot be solved analytically, and, moreover, its complexity grows so fast with the number of the nuclear spins N_b that only systems with limited $N_b \lesssim 50$ have been studied by combining the Bethe ansatz and statistical methods (Faribault and Schuricht, 2013*a*). Alternative approaches are based on consideration of particular initial states of nuclear spin systems, application of high-order perturbation theory (both in the absence or presence of an external magnetic field) (Khaetskii *et al.*, 2002; Coish and Loss, 2004; Yao *et al.*, 2006; Liu *et al.*, 2007; Cywiński *et al.*, 2010; Barnes *et al.*, 2011) or by using various numerical techniques such as the expansion of the time-evolution operator $\exp(-i\mathcal{H}_{CSM}t)$ over the Chebyshev polynomials (Hackmann and Anders, 2014), density-matrix renormalization group method (Stanek *et al.*, 2013, 2014) or by using the Glauber coherent P-states representation (Al-Hassanieh *et al.*, 2006).

The calculations performed in a number of works (Al-Hassanieh *et al.*, 2006; Zhang *et al.*, 2006; Coish *et al.*, 2007; Stanek *et al.*, 2014) demonstrate that on the time scales $t \ll T_N = \sqrt{N_n}/\delta_e$ the semiclassical model of (Merkulov *et al.*, 2002) formulated in Section 7.2 is in agreement with various quantum-mechanical approaches. This corroborates the model of frozen nuclear spins, Eq. (7.13), for wide range of distributions of hyperfine coupling constants A_n.

Somewhat more subtle situation is related to the long-term asymptotic, particularly, with a nondecaying part of the spin polarization, see Eqs. (7.21) and (7.22). The arguments and calculations presented in (Khaetskii *et al.*, 2002; Coish and Loss, 2004; Erlingsson and Nazarov, 2004; Chen *et al.*, 2007*a*) show that for $t \to \infty$ a very slow decay of the electron spin in the form

$$S_z(t) \propto \frac{1}{\ln^{\alpha}(t/T_N)}, \tag{7.33}$$

takes place. Here, $\alpha > 0$ is a constant that depends on the particular model and approximations used. The value of α as well as the applicability conditions of Eq. (7.33) are still debated in the literature (Faribault and Schuricht, 2013*a*,; Hackmann and Anders, 2014; Uhrig *et al.*, 2014). Numerical simulations and the analysis based on general Mazur inequalities on correlations (Faribault and Schuricht, 2013*a*,; Hackmann and Anders, 2014) show that for a finite N_n there are strong arguments in favor of $\gamma \neq 0$ in Eq. (7.22) and $\alpha = 0$ in Eq. (7.33). Correspondingly, the nondecaying fraction of the initial spin polarization should be present, although these works do not exclude the dependence similar to Eq. (7.33) or a power-law asymptotic on the intermediate time scales, $\delta_e^{-1} \ll T \ll T_N$. However, the nondecaying fraction may not be realized in the thermodynamic limit $N_n \to \infty$ if a wide distribution of the hyperfine coupling constants A_n is present. Particularly, in (Chen *et al.*, 2007*a*), the following argument based on the distribution function of the constants A_n has been put forward, which demonstrates the asymptotic (7.33) with $\alpha = 3/2$ for Gaussian envelope function $\varphi_{c.b}(\boldsymbol{r}) \propto \exp(-r^2/2a^2)$: For a nuclear spin coupled with an electron spin via the constant $A_n \ll A_0$, where A_0 is the hyperfine coupling constant in the quantum dot center where $r = 0$, the transfer of the spin angular moment from the electron occurs on the time scale $t \gtrsim \hbar/A_n$. Hence, only the nuclear spins within the sphere of a radius $R(t) \sim a \ln^{1/2}(A_0 t/\hbar)$ contribute to the spin decay by the time t. Assuming that the remaining electron spin is inversely proportional to the number of involved nuclei we have $S_z(t) \propto R^{-3}(t) \propto 1/\ln^{3/2}(A_0 t/\hbar)$ in agreement with Eq. (7.33).

The presence or absence of the nondecaying part of the initial electron spin polarization within the central spin model is still a very active topic in the field. However, in the majority of practical cases the interactions within the nuclear spin systems, such as quadrupole couplings, dipole–dipole interactions, or the possibility of electrons hopping between the localization sites, results in the decay of $S_z(t) \to 0$ in a long time limit.

7.4 Experimental manifestations

The first direct observation of the electron spin decoherence induced by the hyperfine interaction with the lattice nuclei spins in quantum dots has been performed in (Braun *et al.*, 2005) on the ensemble of the *p*-type InAs/GaAs quantum dots grown on a (001) GaAs substrate by means of the time-resolved optical orientation experiments. In this work, the sample was excited by the 1.5 ps duration circularly polarized pulses with the repetition rate of about 80 MHz. The time-resolved photoluminescence of the X^+ trion has been detected in σ^+ (I^+) and σ^- (I^-) polarizations, see the inset in Figure 7.5. In accordance with the optical selection rules, see Sections 2.4 and 5.3, in (001)-grown quantum dots the circular polarization of emission $P_c = (I^+ - I^-)/(I^+ + I^-)$ is directly proportional to the electron spin projection S_z on the growth axis. The circular polarization

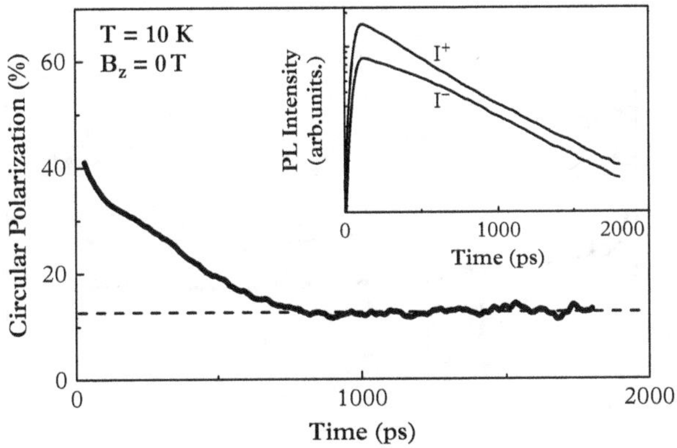

Figure 7.5 *Circular polarization dynamics of the QD luminescence after a circularly polarized σ^+ laser excitation. Inset: Photoluminescence intensity copolarized, I^+, and counter-polarized, I^-, with the laser (semilogarithmic scale). The detection energy is $E_{det} = 1.11$ eV. From (Braun et al., 2005).*

degree is plotted in Figure 7.5 as a function of time. It demonstrates remarkable behavior: Within the first 800 ps after the pump pulse the circular polarization degree drops from its initial value of about 40% down to 12% and then remains constant during the time when the photoluminescence signal is still measurable. From these experimental data one may infer that the relaxation time of the "tail" exceeds 10 ns. Overall, the observed behavior is consistent with the predictions of the central spin model presented in Sections 7.2 and 7.3: The electron spin polarization drops down to one-third of its initial value and then remains constant until the nuclear spin dynamics comes into play.

Further proof of the hyperfine-induced spin decoherence follows from the magnetic field dependence of the circular polarization degree temporal dynamics shown in Figure 7.6. Here, in agreement with Eq. (7.18) and Figure 7.4 a small magnetic field in the range of 100...400 mT applied in the Faraday geometry, that is, $\boldsymbol{B} \parallel z$, strongly suppresses the spin decoherence. The experimental results are in reasonable agreement with the semiclassical model calculations, see the inset of in Figure 7.6, where, additionally, a spread of quantum dot sizes and, hence, of the nuclear fluctuations δ_e has been taken into account. The variation of δ_e in the quantum dot ensemble leads to the suppression of the minimum in the $\boldsymbol{S}(t)$ dependence, see Eq. (7.14), and makes the calculated curves closer to the measured ones. The average decoherence time $1/\delta_e \approx 500$ ps is very close to the estimate made after Eq. (7.11) where the contributions of In and As isotopes are included.

The suppression of the nuclei-induced spin decoherence in the longitudinal magnetic field is most pronounced in the steady-state or time-integrated response of the quantum dots, where one can directly observe the electron spin stabilization by the magnetic field discussed before, see Eq. (7.18), and shown in Figure 7.4 by

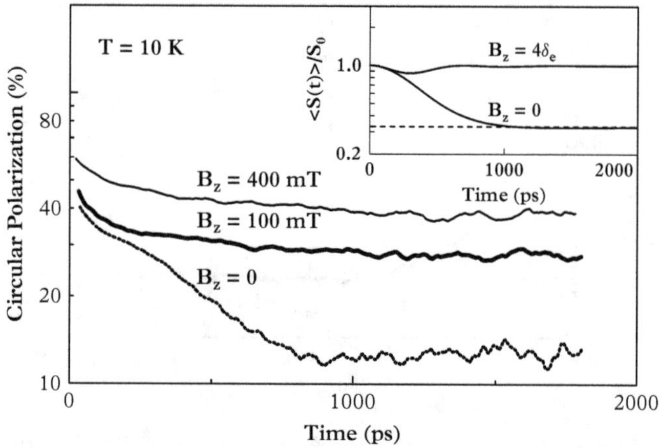

Figure 7.6 *Circular polarization dynamics of the QD ground state luminescence (semilogarithmic scale) for $B_z = 0$, $B_z = 100$ mT, and $B_z = 400$ mT. The inset displays the calculated time dependence of the average electron spin within the semiclassical model with allowance for the variation in quantum dot size in the ensemble, see text for details. From (Braun et al., 2005).*

the solid curve. Corresponding measurements have been performed both on the quantum dot ensembles, see for example (Petrov *et al.*, 2008), and on individual quantum dots (Braun *et al.*, 2006*a*). As an illustration we show in Figure 7.7 the experimental data from (Petrov *et al.*, 2008) (circles) together with the numerical calculation of the nondecaying fraction of the electron spin (solid line) and the approximate analytical formula, the last equality in Eq. (7.18). In this work, the electron spin polarization was extracted from the time-integrated photoluminescence experiments under the nonresonant excitation on *n*-type quantum InGaAs quantum dots, see (Cherbunin *et al.*, 2008) for details on sample and experimental technique. The experimental data shown in Figure 7.4 strongly support the semiclassical model of the spin dynamics presented in Section 7.2. From these data one can estimate the fluctuation of the Overhauser field $B_f = \hbar\delta_e/(g\mu_B)$ to be about 30 mT. The experimental results on single InAs/GaAs quantum dots presented in (Braun *et al.*, 2006*a*) are also in good agreement with the semiclassical model. We note that in these works the modulation of the excitation polarization has been used in order to avoid the dynamical nuclear polarization effects discussed in Chapter 5.

The electron spin decoherence caused by the hyperfine coupling with nuclear spins can be also studied by transport methods in gate-defined GaAs quantum dots (Petta *et al.*, 2005; Johnson *et al.*, 2005), see (Hanson *et al.*, 2007) for a review. In such structures the two-electron spin states can be controlled by applying gate voltages. In particular, by means of gates a double-well potential such as that shown in Figure 7.8(a,b) is formed. Depending on the gate voltages the asymmetric confinement can be realized and the two electrons can be accumulated in one

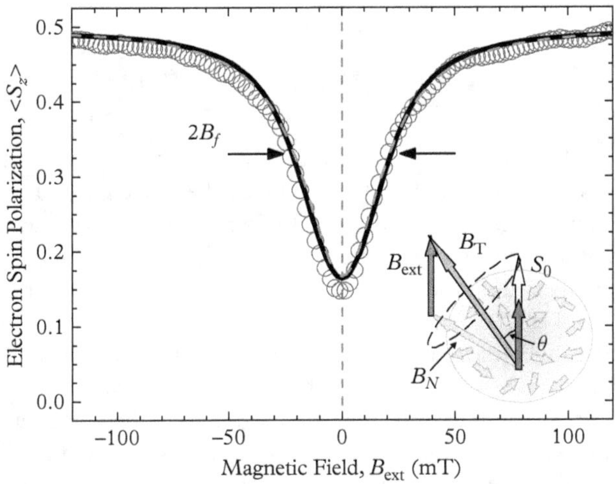

Figure 7.7 *The calculated external magnetic field dependence of the steady-state resident electron spin polarization in the quantum dot ensemble (solid line) compared with the experiment (circles). The dashed line is obtained by approximating the calculations by a Lorentzian function Eq. (7.18). The inset shows schematically the mechanism of suppression of the nuclear spin fluctuations effect by an external magnetic field. From (Petrov et al., 2008).*

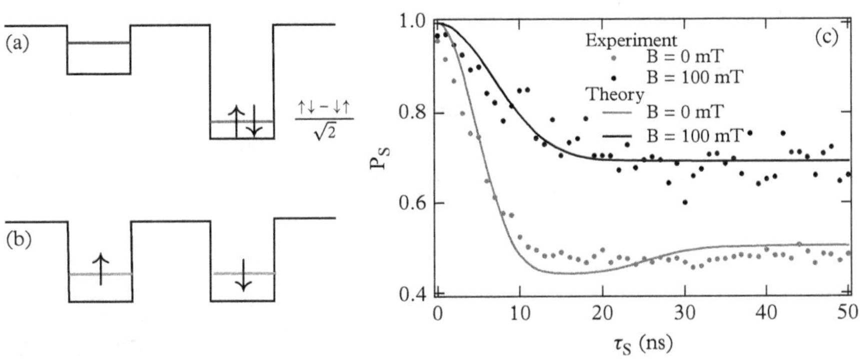

Figure 7.8 *(a) Schematics of the double quantum dot structure in the regime of the two-electron spin singlet initialization and detection. (b) Schematics of the double quantum dot structure in the regime where two electrons are in different dots and their spins evolve independently. (c) Singlet state probability, P_S, measured as a function of separation time, τ_S. Data points are acquired at $B = 0$ mT and 100 mT, see the legend. Solid lines are best fits to the data using the semiclassical model, see Eq. (7.36). From (Hanson et al., 2007).*

quantum dot; for example, the right one in Figure 7.8(a). Such a regime allows one to prepare a two-electron spin singlet state, see Eq. (2.16),

$$\Psi_S = \frac{1}{\sqrt{2}}(|\uparrow\rangle_1|\downarrow\rangle_2 - |\uparrow\rangle_1|\downarrow\rangle_2). \tag{7.34}$$

Such a spin singlet state where the orbital envelope functions are the same for both electrons is quite robust, both to the nuclear field and external field, because the spin flip, that is the transition from the singlet to one of the triplet states, Eq. (2.17), should be accompanied by a change of the orbital functions. Then the gate voltages are adjusted in such a way that the confinement becomes symmetric, Figure 7.8(b), and each electron occupies the corresponding dot. For sufficiently large dots and a high barrier, the exchange interaction between the electrons, Eq. (2.18), is negligible, and the spins of the carriers evolve in the magnetic and nuclear fields. After a certain time delay, τ_S, also termed the pulse separation time, the gate voltage is changed in such a way, that the state shown in Figure 7.8(a), where the two charge carriers are in the same dot, is energetically preferable. The transition to such a state is possible if and only if the two-electron wavefunction is still in the singlet state, see Eq. (7.34). Therefore, by measuring the electrical response of such a double dot system the probability of see finding the electrons in a spin singlet state, P_S, can be studied as a function of time and external magnetic field (Petta *et al.*, 2005). The experimental data obtained in (Petta *et al.*, 2005), see also (Hanson *et al.*, 2007), are shown in Figure 7.8(c) by dots.

Since two-electron spin evolution takes place where the carriers are in different dots and do not interact with each other, the spin decoherence of the pair and, correspondingly, the probability of remaining in the singlet state, $P_S(\tau_S)$, is described by the central spin model, see (Schulten and Wolynes, 1978; Taylor *et al.*, 2007). Here, we are interested only at the spin dynamics on sufficiently short time scales where the nuclear spin dynamics is of no importance. In such a case it is convenient to obtain the probability, P_S, quantum mechanically by explicitly evaluating the action of the two spin evolution operator on the singlet wavefunction Ψ_S:

$$\Psi(t) = \exp\left\{-\mathrm{i}\left[(\boldsymbol{\Omega}_1\cdot\hat{\boldsymbol{s}}_1)+(\boldsymbol{\Omega}_2\cdot\hat{\boldsymbol{s}}_2)\right]t\right\}\Psi_S. \tag{7.35}$$

Here $\hat{\boldsymbol{s}}_i$ $(i = 1,2)$ are the spin operators of the first and second electron, see Eq. (2.18),

$$\boldsymbol{\Omega}_i = \boldsymbol{\Omega}_{N,i} + \boldsymbol{\Omega}_B,$$

are the spin precession frequencies of the first and second electron in the fields of nuclear fluctuations, $\boldsymbol{\Omega}_{N,i}$, and in the external magnetic field $\boldsymbol{\Omega}_B$. Making use of Eq. (7.35) and averaging over the nuclear fields we obtain in accordance with (Schulten and Wolynes, 1978; Taylor *et al.*, 2007)

$$P_S(\tau_S) = \int d\mathbf{\Omega}_{N,1} d\mathbf{\Omega}_{N,2} \mathcal{F}_e(\mathbf{\Omega}_{N,1}) \mathcal{F}_e(\mathbf{\Omega}_{N,2}) \, |\langle \Psi(\tau_S) \Psi_S \rangle|^2$$
$$= \int d\mathbf{\Omega}_{N,1} d\mathbf{\Omega}_{N,2} \mathcal{F}_e(\mathbf{\Omega}_{N,1}) \mathcal{F}_e(\mathbf{\Omega}_{N,2})$$
$$\times \left| \cos(\Omega_1 \tau_S) \cos(\Omega_2 \tau_S) + \frac{(\mathbf{\Omega}_1 \cdot \mathbf{\Omega}_2)}{\Omega_1 \Omega_2} \sin(\Omega_1 \tau_S) \sin(\Omega_2 \tau_S) \right|^2 . \quad (7.36)$$

Here, $\mathcal{F}_e(\mathbf{\Omega}_N)$ is the distribution function of the nuclear spin fluctuations, Eq. (7.9).

The analytical expressions for $P_S(\tau_S)$ can be readily obtained in the limits of a zero magnetic field, $\Omega_B = 0$, and of the strong magnetic field $\Omega_B \gg \delta_e$ (Taylor *et al.*, 2007). Here, let us analyze just the most important limiting cases. In absence of the external magnetic field, $\Omega_B = 0$, for $\tau_S \to \infty$ one obtains from Eq. (7.36) that $P_S(\tau_S \to \infty) \to 1/3$. Note that if the electron spins are fully disordered the probabilities to find the electron pair in the singlet and any of three triplet states are equal and the $P_S(\tau_S \to \infty)$ would be 1/4. The higher value of $P_S(\tau_S \to \infty)$ is related to the nondecaying fraction of the initial spin polarization. In a strong enough magnetic field, $\Omega_B \gg \delta_e$, and for sufficiently long separation times we obtain $P_S(\tau_S \to \infty) = 1/2$. This is because the magnetic field mixes the singlet state with the triplet one with the total spin projection on the field being zero, while other triplet states remain unoccupied. This qualitative analysis is corroborated by the results of the calculations shown in Figure 7.8(c) by solid lines. The calculated $P_S(\tau_S)$ dependence also reproduces the experimental data. The comparison of the experiment and the theory makes it possible to estimate the electron decoherence time by the nuclei $\delta_e^{-1} \sim 10$ ns. Somewhat longer values of the decoherence time measured on the gate-defined quantum dots, compared with the ~1 ns decoherence times reported in Figure 7.5 in self-assembled quantum dots, are mainly related to the difference in the electron confinement. Weaker confinement in the gate-defined quantum dots results in smaller fluctuating Overhauser fields.

The combination of transport and optical methods provides additional insights in the long-time electron spin dynamics interacting with a nuclear environment. In (Bechtold *et al.*, 2015) the authors have studied the electron spin decoherence in a single self-assembled InGaAs quantum dot incorporated into the photodiode structure. The application of the electric field to the structure allows one to control the electron and hole tunneling rates from the quantum dot to the electrodes. After the discharging of the quantum dot (reset stage of the spin control protocol) the electron spin coherence was excited by a short, ~1 ps, circularly polarized pulse (pump stage). In the chosen operation mode the hole tunneling time from the quantum dot was significantly shorter compared with the radiative lifetime of the electron-hole pair and with the time scales related to the electron-hole exchange interaction, ~ 150 ps, which enables quite robust electron spin initialization. After the hole leaves the dot, the electron remains in the quantum dot for quite a long time, ~ 10 μs. After the adjustable storage time, T_{store}, the spin detection is

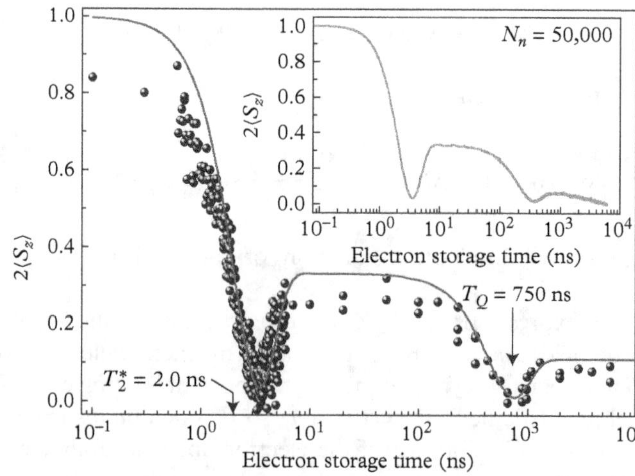

Figure 7.9 *The electron spin dynamics as a function of the* $T_{\rm store}$. *The initial dynamics on the time scale of* $\lesssim 20$ *ns is due to hyperfine interaction with frozen nuclear spins. The* $T_2^* \approx 2$ *ns is related to the inverse spread of nuclear fields,* δ_e^{-1}. *The quadrupole coupling of nuclear spins induces oscillatory fluctuations of the Overhauser field with the time scale* T_Q, *which further reduces the electron spin polarization* $2\langle S_z\rangle$. *The line compares experimental results with analytical predictions based on the semiclassical model. The inset presents results of more sophisticated numerical simulations based on the quantum mechanical model of central spin decoherence by a nuclear spin bath. From (Bechtold et al., 2015).*

performed by another short optical pulse copolarized with the pump (read stage), which is resonantly absorbed in the dot resulting in the X^- trion formation, if the electron flipped its spin during $T_{\rm store}$. If the electron remains in the initial spin state the trion is not formed, see Chapter 6 and Figure 6.2 for details. The electron spin polarization can thus be directly mapped to the X^- trion emission rate.

Figure 7.9 demonstrates the electron spin polarization dynamics, $2\langle S_z\rangle$, as a function of the electron storage time measured in (Bechtold *et al.*, 2015). The first nonmonotonic feature of $\langle S_z\rangle(T_{\rm store})$ is caused by the electron spin precession in frozen nuclear fields being in good agreement with the semiclassical model, see Eq. (7.14) and Figure 7.3. Surprisingly, the authors of (Bechtold *et al.*, 2015) observed a second nonmonotonic feature in the electron spin polarization dynamics at about 1 μs. Further, only a weak decay of the electron spin polarization is observed up to 10 μs. This second feature of electron spin dynamics was related to the spin dynamics of the nuclei induced by the quadrupole splittings in the strained InGaAs quantum dot, see Eqs. (2.24) and (3.33). In the model description presented in (Bechtold *et al.*, 2015) it was assumed that the quadrupole splittings of the nuclei (i) exceed by far both the Knight field exerted by the electron on a given nucleus and the dipole–dipole interactions between the nuclei, and (ii) the quadrupole axes are isotropically distributed in the space. Since a typical time scale related to the quadrupole splitting, $T_Q \sim \hbar/E_Q$, Eq. (3.35), is long compared with the electron spin precession in the field of nuclear fluctuations, the electrons experience quasi-static nuclear fields and Eq. (7.20), where $\boldsymbol{\Omega}_N(t)$ is given by the expression similar to Eq. (7.14) with the replacement of $\mathcal{F}_e(\boldsymbol{\Omega}_N)$ by the distribution function of the quadrupole couplings (Bechtold *et al.*, 2015). The analytical and numerical calculations presented in (Bechtold *et al.*, 2015) and shown, respectively, by the solid curves in the main panel and in the inset of Figure 7.9 are in good agreement with the experimental data. Fitting of the experimental data by the model allowed the authors of (Bechtold *et al.*, 2015) to evaluate the time scale related to the quadrupole coupling $T_Q = 750$ ns in the studied quantum dot. It is worth noting that the particular behavior of the electron spin polarization tail at $T_{\text{store}} \sim T_Q$ is very sensitive to the distribution function of the quadrupole couplings.

Before, we discussed just a small part of extensive experimental activities devoted to the studies of the central spin model in nanosystems. In the standard time-resolved photoluminescence or pump-probe experiments it is not so easy to study the long-time evolution of the electron spin polarization because the finite repetition rate of the lasers is usually about 75...80 MHz limiting the time window by about ~13 ns. That is why, in addition to the experimental methods outlined before, various other crafty techniques based, for example, on the single- and multi-pulse spin-echo method, were employed to study long-time spin evolution, see (Press *et al.*, 2010; Bluhm *et al.*, 2011) and (Urbaszek *et al.*, 2013) for a review. Another option is to use special trains of pulses with much longer repetition periods as realized in (Belykh *et al.*, 2016). In all these cases the initial electron or hole spin state is prepared and its evolution is monitored. Another approach to addressing the slow spin dynamics is to study the spectrum of low-frequency spin fluctuations. In this case, any deliberate excitation of spin degrees of freedom is absent. The spin noise spectroscopy is presented in detail in Chapter 10.

Note that, before, we focused on the electron spin dynamics in the absence of magnetic field or in the magnetic field applied in the Faraday geometry. The spin dynamics in the transverse magnetic field are discussed in Chapter 9 because the interplay of the periodic pumping of the electron spin and the electron spin precession in the magnetic field brings about novel regimes of the spin dynamics of the interacting electron-nuclear spin system.

In conclusion of this chapter, let us reiterate the main features of electron spin dynamics interacting with nuclear spins:

- The charge carrier spin dynamics is characterized by a relatively fast, ~ns scale, decoherence due to randomly distributed "frozen" nuclear spin fluctuations.
- The decoherence in the field of frozen nuclear fluctuations is not complete. The magnitude of the nondecaying fraction depends on the distribution

function of the nuclear fields, strength and orientation of the external magnetic field. It amounts, in absence of an external field, to one third in the case of isotropic distribution function of hyperfine fields, realized in electron-doped quantum dots.

- The long-time spin dynamics within the central spin model is quite sensitive to the inhomogeneity of hyperfine coupling distribution and nuclear spin dynamics controlled by the quadrupole and dipole–dipole interactions.

8

Electron Spin Relaxation Beyond the Hyperfine Interaction

8.1 Phonon-induced electron spin relaxation

In the framework of the central spin model, a sufficiently strong external magnetic field $\boldsymbol{B}$ fully quenches the hyperfine interaction induced spin relaxation of a localized electron. Indeed, the energy difference between the Zeeman sublevels, $|g\mu_B B|$, can be neither absorbed nor taken from the reservoir of nuclear spins interacting with the quantum dot provided that

$$|g\mu_B B| \gtrsim \hbar\delta_e,$$

where $\hbar\delta_e$ is the effective Zeeman splitting caused by the nuclear spin fluctuations, see Eq. (7.10). Recall that g is the electron g-factor and μ_B is the Bohr magneton. In particular, as we have seen in Chapter 7, moderate magnetic fields applied in the Faraday geometry result in the suppression of the electron spin decoherence and saturation of the optically induced electron spin polarization at its maximal value, $|\langle S_z \rangle| = 1/2$. In strong fields, the deviation of the electron spin from the maximal value is controlled by a small parameter $[\hbar\delta_e/(g\mu_B B)]^2 \ll 1$. In fact, electron spin relaxation is fully suppressed in this regime.

This result holds as long as we neglect other interactions in the electron-nuclear spin system. The account of the finite electron correlation time at the localization site or allowance for finite nuclear field correlation time (due to, e.g., the quadrupole couplings or dipole–dipole interaction between the neighboring nuclei) gives rise to a quadratic in the magnetic field slow-down of the electron spin relaxation described by the expression (Meier and Zakharchenya, 1984; Abragam, 2002; Glazov, 2015; Linpeng *et al.*, 2016)

$$T_{1,hf} = \frac{\tau_c \Omega_B^2}{\delta_e^2} \propto \tau_c B^2. \tag{8.1}$$

Electron and Nuclear Spin Dynamics in Semiconductor Nanostructures. M. M. Glazov, Oxford University Press (2018). © M. M. Glazov. DOI: 10.1093/oso/9780198807308.001.0001

Here, $\Omega_B = g\mu_B B/\hbar$ is the electron Larmor frequency in the external field, τ_c is the correlation time, and in derivation of Eq. (8.1) we have assumed that $|\Omega_B| \gg \delta_e$. The suppression of the spin relaxation is related to the mismatch of the electron and nuclear Zeeman splittings. Indeed, as one can check, Eq. (8.1) agrees with (5.14), see also Eqs. (5.15) and (2.52) for the electron nuclear spin-flip rate derived quantum mechanically in Chapter 5. The observation of the quadratic increase of electron spin relaxation time as a function of magnetic field makes it possible to determine the electron correlation time in systems with localized charge carriers (Dyakonov, 2017; Dzhioev *et al.*, 2002*b*,; Linpeng *et al.*, 2016; Belykh *et al.*, 2016; Meier and Zakharchenya, 1984). Schematically, the spin relaxation of the donor-bound electron in a bulk GaAs-like semiconductor in relatively low magnetic fields is illustrated in Figure 8.1(a). On a semiclassical level, Eq. (8.1) can be understood taking into account that in a relatively strong field $|\Omega_B| \gg \delta_e$, the electron spin rotation angle during the correlation time τ_c is random and can be estimated as $\delta\Phi \sim |\delta_e/\Omega_B|$. The relaxation takes place when the mean square of the spin rotation angle is on the order of unity, that is at $\tau_s/\tau_c \sim (\Omega_B/\delta_e)^2$, in agreement with qualitative arguments leading to Eq. (2.42).

The increase of the longitudinal electron spin relaxation time with the magnetic field, Eq. (8.1), is limited by other spin relaxation mechanisms. Among various effects the combination of the electron-phonon interaction and the spin-orbit coupling serves as the prime candidate for the mechanism responsible for the limitation. Note that, as discussed in Section 2.6, the spin-orbit coupling alone cannot be responsible for the spin-flip of the localized electron or hole due to the Kramers degeneracy of the localized carrier states. The rate of the acoustic phonon-assisted transitions between the Zeeman-split electron spin sublevels increases with the magnetic field, particularly because the density of available states for the emission or absorption of a phonon with the energy $\hbar sq = |g\mu_B B|$ increases as q^2, where s is the sound velocity and q is the wavevector of the involved phonon. As a result, the electron spin relaxation time in sufficiently strong magnetic fields scales as

$$T_{1,ph} \propto B^{-\nu}, \tag{8.2}$$

where the positive power ν is, as a rule, larger than 2. Hence, in high magnetic fields the electron spin relaxation time decreases with an increase in the magnetic field. Overall, due to the interplay of the hyperfine interaction and the electron-phonon interaction, the spin relaxation rate of the electron is a non-monotonous function of the magnetic field.

We note that the electron–phonon interaction via, for example, the deformation-potential or the piezoelectric coupling (Gantmakher and Levinson, 1987) is insufficient to cause electron spin-flip by itself. The important ingredient in the phonon-induced spin-flips is either the spin-orbit coupling or the hyperfine interaction (Pines *et al.*, 1957). For example, in the latter case, the phonons slightly shift nuclei from their equilibrium positions and, therefore, cause nuclear fluctuations

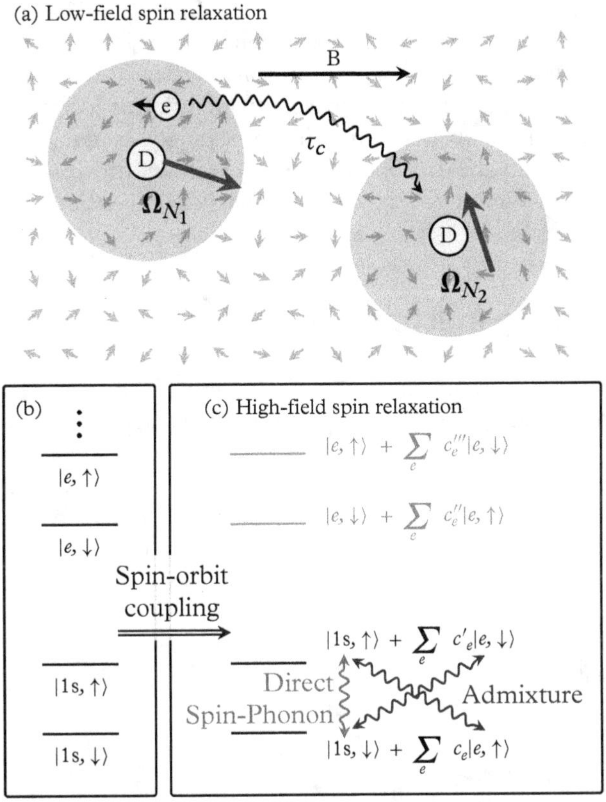

Figure 8.1 *Schematics of the spin relaxation mechanisms. (a) At low magnetic fields, the spin relaxation is dominated by the interaction of the electron spin with lattice nuclear spins. Panels (b, c) are relevant for the high field spin relaxation. (b) Energy level structure for unperturbed localized electron in magnetic field with $|1s\rangle$ denoting the orbital ground state and $|e\rangle$ denoting the excited states, $\uparrow, \downarrow$ are the electron spin orientations. (c) The spin-orbit coupling mixes states with opposite spins. In the admixture mechanism, phonons cause relaxation between the two eigenstates via the components with like spin. The direct spin-phonon interaction causes spin-relaxation via the components with opposite spin.*

$\mathbf{\Omega}_N$ to vary in time. Such a time-dependent magnetic field results in the spin-flip transition of the electron just like the real time-dependent magnetic field, for example, under the conditions of the electron spin resonance, see Chapter 3. However, the modulation of the hyperfine interaction is not the main contribution to the phonon-induced $T_{1,ph}$, Eq. (8.2), although such a process has been observed in recent work (Camenzind *et al.*, 2017). The main channel of the spin-flip at the electron-phonon interaction is due to the spin-orbit coupling (Khaetskii, 2001; Khaetskii and Nazarov, 2001; Woods *et al.*, 2002). The spin-orbit interaction can manifest itself in the spin-flip process in a two-fold way: The spin-dependent terms may explicitly be present in the electron-phonon interaction Hamiltonian, such process is known as the *direct spin-phonon interaction*. Alternatively, the spin-orbit interaction can mix opposite spin states with different orbital wavefunctions, while the electron-phonon interaction involves purely orbital terms. The latter mechanism is termed the *admixture mechanism*. These processes are schematically illustrated in Figure 8.1(c). In what follows, we consider these effects in more detail.

8.2 Admixture mechanism

The detailed theoretical investigation of spin-flip mechanisms has been performed for the related GaAs quantum dot system (Khaetskii and Nazarov, 2001; Woods *et al.*, 2002; Marquardt and Abalmassov, 2005) and for donor-bound electrons (Linpeng *et al.*, 2016). In quantum dots with strong confinement along the growth axis z, and a weaker confinement in the dot plane, the admixture mechanism is the dominant relaxation process. For this mechanism, the spin-orbit interaction modifies the ground state Zeeman sublevels by the admixture of the excited sublevels with the opposite spin component. Hence, the spin-independent electron-phonon coupling causes electron spin relaxation through the components of the states with the same spin, as depicted in Figure 8.1(b)–(c). The interaction Hamiltonian for the admixture mechanism is

$$\mathcal{H}_{adm} = \mathcal{U}_{ph} + \mathcal{H}_{so}, \tag{8.3}$$

where $\mathcal{U}_{ph}$ is the spin-conserving electron-phonon interaction Hamiltonian and $\mathcal{H}_{so}$ is the spin-orbit Hamiltonian. In the high-field limit, the Zeeman splitting can be comparable or even exceed the thermal energy. In such a case, the transition rates from the Zeeman sublevel $\downarrow$ (corresponding to the electron spin component of $-1/2$ onto the field axis) to $\uparrow$ (with spin component $+1/2$), $\Gamma_{\uparrow\downarrow}$, and back, $\Gamma_{\downarrow\uparrow}$, differ. It is convenient to introduce the effective longitudinal spin relaxation time in accordance with

$$T_1 = \frac{1}{\Gamma_{\uparrow\downarrow} + \Gamma_{\downarrow\uparrow}}. \tag{8.4}$$

This time T_1 is observed in the experiments similar to the one reported in (Linpeng *et al.*, 2016), see Section 8.4 for details. The individual rates are found using Fermi's golden rule, namely,

$$\Gamma_{\downarrow\uparrow} = \frac{2\pi}{\hbar} \sum_{q,\alpha} |M_{\downarrow\uparrow}|^2 \delta(\hbar q s_\alpha - |g\mu_B B|), \tag{8.5}$$

where $\boldsymbol{q}$ is the phonon wavevector, s_α is the speed of sound in phonon branch α and $\alpha = t_1, t_2, l$ stands for the two transversal and longitudinal acoustic phonon modes, respectively, and $M_{\downarrow\uparrow}$ is the matrix element of the spin-flip process from the spin-up state, $\uparrow$, to the spin-down state, $\downarrow$. Hereafter, the states are denoted by the spin projection onto the magnetic field direction and we assume for convenience that the spin-up state has higher energy than the spin-down one, that is $g\mu_B B > 0$, as illustrated in Figure 8.1, so that the rate in Eq. (8.5) corresponds to the phonon emission process.

Our next aim is to calculate the matrix element $M_{\downarrow\uparrow}$ within the admixture mechanism. To that end, we account for the electron-phonon interaction and the spin-orbit coupling in Eq. (8.3) each in first-order perturbation theory. Particularly, we assume that the spin-orbit coupling energies are much smaller compared with the size-quantization energy of the localized electron. As a result, the electron spin-relaxation occurs via a second order process due to the quantum interference of $\mathcal{U}_{ph}$ and $\mathcal{H}_{so}$ in the Hamiltonian (8.3), see (Khaetskii and Nazarov, 2001):

$$M_{\downarrow\uparrow,adm} = -\sum_e \left[\frac{\langle 1s, \downarrow |\mathcal{U}_{ph}| e, \downarrow\rangle\langle e, \downarrow |\mathcal{H}_{so}| 1s, \uparrow\rangle}{E_e - E_{1s} + g\mu_B B} + \frac{\langle 1s, \downarrow |\mathcal{H}_{so}| e, \uparrow\rangle\langle e, \uparrow |\mathcal{U}_{ph}| 1s, \uparrow\rangle}{E_e - E_{1s} - g\mu_B B} \right], \tag{8.6}$$

where $|1s\rangle$ is the ground orbital state of the localized electron, $|e\rangle$ denotes the excited orbital states, and E_e, E_{1s} are the energies of the corresponding orbitals. Generally, the states $|e\rangle$ include both the discrete and the continuum states, see Figure 8.1(c).

It is noteworthy that, at $\boldsymbol{B} = 0$, the two contributions in Eq. (8.6) exactly compensate each other so that $M_{\downarrow\uparrow,adm} \to 0$ at $B \to 0$. This behavior has a deep physical meaning related to the time reversal symmetry: In absence of magnetic field the electron states are doubly degenerate even in the presence of the spin-orbit interaction (see Section 2.6 for details). In order to clarify the effect, let us present the wavefunctions of the spin doublet in all orders in $\mathcal{H}_{so}$ in the form [cf. Eq. (2.57)]

$$|\Uparrow\rangle = \varphi_1(\boldsymbol{r})|\uparrow\rangle + \varphi_2(\boldsymbol{r})|\downarrow\rangle, \quad |\Downarrow\rangle = \varphi_2^*(\boldsymbol{r})|\uparrow\rangle - \varphi_1^*(\boldsymbol{r})|\downarrow\rangle,$$

where the $\varphi_1(\boldsymbol{r})$ and $\varphi_2(\boldsymbol{r})$ are the envelope functions of the localized state, the envelope $\varphi_2(\boldsymbol{r})$ describes the admixture of the opposite spin and excited orbital state due to the spin-orbit interaction. The functions $|\Uparrow\rangle, |\Downarrow\rangle$ are orthogonal and, moreover, any spin-independent perturbation, like spin-independent electron-phonon interaction $\mathcal{U}_{ph}$, does not mix these states as it can be readily checked by explicit calculation. The absence of mixing ensures that the $M_{\downarrow\uparrow,adm}$ vanishes without the magnetic field, the effect known as the *van Vleck cancellation* (Van Vleck,

1940; Abrahams, 1957). In fact, one can readily check that the time-reversal symmetry implies:

$$\langle 1s,\uparrow|\mathcal{U}_{ph}|e,\uparrow\rangle = \langle e,\downarrow|\mathcal{U}_{ph}|1s,\downarrow\rangle, \quad \langle 1s,\uparrow|\mathcal{H}_{so}|e,\uparrow\rangle = -\langle e,\downarrow|\mathcal{H}_{so}|1s,\downarrow\rangle.$$

The account for the Zeeman effect of the magnetic field breaks the cancellation and allows the spin-flip process. Usually, it is sufficient to take into account only the lowest-order in $g\mu_B B$ contribution to the spin-flip matrix element, namely,

$$M_{\downarrow\uparrow,adm} = \sum_e \frac{2g\mu_B B}{(E_e - E_{1s})^2}\langle 1s,\downarrow|\mathcal{U}_{ph}|e,\downarrow\rangle\langle e,\downarrow|\mathcal{H}_{so}|1s,\uparrow\rangle. \tag{8.7}$$

The matrix elements in Eq. (8.7) can be evaluated on the wavefunctions which are not perturbed by the magnetic field or, if the orbital effect of the magnetic field is strong enough, on the wavefunctions accounting for the diamagnetic effect of the field.

Before, we established the general form of the spin-flip matrix element in the admixture mechanism. Let us now specify the Hamiltonians for the electron-phonon and spin-orbit interactions in Eq. (8.3). We consider a nanosystem formed of GaAs-like semiconductor and start with the analysis of the electron–phonon interaction. The piezoelectric electron–phonon interaction reads

$$\mathcal{U}_{ph}^{(pz)} = \sqrt{\frac{\hbar}{2\rho\omega_{\boldsymbol{q},\alpha}}}e^{-\mathrm{i}(\boldsymbol{qr}-\omega_{\boldsymbol{q},\alpha}t)}(eA_{\boldsymbol{q},\alpha})b_{\boldsymbol{q},\alpha}^{\dagger} + \text{c.c.}, \tag{8.8}$$

where

$$A_{\boldsymbol{q},\alpha} = h_{14}\sum_{ijk}\beta_{ijk}\xi_i\xi_j\hat{e}_k^{(\boldsymbol{q},\alpha)}, \tag{8.9}$$

ρ is the mass density of the material, $\omega_{\boldsymbol{q},\alpha} = s_\alpha q$ is the phonon frequency, $b_{\boldsymbol{q},\alpha}^{\dagger}$ is the creation operator for a phonon, $\boldsymbol{\xi} = \boldsymbol{q}/q$ is the unit vector along the phonon wavevector, $\hat{\boldsymbol{e}}$ is the phonon polarization vector, and the only nonzero Cartesian components of β_{ijk} are those with different subscripts, $\beta_{xyz} = \ldots = \beta_{zyx} = 1$, and h_{14} is the piezoelectric constant (Gantmakher and Levinson, 1987). Here, we use cubic set of axis, $x \parallel [100]$, $y \parallel [010]$, and $z \parallel [001]$. The deformation potential interaction is described by the Hamiltonian

$$\mathcal{U}_{ph}^{(dp)}(\boldsymbol{r}) = \sqrt{\frac{\hbar}{2\rho\omega_{\boldsymbol{q},\alpha}}}e^{-\mathrm{i}(\boldsymbol{qr}-\omega_{\boldsymbol{q},\alpha}t)}q(\boldsymbol{\xi}\cdot\hat{\boldsymbol{e}})Db_{\boldsymbol{q},\alpha}^{\dagger} + \text{c.c.}, \tag{8.10}$$

where D is the deformation potential constant in the conduction band. It involves longitudinal phonons only. For completeness, we present the expressions for the components of the polarization vectors $\hat{\boldsymbol{e}}$ for the longitudinal, l, and two transversal, t_1 and t_2, phonon modes:

$$\hat{\boldsymbol{e}}_l = q^{-1}[q_x, q_y, q_z], \tag{8.11a}$$

$$\hat{\boldsymbol{e}}_{t_1} = (q_x^2 + q_y^2)^{-1/2}[q_y, -q_x, 0], \tag{8.11b}$$

$$\hat{\boldsymbol{e}}_{t_2} = q^{-1}(q_x^2 + q_y^2)^{-1/2}[q_x q_z, q_y q_z, -(q_x^2 + q_y^2)]. \tag{8.11c}$$

For $\boldsymbol{q} \parallel 0$ Eqs. (8.11b) and (8.11c), formally, do not provide definite values of $\hat{\boldsymbol{e}}_{t_{1,2}}$, in this case one has to take the limit of $q_x, q_y \to 0$ in these expressions. We also note that even in cubic crystals the sound velocities depend on the phonon propagation direction; hereafter, we employ widely used isotropic approximation with effective longitudinal and transversal velocities appropriately averaged over the directions of $\boldsymbol{q}$ (Zook, 1964; Gantmakher and Levinson, 1987).

Importantly, the energy of a phonon involved in the spin-flip transition is the Zeeman splitting between the spin sublevels. Therefore, the phonon wavevector $q_\alpha = g\mu_B B/(\hbar s_\alpha) \to 0$ as $B \to 0$. Thus, at moderate magnetic fields in piezoelectric crystals such as GaAs and in the nanostructures on their basis, the piezoelectric electron-phonon interaction with $\mathcal{U}_{ph}^{(pz)} \propto q^{-1/2}$ dominates over the deformation potential interaction, where $\mathcal{U}_{ph}^{(dp)} \propto q^{1/2}$.

In quantum dot structures, in addition to the deformation potential interaction and piezoelectric coupling the motion of quantum dot heterointerfaces due to the phonons results in a variation of system parameters such as effective mass, g-factor, or the spin-orbit coupling constant. Such an interaction is known as "ripple coupling" (Mishra and Bray, 1977; Knipp and Reinecke, 1995). Its effect on the spin flip transitions is analyzed in detail in (Woods *et al.*, 2002). Here, we resort to the detailed quantitative analysis of the piezoelectric interaction only.

Now we turn to the discussion of the relevant spin-orbit interaction terms. The bulk GaAs-type semiconductors are characterized by the noncentrosymmetric T_d point symmetry group (see Appendix A), and the only relevant spin-orbit coupling comes from the cubic-in-the-electron-wavevector Dresselhaus spin-orbit term, $\mathcal{H}_{so}$, cf. Eq. (2.14). It arises from the lack of inversion symmetry in zinc-blende crystals and takes the form

$$\mathcal{H}_D = \gamma_c \sum_i \sigma_i \hat{k}_i (\hat{k}_{i+1}^2 - \hat{k}_{i+2}^2), \tag{8.12}$$

where γ_c is the Dresselhaus spin-orbit coupling constant, the subscript i cycles through the cubic axes x, y, z, and $\hat{k}_i = -\mathrm{i}\partial/\partial x_i$ are the Cartesian components of the electron wavevector operator. The situation is more involved in bulk strained semiconductors and in quantum well-quantum dot systems where the effective Hamiltonian contains $\boldsymbol{k}$-linear terms in the form (Ivchenko, 2005; Dyakonov, 2017)

$$\mathcal{H}_1 = \sum_{ij} \gamma_{ij} \sigma_i \hat{k}_j,$$

where γ_{ij} are the constants. To be specific we only consider here for low-dimensional systems, following (Khaetskii and Nazarov, 2001), the bulk inversion asymmetry term in the form

$$\mathcal{H}_{BIA} = \gamma(-\sigma_x \hat{k}_x + \sigma_y \hat{k}_y). \tag{8.13}$$

This term appears as a result of the averaging of Eq. (8.12) over the electron size-quantization wavefunction along the growth axis z under the assumption that the in-plane confinement in the quantum well quantum dot is quite shallow. Similar results can be obtained for the structure inversion asymmetry term in the form (Ohkawa and Uemura, 1974; Vas'ko, 1980; Bychkov and Rashba, 1984)

$$\mathcal{H}_{SIA} = \gamma'(\sigma_x \hat{k}_y - \sigma_y \hat{k}_x). \tag{8.14}$$

It is instructive to outline the calculation for the spin-flip rate for the case of such quantum dot assuming that the in-plane confinement is provided by the parabolic potential such that the ground and excited states wavefunctions take the form of the harmonic oscillator functions (Landau and Lifshitz, 1977). Within such a model the spin-orbit interaction, Eq. (8.13), admixes to the ground state with the in-plane envelope function

$$|1s\rangle = \varphi_1(x,y) = \frac{1}{\sqrt{\pi l_x l_y}} \exp\left(-\frac{x^2}{2l_x^2} - \frac{y^2}{2l_y^2}\right), \tag{8.15}$$

only two excited states that are characterized by the envelopes

$$\varphi_{2,x}(x,y) = \frac{\sqrt{2}x}{l_x}\varphi_1(x,y), \quad \varphi_{2,y}(x,y) = \frac{\sqrt{2}y}{l_y}\varphi_1(x,y). \tag{8.16}$$

Here, l_x and l_y are the effective confinement lengths along the main axes x,y of the parabolic potential. In calculation of the spin-flip rate one usually employs the long-wavelength approximation assuming that photon wavelength $2\pi/q$ exceeds by far the confinement lengths l_x, l_y. This assumption usually holds for not too high magnetic fields where

$$\frac{g\mu_B B}{\hbar s_\alpha} \ll l_x^{-1}, l_y^{-1}. \tag{8.17}$$

Within the long-wavelength approximation one can decompose the exponent $\exp(\mathrm{i}\boldsymbol{qr})$ in $\mathcal{U}_{ph}^{(pz)}$, Eq. (8.8), into the power series in $\boldsymbol{qr}$ and retain only a $\boldsymbol{q}$-linear contribution. It is noteworthy that the first, coordinate independent term in the decomposition of exponent does not play a role because the orbital functions of the ground and excited states are orthogonal. The second term can be written in the from $-e\boldsymbol{E}_{ind} \cdot \boldsymbol{r}$ with $\boldsymbol{E}_{ind}$ being the electric field induced by the acoustic phonon, which causes the mixing of the ground and excited states. As a result in $\boldsymbol{B} \parallel z$ (Khaetskii and Nazarov, 2001)

$$|M_{\downarrow\uparrow,adm}|^2 = \frac{(g\mu_B B\gamma)^2}{(E_{1s} - E_e)^4} \frac{\hbar}{2\rho\omega_{\boldsymbol{q},\alpha}} (eA_{\boldsymbol{q},\alpha})^2 (q_x^2 + q_y^2)[1 + N_{ph}(g\mu_B B)]. \tag{8.18}$$

Here, $N_{ph}(\hbar\omega) = [\exp(\hbar\omega/k_B T) - 1]^{-1}$ is the occupancy of the phonon mode with the energy $\hbar\omega$ at the lattice temperature T, the last factor $1 + N_{ph}(\hbar\omega)$ takes into account both the spontaneous and stimulated phonon emission processes. The matrix element of the inverse process, $|M_{\uparrow\downarrow,adm}|^2$ accompanied by the phonon absorption is proportional to $N_{ph}(\hbar\omega)$ and vanishes at $T \to 0$. In derivation of Eq. (8.18) we neglected the splitting between the excited states $\varphi_{2,x}(x,y)$ and $\varphi_{2,y}(x,y)$ compared with the energy gap between the ground and excited states.

Substituting the matrix element into the Fermi's golden rule expression for the transition rate, Eq. (8.5), averaging over the directions of the phonon wavevector $\boldsymbol{q}$ at a fixed absolute value $q = g\mu_B B/(\hbar s_\alpha)$, and summing over the phonon polarizations, we obtain (Khaetskii and Nazarov, 2001)

$$\Gamma_{\downarrow\uparrow} = \frac{2(g\mu_B B)^5}{\hbar(E_{1s} - E_e)^4} \Lambda[1 + N_{ph}(g\mu_B B)], \tag{8.19}$$

where the constant Λ is given by

$$\Lambda = \frac{2}{35\pi} \frac{(eh_{14}\gamma)^2}{\rho\hbar^2} \left(\frac{1}{s_l^5} + \frac{4}{3s_t^5} \right). \tag{8.20}$$

At low temperatures where the thermal phonon occupancy N_{ph} can be neglected, Eq. (8.19) demonstrates the B^5 scaling of the electron spin-flip transition rate. A similar expression for the spin-flip rate with the replacement $\gamma \to \gamma'$ holds for the structure inversion asymmetry induced spin splitting given by Eq. (8.14). The estimates in (Khaetskii and Nazarov, 2001; Woods *et al.*, 2002) show that in the magnetic field of $B = 1$ T, typical spin relaxation rates $\Gamma_{\downarrow\uparrow}$ in quantum dots are in the order of 10^3 s^{-1}, and vary strongly with the quantum dot geometry and material parameters.

For the spin-orbit interaction in the form of either (8.13) or (8.14) the dependence of the spin-flip rate on the magnetic field orientation is described by the factor

$$\frac{1 + \cos^2\theta_B}{2},$$

where θ_B is the angle between the magnetic field and the quantum dot growth axis z and we disregarded possible g-factor anisotropy. The spin-flip rate is maximal at $\boldsymbol{B} \parallel z$ and decreases by a factor of 2 for the in-plane field orientation. Such an anisotropy clearly follows from Eqs. (8.13) and (8.14): The transitions where the spin-z component changes are activated by both contributions to the spin-orbit coupling with σ_x or σ_y Pauli matrices while the flip of x- or y-spin components is activated only by one term in the spin splitting Hamiltonian. For the spin-orbit interaction in the form of a linear combination of the bulk and structure inversion asymmetry terms, the spin-flip rate depends also on the field orientation in the quantum dot plane. Such a spin relaxation anisotropy is typical for the processes related to the $\boldsymbol{k}$-linear terms in the Hamiltonian as demonstrated for quantum well and lateral quantum dot structures both experimentally and theoretically (Dyakonov and Kachorovskii, 1986; Averkiev and Golub, 1999; Averkiev *et al.*, 2006; Fal'ko *et al.*, 2005; Scarlino *et al.*, 2014; Segarra *et al.*, 2015).

We also note that in axially symmetric quantum dots where $l_x = l_y$ and two excited states $\varphi_{2,x}(x,y)$ and $\varphi_{2,y}(x,y)$ are degenerate, the matrix element of the spin-flip process for the $\boldsymbol{B} \parallel z$ can be strongly enhanced due to the orbital paramagnetic effect of the field. Indeed, the linear combinations of the orbital basis functions

$$\varphi_{2p,\pm 1}(x,y) = \mp \frac{\varphi_{2,x}(x,y) \pm \mathrm{i}\varphi_{2,y}(x,y)}{\sqrt{2}} \tag{8.21}$$

transform exactly as the functions with the orbital angular momentum z-components of ± 1 [cf. Eqs. (2.12)]. The paramagnetic effect of the magnetic field is described by the Hamiltonian (Landau and Lifshitz, 1977)

$$V_B = -\frac{e\hbar}{2mc}\boldsymbol{B}[\boldsymbol{r} \times \hat{\boldsymbol{k}}], \tag{8.22}$$

where $\boldsymbol{\rho} = (x,y)$, $e = -|e|$, and m is the electron effective mass. The perturbation (8.22) results in the splitting of the orbital states $\varphi_{2p,\pm 1}$ by

$$\Delta_1 = 2\frac{m_0}{m}\mu_B B_z, \tag{8.23}$$

where m_0 is the free electron mass. Hence, for axially symmetric quantum dots the factor $g\mu_B B$ in Eq. (8.7) should be replaced by $\Delta_1/2$ and, correspondingly, the spin-flip transition rate, Eq. (8.19), acquires a large multiplier $\Delta_1^2/(g\mu_B B)^2 \sim m_0^2/(mg)^2 \sim 10^3$ in GaAs.

With an increase in magnetic field, the spin-flip rate in Eq. (8.19) increases as B^5. For sufficiently strong magnetic fields, deviations from Eq. (8.19) are expected. First, with an increase in B the involved phonon wavevectors increase and the deformation potential interaction, $U_{ph}^{(dp)} \propto q^{1/2}$, Eq. (8.10) may start to play a role in the spin-flip processes (Sherman and Lockwood, 2005). As a result, for the spin-flip rate related to the deformation potential interaction one has

$$\Gamma_{\downarrow\uparrow}^{(dp)} = \frac{4}{3\pi}\frac{(g\mu_B B)^7\gamma^2}{(E_{1s}-E_e)^4}\frac{D^2}{\rho\hbar^6 s_l^7}[1+N_{ph}(g\mu_B B)]. \tag{8.24}$$

The spin-flip rate due to deformation potential scales as B^7 at low temperatures.

Second, with an increase in the field B, the phonon wavelength becomes comparable with the electron localization length. Hence, the long-wavelength approximation (8.17) breaks and the matrix element of the electron–phonon interaction becomes suppressed. For Gaussian envelope functions the suppression is described by the factor $\exp[-(l_x^2 q_x^2 + l_y^2 q_y^2)/4]$, where q_x, q_y are the in-plane components of the phonon wavevector.

Third, the magnetic field affects the envelope functions of the ground and excited states and their energies via its diamagnetic effect. For the quantum dots with parabolic potential the diamagnetic effect simply results in the replacement of the energy difference $E_{1s}-E_e$ in the denominator of Eq. (8.19) by $\sqrt{(E_{1s}-E_e)^2+\hbar^2\omega_c^2}$, where $\omega_c = eB/mc$ is the cyclotron frequency associated with the magnetic field. This results in the suppression of the spin-flip rate with an increase of magnetic field as well. Naturally, in quantum dots with arbitrary potential the diamagnetic effect is more complex.

The orbital effect of the magnetic field is particularly pronounced for the admixture mechanism induced spin-flips for donor-bound electrons in bulk III-V and II-VI semiconductors (Linpeng *et al.*, 2016). Here, depending on the relation between the magnetic length, $l_b = \sqrt{\hbar c/|eB|}$, and the donor-bound electron Bohr radius, $a_B = \varkappa\hbar^2/me^2$, where $\varkappa$ is the static dielectric constant, Eq. (2.54), various regimes of the spin-flip process can be realized.

In particular, at sufficiently weak magnetic fields, where $l_b \gg a_B$, the magnetic field does not affect the hydrogen-like states of the donor-bound electron. In this case, the Dresselhaus spin-orbit interaction, Eq. (8.12), admixes nf-shell hydrogen-like states of the donor-bound electron with principal quantum numbers $n = 4, 5, \ldots$ and orbital momentum $l = 3$ to the $1s$-shell state. This is because the spin-orbit term in Eq. (8.12) contains only third angular harmonics being in stark contrast with quantum dot structures discussed previously, where due to the symmetry reduction the admixture of lowest excited states (with the z component of the angular momentum of ± 1 in the axial approximation) is possible. As a result, within the long-wavelength approximation the matrix element of the electron–phonon interaction contains two extra powers of the phonon wavevector because, in the decomposition of $\exp(\mathrm{i}\boldsymbol{q}\boldsymbol{r})$ only cubic in $\boldsymbol{q}$ terms provide a nonvanishing matrix element $\langle 1s|U_{ph}|nf\rangle$ (Linpeng *et al.*, 2016). Hence, we obtain the longitudinal spin relaxation time

$$\Gamma_{\downarrow\uparrow}^{(low,B)} \propto B^9[1+N_{ph}(g\mu_B B)], \tag{8.25}$$

Hence, at low temperatures, the spin-flip rate is inversely proportional to B^9, while for $k_B T \gtrsim g\mu_B B$, $\Gamma_{\downarrow\uparrow}^{(low)} \propto B^8$. The estimates performed in (Linpeng *et al.*, 2016) demonstrate that in such a low-field regime the donor-bound electron spin relaxation is dominated by the hyperfine interaction with the host lattice nuclei rather than by the admixture mechanism.

In the opposite limit, where $l_b \ll a_{\rm B}$, $\hbar\omega_c \gg E_b$, with the the donor-bound-electron binding energy $E_b = me^4/(2\varkappa^2\hbar^2)$, Eq. (2.55), the magnetic field shrinks the wavefunctions of the ground and excited states of the donor-bound electron. This situation is similar to the case of an electron localized in the (xy) plane by a parabolic potential, as in the quantum dot system discussed previously. The spin-flip rate scales as

$$\Gamma_{\downarrow\uparrow}^{(high,B)} \propto B^3[1+N_{ph}(g\mu_B B)], \tag{8.26}$$

where the reduction of the power from B^5 in Eq. (8.19) to B^3 is related to the fact that the energy distance between the ground and excited states (which contribute to $\Gamma_{\downarrow\uparrow}^{(high,B)}$) becomes equal to $\hbar\omega_c$ (Linpeng *et al.*, 2016). The analysis performed in (Linpeng *et al.*, 2016) shows that in this regime the long-wavelength approximation is no longer valid for the GaAs, CdTe, and InP bulk semiconductors. For this reason, actual dependence of $\Gamma_{\downarrow\uparrow}^{(high)}$ on the field B can be even smoother than that predicted by Eq. (8.26).

The intermediate field behavior is quite involved and can be accessed only numerically. The analysis of the calculated wavefunctions in (Linpeng *et al.*, 2016), motivates (i) restricting the semi-quantitative analysis by consideration of the ground, $1s$, and two lowest excited, $2p$ states with the angular momentum components of ± 1, and (ii) using Gaussian shapes of the ground and excited state wavefunctions,

$$\varphi_{1s} = \frac{1}{\pi^{3/4} l_{\rho,1s}\sqrt{l_{z,1s}}} \exp\left(-\frac{x^2+y^2}{2l_{\rho,1s}^2} - \frac{z^2}{2l_{z,1s}^2}\right), \tag{8.27a}$$

$$\varphi_{2p_\pm} = \frac{(x \pm \mathrm{i}y)}{\pi^{3/4} l_{\rho,2p}^2\sqrt{l_{z,2p}}} \exp\left(-\frac{x^2+y^2}{2l_{\rho,2p}^2} - \frac{z^2}{2l_{z,2p}^2}\right). \tag{8.27b}$$

to obtain an analytic solution for further insight into the intermediate field behavior. In Eqs. (8.27) the magnetic field induced shrinking is taken into account by assuming different characteristic lengths for the motion along and perpendicular to the field, $l_{k,\rho}$, $l_{k,z}$ with $k = 1s$ or $2p$. After some transformations, we obtain for the spin-flip transitions rate

$$\Gamma_{\downarrow\uparrow} = \frac{256\chi^{10}}{35(1+\chi^2)^{12}} \frac{\gamma_c^2 e^4 h_{14}^2 |g\mu_B|^3 B^5}{\pi\rho\hbar^6} \times \\ \times \left(\frac{1}{\Delta E} - \frac{1}{\Delta E + \hbar\omega_c}\right)^2 \left(\frac{f_l}{s_l^5} + \frac{4f_t}{3s_t^5}\right)[1+N_{ph}(g\mu_B B)]. \tag{8.28}$$

Here, $\Delta E = E_{2p_-} - E_{1s}$ is the energy difference between the hydrogen-like ground $1s$ and excited $2p_-$ state, which, in accordance with the numerical calculations, does not vary significantly with field in the intermediate field range. The factors $f_{l,t} = \exp\{-(\chi g \mu_B B l_\rho)^2/[(1+\chi^2)\hbar^2 s_{l,t}^2]\}$ take into account the fact that the phonon wavelength can be comparable with the donor-bound electron state size. These factors are particularly sensitive to the wavefunction shape. Finally, the parameter $\chi = l_{2p,\rho}/l_{1s,\rho} = l_{2p,z}/l_{1s,z}$ is a parameter of the wavefunctions that characterizes the ratio of the effective radii for the excited and the ground states, see Eqs. (8.27). Taking $l_{\rho,1s} = [1/a_B^2 + 1/2l_b^2]^{-1/2}$ and $l_{z,1s} = a_B$ and comparing the approximated wavefunctions Eq. (8.27) to those evaluated numerically, reasonable choices for χ of 1.5, 1.7, and 2.2 for GaAs, InP, and CdTe over the relevant magnetic field range have been found in (Linpeng *et al.*, 2016).

8.3 Direct spin–phonon interaction

The Hamiltonian of the electron phonon interaction in bulk semiconductors with a T_d point symmetry group also contains spin-dependent terms (Pavlov, 1966; Pavlov and Firsov, 1966, 1967) that can be recast in the form (Meier and Zakharchenya, 1984; D'yakonov *et al.*, 1986; Pikus *et al.*, 1988; Khaetskii and Nazarov, 2001; Woods *et al.*, 2002)

$$\mathcal{U}_{dir} = \frac{\hbar V_0}{2}\boldsymbol{\sigma}\cdot\boldsymbol{\varphi} + \frac{\hbar V_0'}{2}\boldsymbol{\sigma}\cdot\boldsymbol{\psi}, \tag{8.29}$$

where the pseudovectors $\boldsymbol{\varphi}$ and $\boldsymbol{\psi}$ can be expressed via the symmetrized combinations of the wavevector operator $\hat{k}_i$ and the deformation tensor components u_{ij} $(i,j = x,y,z)$ as

$$\varphi_x = \{u_{xy}\hat{k}_y - u_{xz}\hat{k}_z\}_s,\ \varphi_y = \{u_{yz}\hat{k}_z - u_{yx}\hat{k}_x\}_s,\ \varphi_z = \{u_{zx}\hat{k}_x - u_{zy}\hat{k}_y\}_s, \tag{8.30a}$$

$$\psi_x = \{\hat{k}_x(u_{yy} - u_{zz})\}_s,\ \psi_y = \{\hat{k}_y(u_{zz} - u_{xx})\}_s,\ \psi_z = \{\hat{k}_z(u_{xx} - u_{yy})\}_s. \tag{8.30b}$$

Here, the deformation tensor u_{ij} due to the presence of the phonon in the state $\boldsymbol{q},\alpha$ reads [cf. Eq. (8.10)]

$$u_{ij}^{\boldsymbol{q},\alpha} = \sqrt{\frac{\hbar}{2\rho\omega_{\boldsymbol{q},\alpha}}}\, e^{-\mathrm{i}(\boldsymbol{q}\boldsymbol{r} - \omega_{\boldsymbol{q},\alpha}t)}\frac{\mathrm{i}(\hat{e}_i^{(\boldsymbol{q},\alpha)}q_j + \hat{e}_j^{(\boldsymbol{q},\alpha)}q_i)}{2}b_{\boldsymbol{q},\alpha}^\dagger + \text{c.c.}, \tag{8.31}$$

and the curly brackets denote symmetrization of two operators $\{AB\}_s = (AB + BA)/2$. Microscopically, the contribution to $\mathcal{U}_{dir}$ described by the parameter V_0 can be derived within the two-band model accounting only for the conduction Γ_6, and valence Γ_7, Γ_8 bands of the energy spectrum of GaAs-like semiconductors, see Figure 2.2, taking into account the linear in u_{ij} mixing of the conduction and valence band Bloch amplitudes at the Γ-point. By contrast, the term $\propto V_0' \boldsymbol{\sigma} \cdot \boldsymbol{\psi}$ appears only with account for higher bands, which are not shown in Figure 2.2. On the basis of the $\boldsymbol{k} \cdot \boldsymbol{p}$ calculations (Pikus *et al.*, 1988) and atomistic tight-binding model (Nestoklon *et al.*, 2016) it is expected that $|V_0| \gg |V_0'|$ and the contribution $\propto V_0' \boldsymbol{\sigma} \cdot \boldsymbol{\psi}$ is disregarded in what follows.

The spin-flip rate $\Gamma_{\downarrow\uparrow}$ can be conveniently expressed by means of the Fermi golden rule via the matrix element

$$M_{\downarrow\uparrow} = \langle 1s, \downarrow |\mathcal{U}_{dir}| 1s, \uparrow \rangle. \tag{8.32}$$

Due to the symmetrization of the coordinate-dependent deformation and wavevector operator in Eqs. (8.30) and the fact that the ground state is a localized state, the matrix element (8.32) is nonzero only if the vector potential of the magnetic field $\boldsymbol{A}$ is taken into account in the electron wavevector operator

$$\hat{\boldsymbol{k}} = -\mathrm{i}\frac{\partial}{\partial \boldsymbol{r}} - \frac{e}{\hbar c}\boldsymbol{A}.$$

The cancellation of Eq. (8.32) for a zero magnetic field has the same origin as Kramers degeneracy of the localized states, see Section 2.6 for details. For the magnetic field $\boldsymbol{B} \parallel z$ one has for the matrix element describing the phonon emission process at $T = 0$

$$\langle 1s, \downarrow |U_{dir}| 1s, \uparrow \rangle = i\frac{\hbar V_0}{4}\sqrt{\frac{\hbar}{2\rho\omega_{\boldsymbol{q},\alpha}}}\left(\hat{e}_x^{(\boldsymbol{q},\alpha)} q_y + \hat{e}_y^{(\boldsymbol{q},\alpha)} q_x\right)$$
$$\times \left[\langle 1s|\{\exp(\mathrm{i}\boldsymbol{q}\boldsymbol{r}), k_y\}|1s\rangle - i\langle 1s|\{\exp(\mathrm{i}\boldsymbol{q}\boldsymbol{r}), k_x\}|1s\rangle\right], \tag{8.33}$$

where

$$\langle 1s|\{\exp(\mathrm{i}\boldsymbol{q}\boldsymbol{r}), k_x\}|1s\rangle = \frac{eB}{2\hbar}\langle 1s|\exp(\mathrm{i}\boldsymbol{q}\boldsymbol{r})y|1s\rangle,$$
$$\langle 1s|\{\exp(\mathrm{i}\boldsymbol{q}\boldsymbol{r}), k_y\}|1s\rangle = -\frac{eB}{2\hbar}\langle 1s|\exp(\mathrm{i}\boldsymbol{q}\boldsymbol{r})x|1s\rangle.$$

Substituting Eq. (8.32) into Eq. (8.5), and taking the phonon occupancies into account, we obtain the general expression for the spin-relaxation rate (Linpeng *et al.*, 2016)

$$\Gamma_{\downarrow\uparrow} = [1 + N_{ph}(g\mu_B B)]\frac{V_0^2 e^2 B^2}{2^9 \pi^2 \hbar \rho}$$
$$\times \sum_\alpha \frac{q_\alpha^3}{s_\alpha^2} \int d\Omega_q \left| \left(\hat{e}_x^{\boldsymbol{q},\alpha} \xi_y + \hat{e}_y^{\boldsymbol{q},\alpha} \xi_x\right) \langle 1s| \exp(\mathrm{i}\boldsymbol{q}_\alpha \boldsymbol{r})(x - \mathrm{i}y)|1s\rangle \right|^2. \tag{8.34}$$

It follows from Eqs. (8.33) and (8.34) that the direct spin–phonon interaction is sensitive to the shape of the ground state wavefunction only, because an admixture of the excited states is not required in this process.

Within the long wavelength approximation, Eq. (8.17), it is sufficient to retain only a $\boldsymbol{q}$-linear term in the matrix element involving the orbital function $\langle 1s|\exp(i\boldsymbol{q}_\alpha \boldsymbol{r})(x - iy)|1s\rangle$. As a result, we obtain the scaling of spin-flip transition rate related to the direct spin-phonon coupling as B^7 in low magnetic fields. For quantum dot structures the estimates presented in (Khaetskii, 2001; Khaetskii and Nazarov, 2001) demonstrate that the direct mechanism is, at $B = 1$ T, about $10^2 \dots 10^3$ times weaker compared with the admixture mechanism. The situation is different for bulk semiconductors (Linpeng *et al.*, 2016) where, as we already mentioned, due to the $\boldsymbol{k}$-cubic spin splitting, the admixture mechanism is to some degree suppressed. The spin-flip rate via the direct spin-phonon interaction in bulk GaAs-like semiconductor can be recast in a compact form

$$\Gamma^{dir}_{\downarrow\uparrow} = \frac{(eV_0 l^2)^2 |g\mu_B|^5 |B|^7}{560\pi\rho\hbar^6} \left(\frac{F_l}{s_l^7} + \frac{4F_t}{3s_t^7} \right) [1 + N_{ph}(g\mu_B B)], \qquad (8.35)$$

where the suppression factors F_l and F_t equal to $F_\alpha = \exp(-q_\alpha^2 l^2/2)/4$ or $(1 + q_\alpha^2 l^2/4)^6$ for the isotropic Gaussian, $\varphi_{1s} = (\sqrt{\pi}l)^{-3/2} \exp(-r^2/2l^2)$, and exponential, $\varphi_{1s} = \sqrt{\pi l^3} \exp(-r/l)$, donor-bound electron trial functions, respectively, with $q_\alpha = g\mu_B B/\hbar s_\alpha$ being the phonon wavevector. Note that in strong magnetic fields the B^7 scaling is changed to a weaker function of the field due to (i) breakdown of the long wavelength approximation and (ii) field-induced modification of the ground state wavefunction.

8.4 Experiment and theory

Experimentally, spin-flip transition rates between Zeeman levels in quantum dots have been studied in various semiconductor structures including, for example, electrostatic gate-defined GaAs/AlGaAs (Elzerman *et al.*, 2004; Amasha *et al.*, 2008) and InGaAs (Hayes *et al.*, 2009) quantum dots, as well as in self-assembled GaAs (Kroutvar *et al.*, 2004) and InGaAs (Heiss *et al.*, 2005; Lu *et al.*, 2010) quantum dots. The longitudinal spin-flip rates [cf. Eq. (8.4)]

$$1/T_1 \equiv \Gamma = \Gamma_{\downarrow\uparrow} + \Gamma_{\uparrow\downarrow}, \qquad (8.36)$$

as a function of the magnetic field measured in these works are summarized in Figure 8.2. The data presented in Figure 8.2 clearly demonstrate the enhancement of the spin-flip rate with an increase of magnetic field. The detailed theoretical analysis of the data is presented in (Abalmassov, 2014). It shows that depending on the studied sample and magnetic field range the data can be described by

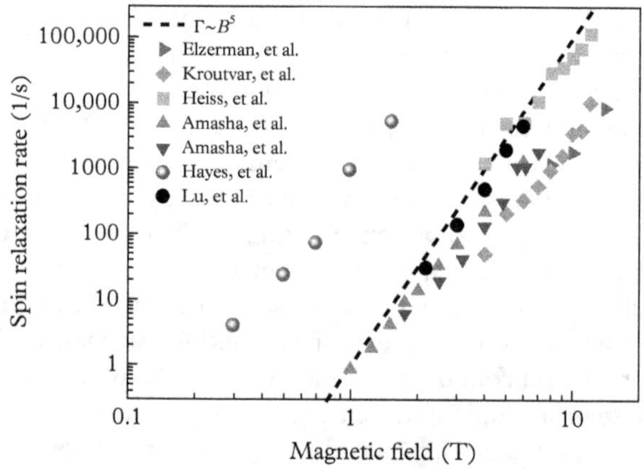

Figure 8.2 *Experimental data on the spin-flip rate measured in different III-V quantum dots on a double logarithmic scale illustrating power law dependence $\Gamma \propto B^{\nu}$. The dashed line shows $\propto B^5$ scaling law, Eq. (8.19). The data are collected from (Elzerman et al., 2004; Kroutvar et al., 2004; Heiss et al., 2005; Amasha et al., 2008; Hayes et al., 2009; Lu et al., 2010).*

approximate scaling relations $\Gamma \propto B^{\nu}$ with the power $\nu = 4\ldots5$. Usually, these data are discussed in the framework of admixture mechanism, see Section 8.2. In fact, given the uncertainty of the data, the B^5 scaling relevant for the admixture mechanism and piezoelectric interaction with acoustic phonons, Eq. (8.19), is in qualitative agreement with the experimental data. This is further supported by the scaling of the spin relaxation rate Γ with the localization energy E_{loc} studied, for example, in (Amasha *et al.*, 2008). In agreement with Eq. (8.19) the experimental data are described by the $\Gamma \propto E_{loc}^{-4}$.

However, a strong spread of the experimental data and significant uncertainties of the sample parameters (mainly, the quantum dot geometry, size quantization energies, and the spin-orbit coupling strength), do not allow one, in general, to describe the $\Gamma(B)$ dependence quantitatively. Moreover, the analysis performed in (Abalmassov, 2014) demonstrates that in some of the studied systems other mechanisms of the spin-flip transitions can be important. Among those is the admixture mechanism where the energy needed for the spin-flip transition is given to or taken from the fluctuating electric fields created by the mobile charge carriers present in the lateral quantum dot structures with electrostatic gates (Marquardt and Abalmassov, 2005). Within this mechanism the scaling $\Gamma \propto B^3$ is expected for low magnetic fields as compared with the B^5 scaling for the phonon-induced transitions. Other spin-flip mechanisms related to the modulation of the hyperfine coupling with lattice nuclei via phonons, spin-spin interactions of the localized, and free electrons and spin-flip transitions related to the current-noise induced fluctuating magnetic field (Pines *et al.*, 1957; Abrahams, 1957; Kaminski *et al.*, 2000; Khaetskii and Nazarov, 2001) are estimated to be very weak for III-V quantum dot systems.

In this regard, donor-bound electrons in bulk semiconductors are quite attractive to the systematic studies of the spin-flip rates because the wavefunctions of the donor-bound states and the material parameters are, to a large extent, well known. The experimental results on the longitudinal spin relaxation of the donor-bound electrons in GaAs, InP, and CdTe bulk samples are reported in (Linpeng *et al.*, 2016). The longitudinal spin relaxation times T_1 as a function of the electron Zeeman splitting $\Delta E = |g\mu_B B|$ for InP, GaAs, and CdTe are shown in Figure 8.3. In these experiments, electron spin polarization and detection have been carried out by optically depleting one of the Zeeman spin sublevels and monitoring the recovery of its thermal population in the course of spin relaxation, see the end of Section 6.2 and (Linpeng *et al.*, 2016) for details.

The data in Figure 8.3 show several notable features. First, all samples demonstrate a $T_1 \sim B^{-\nu}$ dependence, with $3 \lesssim \nu \lesssim 4$, at high magnetic fields. The proportionality constant depends on the semiconductor sample. A B^{-3} dependence, included in Figure 8.3, fits all curves well but reliable determination of the scaling law is difficult, especially for GaAs because the small electron g-factor prevents us from accessing the high-Zeeman-splitting limit, where $|g\mu_B B| \gg k_B T$.

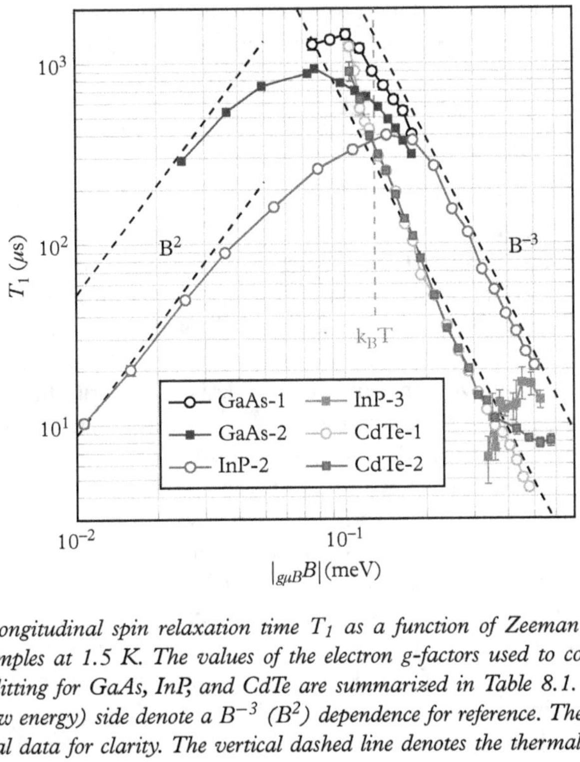

Figure 8.3 *Longitudinal spin relaxation time T_1 as a function of Zeeman splitting for the six different samples at 1.5 K. The values of the electron g-factors used to convert from B to the Zeeman splitting for GaAs, InP, and CdTe are summarized in Table 8.1. The lines in the high energy (low energy) side denote a B^{-3} (B^2) dependence for reference. They are offset from the experimental data for clarity. The vertical dashed line denotes the thermal energy $k_B T$ for reference. From (Linpeng et al., 2016).*

Also, a B^{-4} power-law is reasonable for CdTe, as the magnetic field dependence becomes steeper in CdTe with a decrease in field. The high-field longitudinal spin relaxation time, T_1, appears to be independent of donor concentration. At low fields, T_1 in InP and GaAs approaches a B^2 dependence with a donor-concentration-dependent pre-factor. Finally, the maximum T_1 observed in all three materials is similar: $T_1 = 1.4$, 0.4, and 1.2 ms for GaAs, InP, and CdTe, respectively.

Although spin relaxation in the high-field regime is of main importance for the following, before proceeding to the detailed comparison between the experiment and the theory, let us comment briefly on the $T_1(B)$ dependence in the low-field regime. Here the increase in spin relaxation time with increasing field is related to the suppression of the nuclei-induced relaxation by the magnetic field. The suppression is described by Eq. (8.1). For relevant mechanisms of electron correlation time at a donor, such as electron hopping and the exchange diffusion, see (Kavokin, 2008) for a review, an exponential sensitivity to the donor density (and, in the former case, to the temperature) is expected (Shklovskii and Efros, 1984; Kavokin, 2008). Correspondingly, for these mechanisms T_1 should be strongly affected by these parameters. Such trends are clearly seen in the experiment, see Figure 8.3. The fit of the experimental data to Eq. (8.1) allows one to evaluate the electron correlation time, τ_c, because the nuclear spin fluctuations, δ_e, for the donor-bound electrons can be readily evaluated as

$$\delta_e^2 = \frac{v_0}{12\pi a_{\rm B}^3 \hbar^2} \sum_n A_n^2 I_n(I_n+1), \tag{8.37}$$

where $a_{\rm B}$ is the donor Bohr radius, v_0 is the primitive cell volume, I_n is the spin of nth nucleus in the cell, and A_n is the hyperfine interaction constant, see Eq. (7.11) and (Dyakonov and Perel', 1974). Taking the hyperfine coupling constants from Table 4.1, we determine the correlation time $\tau_c \approx 25$ ns for the GaAs-2 sample. A somewhat longer τ_c of ~ 40 ns is determined for the InP-2 sample. Although, as already noted, the sensitivity of observed T_1 values to temperature and carrier density are consistent with the known mechanisms contributing to the donor electron correlation time, the magnitude of τ_c is orders of magnitude shorter than these mechanisms predict for the low donor densities used in this study. However, these short values of τ_c are consistent with other studies (Dzhioev *et al.*, 2002*a*; Berski *et al.*, 2015) and suggests additional, unknown mechanisms may be at play, such as an inhomogeneous donor distribution resulting in the formation of clusters with a relatively high donor density, and short τ_c.

Now let us turn to the discussion of spin relaxation in a high magnetic field regime where T_1 decreases with an increase of the field. The experimental data together with the results of calculations performed in (Linpeng *et al.*, 2016) are summarized in Figure 8.4. For the GaAs the admixture mechanism [dotted curve in Figure 8.4(a)] is already in order-of-magnitude agreement with the experiment. Moreover, in GaAs the magnitude of the direct-phonon mechanism

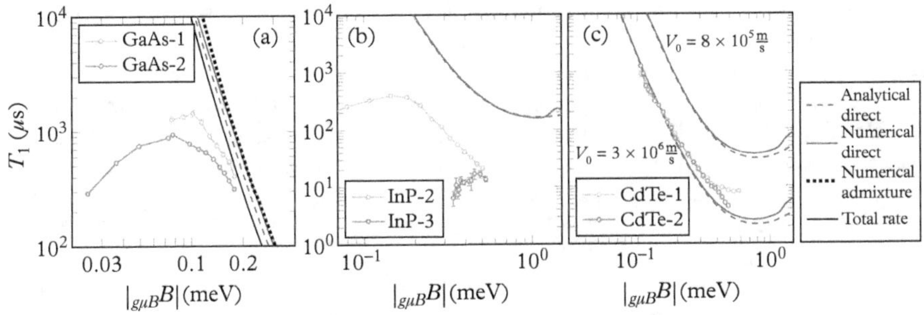

Figure 8.4 *Theoretical and experimental results for the spin-relaxation time T_1 in GaAs (a), InP (b), and CdTe (c). Dots show the experimental data. $T = 1.5$ K. The two dashed lines and two solid lines in (c) represent the analytic and numerical calculation results of T_1 using $V_0 = 8 \times 10^5$ m/s (upper curves) and $V_0 = 3 \times 10^6$ m/s (lower curves). In panel (a) dotted and solid curves correspond to the account for the admixture mechanism and both admixture and direct spin-phonon coupling mechanisms, respectively. The analytical curves are calculated after Eq. (8.4), Eq. (8.28) (admixture mechanism), and Eq. (8.35) (direct mechanism). From (Linpeng et al., 2016).*

is approximately the same as the admixture mechanism. Accounting for both mechanisms for GaAs [solid curve in Fig. 8.4(a)] results in a difference between the theory and the experimental data of approximately a factor of about 2, which can be easily attributed to the uncertainties in the system parameters in Table 8.1.

The numerical and analytical calculations of the spin-flip rate in the admixture mechanism performed for the CdTe and InP semiconductors provide values of T_1 being orders of magnitude larger compared with the experimental data. Hence, in this sample the admixture mechanism is not relevant at all. For these materials the direct spin-phonon mechanism is found to be significantly stronger than the admixture mechanism. For CdTe the magnitude of the direct spin-phonon coupling constant V_0 is not known. Hence, the calculations for two values of V_0: (i) such as in GaAs and (ii) such as in InSb, have been performed. The agreement between the theory and experiment is very good if the direct spin–phonon interaction strength in CdTe is similar to that of InSb. Although in these two materials the valence band spin-orbit splitting is similar (Marple and Ehrenreich, 1962; Cardona *et al.*, 1967; Twardowski *et al.*, 1980; Cardona *et al.*, 1988) the band gaps differ. Thus an independent measurement of V_0, like those performed in (D'yakonov *et al.*, 1986) for GaAs and InP, and its independent first-principles calculation, is needed to corroborate the result.

There is still a significant discrepancy between theory and experiment for InP, where the experimental spin-relaxation time is 15–30 times shorter than the predicted value from the direct spin-phonon coupling. Its origin is not clear and further studies, both experimental and theoretical, are needed to resolve this discrepancy.

Table 8.1 *Material parameters relevant to the donor-bound electron spin-relaxation in GaAs, InP, and CdTe. g is the effective electron g-factor, m is the electron effective mass (m_0 is the free electron mass), h_{14} is the piezoelectric constant, γ_c is the Dresselhaus spin-orbit coupling constant, ρ is the mass-density, s_l is the longitudinal sound velocity, s_t is the transverse sound velocity, $\varkappa$ is the static dielectric permittivity of the material, V_0 characterizes the strength of the direct spin-phonon coupling interaction, and D is the deformation potential interaction constant. The data are taken from (Madelung et al., 1999; Gavini and Cardona, 1970; D'yakonov et al., 1986; Rottner et al., 1993; Madelung, 2004; Jancu et al., 2005; Petrovićfi and Vukmirović, 2012) and www.ioffe.ru/SVA/NSM/Semicond/.*

	GaAs	**InP**	**CdTe**
g	−0.44	1.3	−1.65
m/m_0	0.067	0.08	0.106
h_{14} (V/m)	14.5×10^8	7.4×10^8	3.94×10^8
γ_c (eV·Å^3)	23.7	10.1	11.74
ρ (kg/m^3)	5.32×10^3	4.81×10^3	4.85×10^3
s_l (m/s)	4.73×10^3	4.58×10^3	3.08×10^3
s_t (m/s)	3.35×10^3	3.08×10^3	1.85×10^3
$\varkappa$	12.56	12.5	10.3
V_0 (m/s)	8×10^5	4×10^5	unknown
D (eV)	−5.55	−4.4	−5.45

8.5 Spin lifetime limited by radiative processes

In bulk GaAs-like semiconductors the longitudinal spin relaxation time of the electron reaches ~ 1 ms at optimal magnetic fields where the crossover between the hyperfine-interaction and the phonon-induced spin-flip transitions regimes takes place, see Figure 8.3. In III-V quantum dots the spin lifetimes can be much longer and reach seconds, see Figure 8.2 and (Camenzind *et al.*, 2017). Hence, it is natural to ask a question about the upper limits of the electron spin relaxation time in nanosystems. In this regard, Si-based nanostructures play a special role (Zwanenburg *et al.*, 2013). This is because bulk Si is a centrosymmetric semiconductor characterized by the O_h point symmetry group where any spin-dependent terms in the form of Eq. (2.15) in the electron effective Hamiltonian are forbidden. In Si-based quantum wells the inversion symmetry is broken, as a rule, which gives rise to the linear in the electron wavevector spin-dependent terms like those in Eqs. (8.13) and (8.14), see (Wilamowski *et al.*, 2002; Golub and Ivchenko, 2004; Nestoklon *et al.*, 2006). However, a much weaker spin-orbit coupling in silicon as compared with III-V compounds, results in significantly longer spin lifetimes in Si-based low-dimensional structures (Wilamowski and Jantsch, 2004; Zinovieva *et al.*, 2008).

Correspondingly, in Si-based quantum dots fairly long longitudinal spin relaxation times have been observed (Hayes *et al.*, 2009; Xiao *et al.*, 2010; Yang *et al.*, 2013) reaching 30 s in specially engineered structures (Muhonen *et al.*, 2014; Watson *et al.*, 2017). In these isotopically purified nanostructures, where the amount of ^{29}Si is negligible, the spin relaxation is controlled by the electron–phonon and spin–orbit interactions.

The ultimate limit on the spin relaxation rate is imposed simply by the radiative transitions between the Zeeman-split states (Purcell, 1946). Indeed, as discussed in Chapter 3 the transitions between the spin-up and spin-down states are active in the magneto-dipole approximation: The states $\uparrow$ and $\downarrow$ are mixed by the in-plane magnetic field components. These transitions are extremely weak for a spin in a free space. For example, taking the spin precession frequency $\Omega_B/(2\pi) = 8$ GHz the spontaneous emission rate is about 10^{-12} s^{-1} (Bienfait *et al.*, 2016). Of course, corresponding T_1 time is too long to be reliably detected and, moreover, it is too long compared with times related to other spin-flip mechanisms. By putting the spin in the cavity where the density of photon states can be greatly enhanced in a certain energy range one can also get a strong increase in the the spontaneous emission rate. Correspondingly, the spin-flip rate related to the radiative transition between the Zeeman levels takes the form (Bienfait *et al.*, 2016)

$$\frac{1}{T_1} \equiv \Gamma = \Gamma_{nr} + \frac{\gamma_0}{1 + (2\delta/\kappa)^2}. \tag{8.38}$$

Here, Γ_{nr} is the spin-flip rate related to the nonradiative processes, δ is the detuning between the Zeeman splitting $\hbar\Omega_B$ and the cavity resonance frequency, κ is the damping rate of the photon mode of the cavity, and γ_0 is the spin-flip rate due to the radiative process exactly in the resonance. Equation (8.38) demonstrates the possibility to control the spin-flip rate by changing the detuning δ; for example, by varying the external magnetic field. The enhancement of Γ due to the cavity is known as the *Purcell effect* (Purcell, 1946). In the original paper, E.M. Purcell demonstrated that the rates of magneto-dipole transitions between nuclear spin states are strongly enhanced in the presence of a resonator. Equation (8.38) is derived for the weak coupling regime where the reabsorption of the photon can be disregarded.

Experimentally, the spin-flip process limited by the spontaneous emission of photons has been studied in (Bienfait *et al.*, 2016) on the isotopically purified bulk ^{28}Si crystal where nuclear spins are absent. The bismuth (^{209}Bi) donors have been implanted into the crystalline lattice and superconducting resonators have been patterned onto this silicon chip. The bismuth donor has a nonzero nuclear spin $I = 9/2$, this spin interacts with the donor-bound electron spin via hyperfine coupling in the form of Eq. (2.22)

$$\mathcal{H}_{hf} = A_{Bi}(\hat{\boldsymbol{I}} \cdot \hat{\boldsymbol{s}}), \tag{8.39}$$

with the effective constant $A_{Bi}/h = 1.475$ GHz (Wolfowicz *et al.*, 2012). The hyperfine interaction dominates all other interactions in the spin system of "electron+nucleus" and results, in the absence of a magnetic field, in the splitting between the multiplets with different values of the total spin $\hat{\boldsymbol{F}} = \hat{\boldsymbol{s}} + \hat{\boldsymbol{I}}$; namely, with $F = 5$ and $F = 4$, see Chapter 11 for details. This splitting is about 7.375 GHz (Wolfowicz *et al.*, 2012) and is very close to the cavity resonance frequency of $\omega_0/(2\pi) = 7.305$ GHz used in the reported experiments. The application of the external magnetic field to the system gives rise to the splitting of each multiplet into Zeeman sublevels and allows for a fine tuning of the resonant transition between the multiplet and the cavity frequency, see the scheme in the inset of Figure 8.5.

The spin-flip time T_1 corresponding to the transitions between the levels in the multiplet with $F = 5$ to that of $F = 4$ is measured by means of the "inversion recovery" technique (Bienfait *et al.*, 2016) based on the erasing of the spin polarization by the radio-frequency pulse and monitoring the recovery of the spin polarization via a spin-resonance signal. The detuning δ has been varied by changing the magnetic field applied to the sample. The resulting dependence of the T_1 time is shown by circles in Figure 8.5. It is very well described by the theoretical dependence (8.38) (solid curve) making it possible to evaluate the

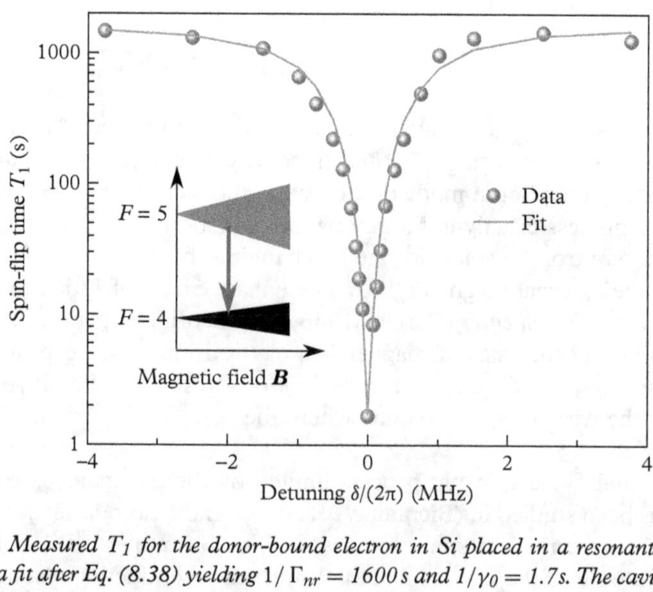

Figure 8.5 *Measured T_1 for the donor-bound electron in Si placed in a resonant cavity. The solid line is a fit after Eq. (8.38) yielding $1/\Gamma_{nr} = 1600$ s and $1/\gamma_0 = 1.7$s. The cavity resonance frequency $\omega_0/(2\pi) = 7.305$ GHz and the damping rate $\kappa = 68$ KHz. Inset schematically shows the multiplets with $F = 5$ and $F = 4$ split by the hyperfine interaction Eq. (8.39) (states within the multiplets are not shown) and vertical arrow shows the spin flip transition resonant with the cavity. The data are extracted from (Bienfait et al., 2016).*

nonradiative contribution to the spin-flip time $\sim 1.6 \times 10^3$ s (i.e., about half an hour), and the T_1 value at the zero-detuning, which corresponds to the radiative limit, $T_1 \approx 1.7$ s. The latter quantity is quite close to the theoretical estimate based on the cavity parameters and spin-photon coupling strength, g_c (Bienfait *et al.*, 2016):

$$\gamma_0 = \frac{4g_c^2}{\kappa} \approx 0.8\ \text{s}^{-1}, \quad 1/\gamma_0 = 1.25\ \text{s}.$$

The results obtained in (Bienfait *et al.*, 2016) not only demonstrate the possibility to reach the regime where the spin relaxation rate in the electronic system is limited by the radiative process, but also show the possibility to control the spin relaxation time over three orders of magnitude by varying the external magnetic field. We also note that the upper limit of the $T_1 = 1/\Gamma_{nr}$ in Figure 8.5 can be related to, for example, two-phonon processes, which provide the donor-bound electron spin-flip rate weakly dependent on the magnetic field (Abrahams, 1957).

9

Electron Spin Precession Mode Locking and Nuclei-Induced Frequency Focusing

9.1 Phenomenology and overview of experimental results

The pump-probe technique introduced in Chapter 6 provides unprecedented access to the electron and nuclear spin dynamics in nanosystems. It is particularly informative if an external magnetic field is applied in the Voigt geometry; that is, perpendicular to the propagation direction of pump and probe beams. In this case, the spin polarization of the charge carriers induced by the circularly polarized pump pulse precesses around the magnetic field. As a result, it gives rise to the temporal oscillations of the Faraday or Kerr rotation angles detected by the linearly polarized probe, see Figure 6.1. The period of oscillations is directly related to the magnetic field strength and the charge carriers g-factor values. The damping of the oscillations reflects transversal relaxation time, T_2^*, related in the electron ensembles to both the spin dephasing/decoherence processes and inhomogeneous broadening caused, for example, by the nuclear fluctuations and spread of the g-factor values (Glazov, 2012; Dyakonov, 2017).

The application of the pump-probe technique to nanosystems with localized electrons, particularly to quantum dot systems, has revealed a number of fascinating phenomena related to the spin dynamics of electrons and nuclei. In this chapter, we address the most prominent ones, the reviews (Glazov, 2012; Yakovlev and Bayer, 2017) contain more details on electron spin dynamics under pump-probe conditions.

Figure 9.1(a) shows typical spin signals of Faraday rotation and ellipticity obtained in (Glazov *et al.*, 2010) on the n-type InGaAs quantum dot ensemble (quantum dots are charged with electrons). The measurements were carried out using the so-called "two-color" or nondegenerate pump-probe technique where pump and probe pulses are generated by different lasers, hence their carrier

Electron and Nuclear Spin Dynamics in Semiconductor Nanostructures. M. M. Glazov, Oxford University Press (2018). © M. M. Glazov. DOI: 10.1093/oso/9780198807308.001.0001

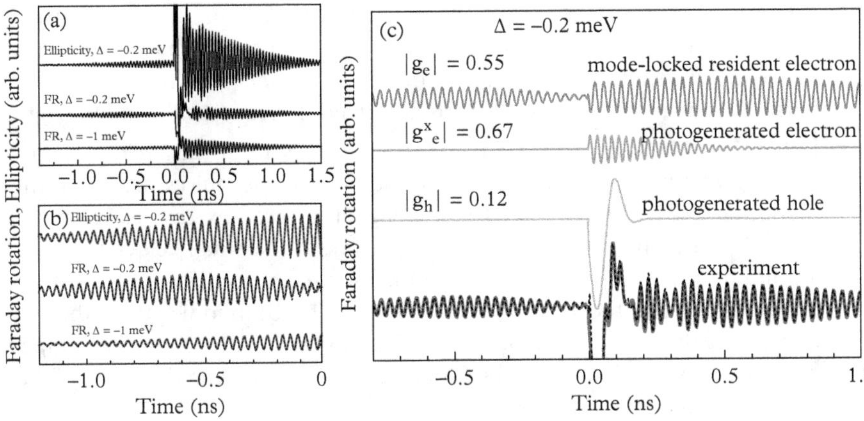

Figure 9.1 *(a) Spin-induced Faraday rotation and ellipticity signals as functions of the temporal delay between pump and probe pulses. The top two curves are the ellipticity and Faraday rotation signals at (almost) spectrally degenerate pump and probe lasers ($\hbar\omega_P - \hbar\omega_{pr} = \Delta = -0.2$ meV), the bottom curve shows the Faraday rotation signal for a detuned probe pulse ($\Delta = -1.0$ meV). (b) Corresponding signals at negative delays. Dotted curves are experimental data, solid ones are the fitting. (c) Faraday rotation signal (bottom curve) at almost degenerate pump and probe pulses. The fitting is superimposed over the experimental curve. The curves presented above (from top to bottom): signal related with the long living spin polarization in charged quantum dots, signal related to electron-in-exciton spin precession in neutral quantum dots, and signal related to hole spin precession (both in neutral quantum dots and in trions). Measurements are carried out on the structure consisting of 20 layers of InGaAs/GaAs quantum dots with the dot density in each layer of 10^{10} cm^{-2}, the structure is doped with one electron per dot on average. Temperature $T = 6$ K, magnetic field $B = 4$ T. Data are reproduced from (Glazov et al., 2010).*

frequencies can be tuned independently. The pulses are synchronized at a high precision of less than 100 fs. Pumping and probing is carried out by a periodic sequence of pulses arriving at the sample with a repetition period of $T_R = 13.2$ ns. The magnetic field is applied in the Voigt geometry, that is $\boldsymbol{B} \parallel x$, where the used set of axes, with the z-axis being the light propagation axis and the x- and y-axes being the axes in the sample plane, as depicted in Figure 6.1.

Figure 9.1 (a) demonstrates key features of the experimental data:

- Signals at positive delays, that is when the probe pulse arrives after the corresponding pump pulse, have a complex character corresponding to the superposition of oscillations with different frequencies. Analysis of the oscillation frequencies and decay times, as shown in Figure 9.1 (c), demonstrates that the observed signal is a superposition of the resident electron spin contribution, as well as the contributions of photocreated (in neutral dots) electrons and holes (Yugova *et al.*, 2007*b*; Glazov *et al.*, 2010).

- Remarkable signals take place also at negative delays; that is, when the probe pulse arrives before the next pump pulse. Effects of spin signal generation at negative delays and spin accumulation at the excitation of nanosystems by periodic sequences of the pulses are described in Sections 9.2 and 9.3.
- In the situation where the pump and probe pulses are spectrally degenerate, that is, where the optical (carrier) frequency of the pulses are the same, the amplitude of the Faraday rotation signal induced by resident electrons behaves nonmonotonically as a function of delay: At small delays the signal amplitude increases with time (i.e., builds-up) and decays at larger pump-probe delays. It is most clearly seen in Figure 9.1 (b), middle curve; the area of negative delays is shown there. The ellipticity signal demonstrates damped oscillations only without any buildup. Section 9.4 is devoted to the buildup of the Faraday rotation signal.

9.2 Resonant spin amplification and electron spin coherence mode locking

The spin-Kerr or -Faraday rotation signals at negative pump-probe delays arise due to the fact that the pumping and probing of the electron spins is carried out by the periodic train of pulses and the resident carrier spin coherence does not fully relax during the pulse repetition period. Depending on the relation between the electron spin precession period and the pump pulse repetition period, the spin polarization in the system can be either accumulated or get suppressed.

Indeed, as can be seen from Figure 9.2, if the pump pulse repetition period, T_R, is a multiple of the electron spin precession period in the external field $T_L = 2\pi/\Omega_B$ [$\Omega_B = g\mu_B B/\hbar$, Eq. (2.10), is the electron spin precession frequency with g being the Landé factor and μ_B the Bohr magneton],

$$T_R = NT_L = \frac{2\pi N}{\Omega_B}, \quad N = 1,2,\dots, \tag{9.1}$$

then the next pump pulse adds the spin in phase with the precessing one. In this case, the spin polarization in the system is enhanced as compared with that formed by a single pulse. This effect is known as *resonant spin amplification.* Equation (9.1) describes the *phase synchronization condition.* If condition (9.1) is not fulfilled, the phase synchronization fails and spin accumulation does not occur.

As a rule, in experiments the spin dynamics of an electron ensemble is studied. Optical excitation of the quantum dot ensemble or the quantum well with localized electrons results in the spin polarization of charge carriers with different energies being spread within the pump pulse spectral width $\sim \hbar/\tau_p$, see Chapter 6 for details on spin coherence generation and detection by optical pulses. Electron g-factor values are different and the nuclear spin fluctuations acting on different electrons

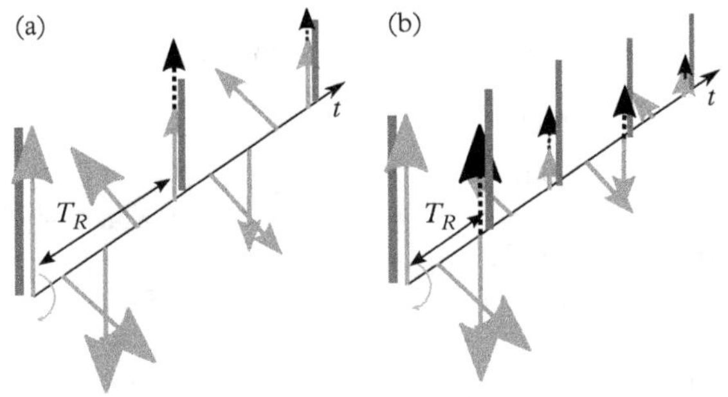

Figure 9.2 *Schematic illustration of the resonant spin amplification. Pump pulse arrival time moments are marked. Solid arrows show the electron spin orientation at different time moments. Dotted arrows show the spin polarization generated by pump pulses. Panel (a) shows the case of equal spin precession and repetition periods, $T_R = T_L$, in panel (b) the pulses arrive twice as often: $T_R = T_L/2$. The sense of spin precession is shown by the curved arrows.*

are random. As a result, the electron spin precession frequencies differ from one localization site to another. Spin beats damping are characterized by the following parameters: T_2 being the relaxation time of transversal to the external field electron spin components, and

$$T_2^* = T_2 T_{\rm inh}/(T_2 + T_{\rm inh}), \tag{9.2}$$

being the dephasing time of electron spin in the ensemble contributed both by the relaxation processes and the spread of Larmor precession frequencies. The latter one is characterized by the time $T_{\rm inh} \sim (\Delta\Omega)^{-1}$, where $\Delta\Omega$ is the spin precession frequency spread. The specifics of spin dynamics under the excitation by a periodic train of pulses is determined by the repetition period of the pulses T_R and the spin beat decay times. Obviously, if $T_2 \ll T_R$, the effects of spin polarization accumulation are unessential, since spin fully vanishes before the next pulse arrival. The accumulation can take place if

$$T_2 \gtrsim T_R. \tag{9.3}$$

Next, we assume that the condition (9.3) is fulfilled and analyze two important cases: (i) of weak dephasing related to the spin precession frequency spread $T_{\rm inh} \gg T_R$, where the resonant spin amplification is realized, and (ii) the regime of strong dephasing, $T_{\rm inh} < T_R$, where the *spin precession mode locking* becomes possible.

We start with a situation where the inhomogeneous spread of the spin precession frequencies is not relevant, $T_{\rm inh} \gg T_2$, and spin beats damping is determined by the spin relaxation processes. We assume for simplicity that the average electron spin, both generated by a single pump pulse and accumulated by a pump pulse train, is small. Therefore, we obtain from Eqs. (6.1) the periodic sequence consisting of a large enough number of pulses:

$$S_z^{\mathrm{tot}}(\Delta t) = \sum_{n=1}^{\infty} S_z^{(0)} \mathrm{e}^{-(\Delta t + nT_R)/T_2} \cos\left[\Omega_B(\Delta t + nT_R)\right]. \tag{9.4}$$

Here, $S_z^{(0)}$ is the electron spin created by a single pulse, Δt is the delay between the probe pulse and the nearest next pump pulse, it can take any negative value in the interval $\Delta t \in (-T_R, 0]$. The summation of the geometric series in Eq. (9.4) yields (Kikkawa and Awschalom, 1998; Beschoten, 2005)

$$S_z^{\mathrm{tot}}(\Delta t) = \frac{S_z^{(0)}}{2}\, \mathrm{e}^{-(T_R+\Delta t)/T_2} \frac{\cos(\Omega_B \Delta t) - \mathrm{e}^{T_R/T_2} \cos\left[\Omega_B(T_R + \Delta t)\right]}{\cos(\Omega_B T_R) - \cosh(T_R/T_2)}. \tag{9.5}$$

It follows from Eq. (9.5) that the dependence of the accumulated spin on the Larmor precession frequency Ω_B (and, correspondingly, on the magnetic field) at a fixed delay Δt consists of a sequence of maxima corresponding to the phase synchronization condition (9.1): $\Omega_B T_R \approx 2\pi N$, where N is an integer. In the vicinity of the maximum where $|\Omega_B T_R - 2\pi N| \ll 1$ and $|\Delta t/T_R| \ll 1$, Eq. (9.5) can be rewritten in the form of Lorentz function

$$S_z^{\mathrm{tot}}(0) = S_z^{(0)} \frac{1 - S_z^{(0)} e^{-\frac{T_R}{T_2}}}{(\Omega_B T_R - 2\pi N)^2 + 2\left[\cosh(T_R/T_2) - 1\right]}. \tag{9.6}$$

In this approximation the peak width (as function of $\Omega_B T_R$) is determined by the quantity

$$\Delta = \sqrt{2\left[\cosh(T_R/T_2) - 1\right]}, \tag{9.7}$$

which, in the limit of long spin relaxation times, $T_R/T_2 \ll 1$, passes to $\Delta \approx T_R/T_2$, and the width is the smaller the longer the spin relaxation time T_2.

Exactly at the resonance where $\Omega_B T_R = 2\pi N$ and for a sufficiently long electron spin relaxation time, $T_2 \gg T_R$, the accumulated value of electron spin can be presented in the form

$$S_z^{\mathrm{tot}}(0) \approx S_z^{(0)} \frac{T_2}{T_R}.$$

It grows with an increase of the relaxation time T_2. The factor T_2/T_R in the expression here has a transparent physical meaning: it describes the number of pulses that contribute to the spin accumulation during the spin relaxation time. In real systems the electron spin polarization is, of course, limited, $|S_z| \leqslant 1/2$. In order to account the saturation of the electron spin polarization the modification of the electron spin state by the circularly polarized pulse described by the quantum mechanical model, Eqs. (6.16), should be taken into account. General expressions describing the resonant spin amplification were derived in (Yugova *et al.*, 2009*a*,; Fokina *et al.*, 2010; Yugova *et al.*, 2012). With an increase of the pumping the peaks become broader and the spin polarization dependence on magnetic field becomes smoother. In order to illustrate the effect, here we consider the resonant situation where Eq. (9.1) is fulfilled and assume that the spin excitation is carried out by the train of σ^+ polarized pulses. From Eq. (6.16a) it follows in the steady state that

$$S_z^{\rm tot}(0) = \frac{Q^2-1}{4} + \frac{Q^2+1}{2} e^{-T_R/T_2} S_z^{\rm tot}(0), \tag{9.8}$$

where the parameter $0 \leqslant Q \leqslant 1$ describes the efficiency of the spin coherence generation by a single pulse, see Chapter 6 for details. Solution of Eq. (9.8) under the condition $T_R/T_2 \ll 1$ reads

$$S_z^{\rm tot}(0) = \frac{Q^2-1}{4} \frac{1}{\frac{1-Q^2}{2} + \frac{1+Q^2}{2}\frac{T_R}{T_2}},$$

and the electron spin polarization is indeed limited, $|S_z^{\rm tot}(0)| \leqslant 1/2$.

The expression for the resonant spin amplification spectrum Eq. (9.5) can be easily extended to allow for the arbitrary spin relaxation anisotropy (Glazov and Ivchenko, 2008; Griesbeck *et al.*, 2012) inherent to semiconductor nanosystems. To analyze the role of spin relaxation anisotropy we consider a system of C_{2v} point group symmetry with the z-axis being the two-fold rotation axis C_2, see Appendix A for details. Such symmetry is relevant for quantum dot or quantum well structures grown on a (001) plane. In the geometry where the $\boldsymbol{B} \parallel x \parallel [1\bar{1}0]$, the spin precession involves only S_y ($y \parallel [110]$) and S_z components. The relaxation times of z and y spin components T_{zz} and T_{yy} are different. In this case instead of Eq. (9.5) one has

$$S_z^{\rm tot}(\Delta t) = \frac{S_z^{(0)}}{2} e^{-(T_R+\Delta t)/\bar{T}} \frac{e^{T_R/\bar{T}} C[\tilde{\Omega}(T_R+\Delta t)] - C(\tilde{\Omega}\Delta t)}{\cosh(T_R/\bar{T}) - \cos(\tilde{\Omega} T_R)}, \tag{9.9}$$

where

$$\tilde{\Omega} = \sqrt{\Omega_B^2 - \frac{1}{4}\left(\frac{1}{T_{zz}} - \frac{1}{T_{yy}}\right)^2}, \quad \frac{1}{\bar{T}} = \frac{1}{2}\left(\frac{1}{T_{zz}} + \frac{1}{T_{yy}}\right)^2, \tag{9.10}$$

and the function $C(x)$ is defined by

$$C(x) = \cos x - \frac{1}{2\tilde{\Omega}}\left(\frac{1}{T_{zz}} - \frac{1}{T_{yy}}\right)\sin x .$$

Note that for the experimentally relevant parameters where $T_R \ll T_{zz}, T_{yy}$ the spin relaxation anisotropy affects the relative amplitude and width of the zero field peak compared with the parameters of other peaks (for $\Omega_B T_R = 2\pi N$ with $N = 1,2,\ldots$). Indeed, for sufficiently strong magnetic fields the difference in spin relaxation times is effectively averaged out and the spin dynamics can be described by a single effective relaxation time $\bar{T}$, Eq. (9.10). In contrast, at small magnetic fields, $\Omega_B T_{zz} \ll 1$, the spin precession is practically unimportant and the spin dynamics is controlled by the T_{zz} relaxation time. The anisotropy of the spin relaxation makes small-field spin dynamics quite involved (Kalevich *et al.*, 1997; Döhrmann *et al.*, 2004). That is why the zero-field peak in the resonant spin amplification spectrum, S_z^{tot}, is different from the remaining ones: Particularly, as follows from Eq. (9.9) for $T_{zz} > T_{yy}$ the zero-field peak is higher than the remaining ones, for $T_{zz} < T_{yy}$ the relation between the peak heights is opposite (Glazov and Ivchenko, 2008).

Figure 9.3(a) shows the resonant spin amplification spectrum, that is the dependence of the Kerr rotation signal, $\vartheta_K \propto S_z^{\text{tot}}$, on the magnetic field obtained at CdTe/$Cd_{0.78}Mg_{0.22}$Te quantum well structure at a small negative delay (Astakhov *et al.*, 2008). Peaks correspond to the commensurability condition of the spin precession period and the pulse repetition period, see Eq. (9.1). Note, that the zero field peak is larger than adjacent ones. This is related to the effect of the nuclear spin fluctuations: In the studied structure, the electrons are localized

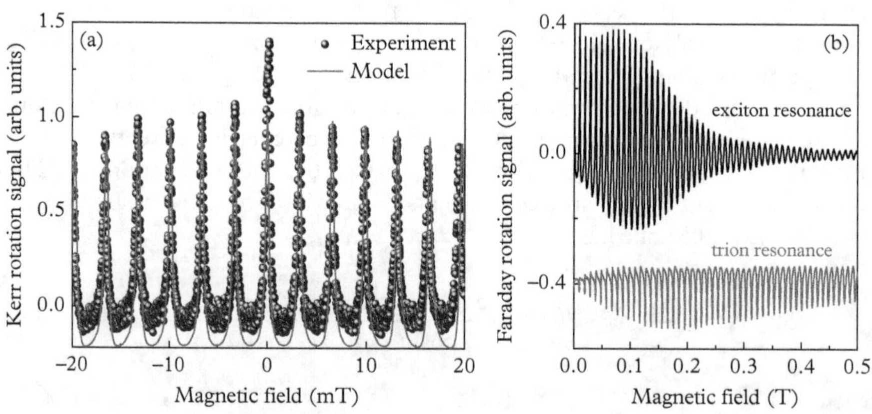

Figure 9.3 *(a) Kerr rotation signal dependence on the external magnetic field. Points are the experimental data obtained at the CdTe/$Cd_{0.78}Mg_{0.22}$Te quantum well structure (five wells of 20 nm width, electron density in the well $N \approx 10^{10}$ cm^{-2}) for a small negative delay $\Delta t = -80$ ps, pulse repetition period is $T_R = 12.5$ ns. Sample temperature $T = 2$ K. Data are reproduced from (Astakhov et al., 2008). The solid curve is the fit of experimental data, transverse spin relaxation time $T_2 = 30$ ns, mean g-factor $|g| = 1.64$, spread of g-factor values $\sigma_g/(2g) =$ 0.4% after (Glazov and Ivchenko, 2008). (b) Faraday rotation signals in the resonant spin amplification regime measured at the exciton (1.4395 eV, top curve) and trion (1.4382 eV, bottom curve) resonances at $T = 1.8$ K on two coupled 8-nm-thick $In_{0.09}Ga_{0.91}$As/GaAs quantum wells separated by a thin 1.7 nm GaAs barrier. Wells are slightly doped with electrons. Data are reproduced from (Fokina et al., 2010).*

in the quantum well plane on the well width fluctuations and experience random fields. The hyperfine interaction effectively stabilizes the spin-z component due to the nondecaying fraction of electron spin discussed in detail in Section 7.2. Peak height decreases monotonously with an increase in the magnetic field and the peaks themselves become slightly broadened. It is caused by the g-factor spread in the localized electron ensemble, which gives rise to an additional spin dephasing. Spin dephasing due to the fluctuations of g-factor can be modeled assuming Gaussian distribution of the g-factor fluctuations, Δg:

$$\mathcal{F}_g(\Delta g) = \frac{1}{\sqrt{\pi}\sigma_g} \exp\left(-\frac{(\Delta g)^2}{\sigma_g^2}\right),$$

where σ_g describes the g-factor spread, see Eq. (7.9). The spin dynamics between the pulses is given by

$$S_z(t) = S_z(0)\cos(\Omega_B t)\exp\left(-\frac{t}{T_2} - \frac{\sigma_g^2 \mu_B^2 B^2 t^2}{4}\right). \tag{9.11}$$

The decoherence rate due to the g-factor spread, $\sim \sigma_g \mu_B B$, is the higher the larger is the external magnetic field. Correspondingly, the peaks in the resonant spin amplification spectrum become smaller with an increase in magnetic field. The experimental data fitting with allowance for the g-factor values spread carried out by using theoretical formulae obtained by averaging of Eqs. (9.5) and (9.9), (Glazov and Ivchenko, 2008; Astakhov *et al.*, 2008) is shown by a solid line in Figure 9.3(a). It describes the experiment rather well. The comparison of the theoretical calculation and the experimental data has made it possible to extract the main parameters of the resident electron spin dynamics (Astakhov *et al.*, 2008): mean g-factor absolute value, $|g| = 1.64$, its spread $\sigma_g/(2|g|) = 0.4\%$, transverse spin relaxation time $T_2 = 30$ ns. These parameters obtained by means of a resonant spin amplification technique agree well with the values extracted from the Hanle effect and temporal dependence of the spin signals on the same structure.

The additional specifics of the resonant spin amplification spectra can be brought about by the unpaired hole-in-trion (in n-type structures) and by the electron-in-trion (in p-type structures) spin dynamics. If the trion spin relaxation is suppressed, then the considerable spin polarization of resident charge carriers appears in the magnetic field only, because in the absence of the field, the spin returning from the trion exactly compensates the photogenerated spin, see Section 6.1 and (Zhukov *et al.*, 2007; Yugova *et al.*, 2009*b*). Hence, the resonant spin amplification peak amplitude first increases with an increase of magnetic field strength and then decreases due to the g-factor spread as illustrated in Figure 9.3(b). As a result the resonant spin amplification spectrum envelope has a characteristic "bat-like" shape. One can extract the spin relaxation times of unpaired charge carrier in the trion from such spectra (Yugova *et al.*, 2009*b*;

Fokina *et al.*, 2010; Korn *et al.*, 2010; Kugler *et al.*, 2011; Yugova *et al.*, 2012). To conclude the discussion on the resonant spin amplification effect, we note that the observation of spin accumulation induced by the train of pump pulses in transverse magnetic field makes it possible to measure long spin decoherence times $T_2^* \gg T_R$, where the direct application of the pump-probe technique is hampered by the finite value of the pulse repetition period. Other techniques to address long-time spin dynamics include the spin inertia method (Heisterkamp *et al.*, 2015*b*), extended pump-probe technique (Belykh *et al.*, 2016), and spin noise spectroscopy; the latter technique is discussed in detail in Chapter 10.

Let us turn now to the opposite limiting case where the spread of electron *g*-factors and the random nuclear fields result in fast electron spin dephasing, that is,

$$T_2^* \approx T_{\text{inh}} < T_R, \tag{9.12}$$

but, as before, the transversal spin relaxation time T_2 is assumed to be much longer than T_R. At first glance, in this situation no remarkable spin signals at small negative delays are expected because the spin of the electron ensemble averages out before the next pulse arrival. However, for localized electrons the spin of a given electron is conserved over a long time, which exceeds by far the pulse repetition period $T_2 \gg T_R \sim 10$ ns in typical experiments. Furthermore, if condition (9.12) holds, a wide spectrum ($\sim 1/T_{\text{inh}}$) of spin precession frequencies is excited. Hence, there are the spins in the whole ensemble of the precessing ones for which the spin precession and the pump pulse repetition frequencies are synchronized. Obviously, the spins of these electrons will always be in phase at the time moments $t = 0,\ T_R,\ 2T_R, \ldots$; that is, when the next pump pulse arrives. The spins of remaining charge carriers in these time moments have random precession phases, and they do not contribute to the observed signal. Therefore, if condition (9.12) is satisfied, the spin signal decays during a time scale in the order of T_2^*, and afterward emerges by the next pump pulse arrival over approximately the same time.

It is shown in Figure 9.4(a) where the results of model calculations are presented. This effect was first observed in the pump-probe experiments on *n*-type InGaAs quantum dot ensembles (Greilich *et al.*, 2006*b*) and was termed spin precession mode locking in analogy with laser physics where laser generation is possible only for certain modes with frequencies being multiples of the cavity mode frequency. Such behavior is indeed typical for quantum dot structures with a significant spread of *g*-factors and relatively strong nuclear spin fluctuations, and it can be clearly seen in the experimental data shown in Figure 9.1.

The spin precession mode-locking effect allows one, to a certain extent, to overcome the electron spin dephasing effects related to an inhomogeneity of the electron ensemble. Roughly 10^6 electron spins precess with commensurable frequencies under experimental conditions (Greilich *et al.*, 2006*b*). Since the spread of spin precession frequencies decreases with a decrease in the magnetic field,

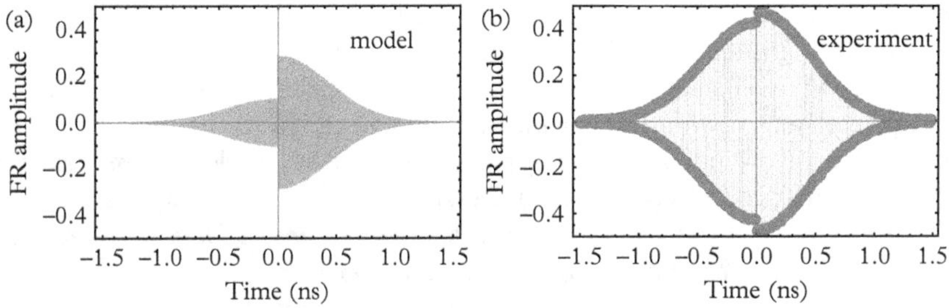

Figure 9.4 *(a) The Faraday rotation signal from a quantum dot array calculated for the experimental conditions of (Greilich et al., 2007) assuming that nuclear spin fluctuations are frozen. (b) The envelope of the experimentally measured electron spin-induced Faraday rotation signal. Measurements were carried out on a structure consisting of 20 layers of InGaAs/GaAs quantum dots with a dot density per layer of 10^{10} cm^{-2}, the structure doped in such a way that there is one electron per dot on average. Magnetic field $B = 6$ T, sample temperature $T = 6$ K. Data are reproduced from (Greilich et al., 2007).*

at small fields the situation where one or two spin precession modes are excited can be achieved, as shown experimentally in (Greilich *et al.*, 2009*b*). Application of spin precession mode locking allowes one to determine experimentally the transverse relaxation time of electron spin, $T_2 \sim 1$ μs, which is three orders of magnitude longer than the $T_2^* \sim 1$ ns under the same conditions. It is also possible to excite the spin echo under pumping by a sequence containing pairs of circularly polarized pulses (Greilich *et al.*, 2006*b*) paving the way for advanced manipulation of spins by complex pulse patterns (Varwig *et al.*, 2016). The hole spin precession mode locking is also observed in *p*-type quantum dot ensembles (Fras *et al.*, 2012; Varwig *et al.*, 2012).

9.3 Nuclei-induced electron frequency focusing effect

From the analysis presented in Section 9.2, it follows that the ratio of the long living (electron) spin signal amplitudes at negative and positive delays $A_{\text{neg}}/A_{\text{pos}}$ should be determined by the fraction of electrons, whose spin precession frequencies satisfy the phase synchronization condition, see Eq. (9.1). Indeed, the contribution to the spin-Kerr, -Faraday, and ellipticity signals at negative delays is given by only those electrons in ensemble, whose spin precession is synchronous with pump pulses, while at small positive delays all electrons from the ensemble contribute to the signal. Analysis shows that this ratio should not exceed, under experimental conditions of (Greilich *et al.*, 2006*b*, 2007), 0.2...0.3 as can also be seen from the model calculation in Figure 9.4(a).

However, the experiments performed in (Greilich *et al.*, 2007) revealed that it is not generally the case: The signal amplitude at negative delays, A_{neg}, is just slightly smaller than A_{pos}, see Figure 9.4(b). It means, that synchronization condition (9.1) is satisfied in almost all quantum dots in ensemble. The conclusion was drawn in (Greilich *et al.*, 2007) that it is the hyperfine interaction of electrons with nuclear spins responsible for the observed effect: In the process of electron spin coherence excitation by the periodic pulse train, nuclear spins orient in such a way that the electron spin precession period becomes a divisor of the pump pulse repetition period. The effect is termed *nuclei-induced electron spin precession frequency focusing*. At first glance this conclusion is surprising because in the Voigt configuration and for a sufficiently strong magnetic field, of several tesla used in the reported experiments, the buildup of a significant dynamical nuclear polarization is not expected; see Section 5.4 for details. However, there is an important difference between the situation studied in Chapter 5 and here: In the discussed experiments on nuclei-induced frequency focusing in quantum dots the electron spin is pumped by a periodic train of light pulses rather than by the continuous wave (*cw*) laser light. An important experimental argument in favor of nuclear effects responsible for the frequency focusing comes from the fact that, if after a long pump period, the quantum dot structure was left in the "dark," the high amplitude of the Faraday rotation at negative delays remains for tens of minutes (Greilich *et al.*, 2007).

There are two theoretical approaches aimed at the description of the frequency focusing process: *quantum-mechanical* and *semiclassical* ones. In the quantum-mechanical model suggested in (Greilich *et al.*, 2007) the random nuclear spin flips are considered, which are caused by hyperfine interaction with the electron spin. In accordance with Eq. (2.52) [see also Eqs. (5.14) and (5.15)] the rate of the nuclear spin-flip processes can be estimated as

$$W_{qnt} \sim \left(\frac{A}{\hbar N_n}\right)^2 \frac{1}{\Omega_B^2 \tau_c}. \tag{9.13}$$

We recall that A is the hyperfine interaction constant (2.22), N_n is the number of nuclei interacting with the localized electron, and τ_c is the electron spin correlation time in the quantum dot. It is assumed here that $\Omega_B \tau_c \gg 1$; that is, electron-nuclear spin flips are suppressed by the electron Zeeman splitting. The key assumption of this approach is that the electron spin correlation time is governed by the processes of the pump pulse interaction with the quantum dot. In more detail, the inverse correlation time is proportional to the probability (per unit time) to excite the trion in the dot. Since the trion is excited by a periodic sequence of the circularly polarized pump pulses arriving each T_R, the following estimate holds (Greilich *et al.*, 2007):

$$\frac{1}{\tau_c} \sim \frac{P_{\text{tr}}}{T_R}. \tag{9.14}$$

Here $P_{tr} = 1 - Q^2$ is the trion formation probability by a single pump pulse and the parameter Q has already been introduced in Eq. (6.15). In those dots where the phase synchronization condition accounting for both the external and nuclear Overhauser field, $\boldsymbol{\Omega}_N$, is fulfilled [cf. Eq. (9.1)]:

$$\Omega T_R = |\boldsymbol{\Omega}_B + \boldsymbol{\Omega}_N| T_R = 2\pi N, \quad N = 1, 2, \ldots \tag{9.15}$$

the electron spin polarization by the moment of the next pump pulse arrival is saturated, $|S_z| = 1/2$. Hence, the trion is not formed in accordance with the selection rules discussed in Chapter 6. Therefore, the effective correlation time $\tau_c \to \infty$ in Eq. (9.13), and the nuclear spin-flips stop. In those dots where the spin precession phase synchronization condition is not fulfilled, the nuclear spin flips take place until Ω changes in random manner in such a way that the condition Eq. (9.15) is reached. Hence, the distribution function of the nuclear spin frequencies $\mathcal{F}(\boldsymbol{\Omega}_N)$ evolves from the Gaussian, Eq. (7.9), to a series of sharp peaks at those Ω_N where the condition (9.15) is satisfied. In the work (Carter *et al.*, 2009), this approach initially formulated to describe the resonant pulses has been extended to allow for the detuning between the pump pulse frequency and the trion resonance frequency in quantum dots, and applied to describe corresponding experimental situations.

The effect of detuning between the pump pulse carrier frequency and the trion resonance on the electron-nuclear spin dynamics in a quantum dot system excited by a periodic train of pump pulses has been also studied in (Korenev, 2011). In this work the generation of the electron spin component S_x parallel to the external magnetic field by detuned pulses, see Eq. (6.16b), has been taken into account. The resulting nonzero product $(\boldsymbol{S} \cdot \boldsymbol{B})$ within the standard model of the dynamic nuclear polarization, Eqs. (5.13) and (5.14), gives rise to dynamical nuclear polarization. The stability analysis of the interacting electron-nuclear spin system performed in (Korenev, 2011) demonstrates the multitude of the stable states of the nuclear spin system whose frequencies are close to the ones given by Eq. (9.15).

An alternative approach to the nuclei-induced electron spin precession frequency focusing was suggested in (Glazov *et al.*, 2012). The description of the coupled dynamics of electron and nuclear spins is carried out within the *semi-classical model.* In this approach the electron spin, $\boldsymbol{S}$, and nuclear spin fluctuation, $\boldsymbol{M} = \sum_i \boldsymbol{I}_i$, with the summation being carried out over the nuclei interacting with the electron, see Eq. (3.20), are treated as classical vectors, which precess around the external field and around each other. The analysis of the corresponding Bloch equations shows that the nuclear spin precession in the external magnetic field with the frequency $\boldsymbol{\omega}_N = -g_n \mu_B \boldsymbol{B}/\hbar$ [g_n and μ_N are the nuclear g-factor and magneton, respectively, see Eq. (2.21)], results, owing to the hyperfine interaction, in additional small oscillations of electron spin with this very frequency. These electron spin oscillations, in turn, affect the nuclear spin dynamics. Effectively, they cause nuclear magnetic resonance. As a result, the projection of nuclear spin onto the external magnetic field, M_x, starts to vary slowly, until the electron spin precession frequency $\Omega \approx \Omega_B + AM_x$ satisfies the mode-locking condition in Eq. (9.15). This process is no more random and its rate can be estimated as (Glazov *et al.*, 2012)

$$W_{class} \equiv \frac{1}{\tau_{\rm nf}} \sim \left(\frac{A}{\hbar N_n}\right)^3 \frac{\sqrt{N_n}}{\omega_N \Omega_B^2 T_R}. \tag{9.16}$$

The comparison of Eqs. (9.13) and (9.16) demonstrates that the rate of such a regular process is, by factor $W_{class}/W_{qnt} \sim A/(\hbar\sqrt{N_n}\omega_N)$ larger than that of the nuclear spin flips in the model of random flips. Typically the factor $W_{class}/W_{qn} \sim 10$ under experimental conditions (Greilich *et al.*, 2007).

Figure 9.5 demonstrates the results of the calculations of the nuclei–induced electron spin precession frequency focusing time within the semiclassical model of (Glazov *et al.*, 2012) as a function of the pump pulse area Θ defined in Eq. (6.20). One can see that $\tau_{\rm nf}$ becomes extremely long for $\Theta \ll 1$ because electron spin orientation is inefficient. An increase in Θ increases the electron spin polarization and consequently the Knight field acting on the nuclear spins, which in turn shortens $\tau_{\rm nf}$. The frequency focusing time $\tau_{\rm nf} \propto \Omega_B^2\omega_N \propto B^3$, see Eq. (9.16), increases significantly with an increase in magnetic field. This explains the rapid increase of $\tau_{\rm nf}$ with Ω_B seen in Figure 9.5. Estimations show that in

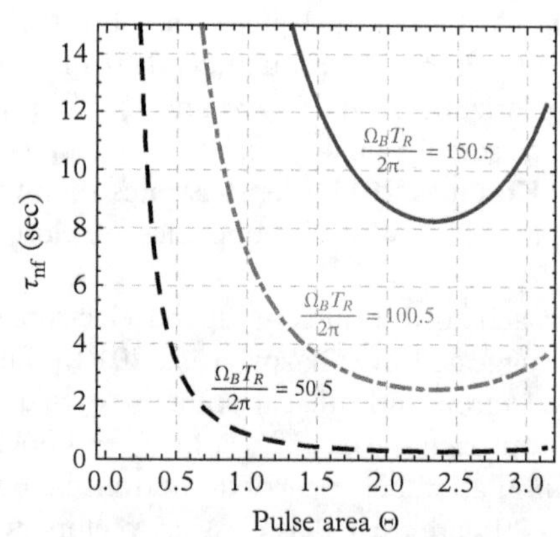

Figure 9.5 *Dependence of the nuclei-induced electrons spin precession frequency focusing time, $\tau_{\rm nf}$, on the pump pulse area, Θ [cf. Eq. (6.20)] calculated within the model of (Glazov et al., 2012). The three curves were calculated at the magnetic fields that give the following electron spin precession frequencies: $\Omega_B T_R/(2\pi) = 50.5$ (dashed), 100.5 (dash-dotted), and 150.5 (solid). The parameters used, $A/(\hbar N_n) = 5 \times 10^6 s^{-1}$, $T_R = 13$ ns, $\omega_N/\Omega_B = 10^{-3}$, and $M = 126$, which could be created by 6×10^4 nuclei with spin 1/2, were selected to keep the calculations relevant to (Greilich et al., 2007).*

experimental conditions of (Greilich *et al.*, 2007) the nuclei spin focusing time ranges from units to tens of seconds, Figure 9.5, which is in satisfactory agreement with the experimental data.

A natural question arises about the comparison between the semiclassical and quantum-mechanical approaches. To that end, more advanced calculations are required that go beyond the "mean-field" and box-model approximations used previously, which account for either average nuclear polarization or the distribution function of nuclear spins. The significant steps in this direction have been performed in (Petrov and Yakovlev, 2012; Economou and Barnes, 2014; Beugeling *et al.*, 2016, 2017; Jäschke *et al.*, 2017) where the intertwined dynamics of the nuclear and electron spins under periodic excitation have been studied by various sophisticated approaches. On the experimental side, further studies are needed; in particular, the analysis of the focusing time as a function of the magnetic field and pump pulse power as well as the measurements of the nuclear spin distribution function (see Section 10.6, Chapter 10 for possible experimental protocol) in the course of the frequency focusing.

9.4 Buildup of Faraday rotation signal

So far, we have discussed the origins of the spin-Faraday, -Kerr, and ellipticity signals at negative delays between the pump and probe pulses. In this section we address another intriguing observation shown in Figure 9.1: The amplitude of the Faraday rotation signal, related to the resident electron, that is the top curve in Figure 9.1(c), increases with time before decaying. It is especially remarkable at negative delays: with an increase of $|\Delta t|$ spin beats amplitude first increases, and decreases afterwards, see Figure 9.1(b). It is evident that nuclear effects cannot be responsible for such a behavior for two reasons. First, the buildup of Faraday rotation signal is fast, approximately during 500 ps. This buildup time is by far shorter compared with any time scale of nuclear spin dynamics. Second, the ellipticity signal behavior is quite standard: its envelope function decays with delay time $|\Delta t|$. Therefore, nonmonotonic behavior of the Faraday signal envelope compared with ellipticity can be related to the specifics of this signal spectral sensitivity only.

In order to describe this effect qualitatively and quantitatively, we note that the electron *g*-factor depends on the localization energy of this charge carrier. Indeed, the renormalization of the *g*-factor in direct band semiconductors is determined mostly by the admixture of the valence band states to the conduction band states, see (Roth *et al.*, 1959; Ivchenko and Kiselev, 1992; Ivchenko, 2005) and Appendix B. Thus, the electron *g*-factor is controlled by the effective band gap energy in the nanostructure. Since the trion excitation energy depends on the electron localization energy, the resident carrier *g*-factor in a quantum dot is a function of the optical transition frequency ω_0^{T}. This dependence can be quite accurately described by a linear law in a narrow frequency range relevant for the experiments (Greilich *et al.*, 2006*b*; Yugova *et al.*, 2007*a*):

$$|g_e(\omega_0^{\mathrm{T}})| = a\hbar\omega_0^{\mathrm{T}} + c\,, \tag{9.17}$$

where a and c are the parameters determined by the material system and quantum dot properties.

Out of a broad distribution of quantum dots over energy in the sample, the pump pulse excites a subensemble, whose spectral width amounts to $\hbar/\tau_p \sim$ 1 meV for typical pulses with a duration of $\tau_p \sim 1$ ps. We introduce a function $S_z^+(\omega_0^{\mathrm{T}}, \Omega_B, \omega_{\mathrm{P}})$, which describes the magnitude of the electron spin z component in the quantum dot with the resonant frequency ω_0^{T} and spin precession frequency $\Omega_B = g\mu_B B/\hbar$ right after the pump pulse arrival with the carrier frequency ω_{P}. Note that the Larmor frequency in a quantum dot is determined, in general, not only by the g-factor value Eq. (9.17), but also the nuclear spin polarization fluctuation as discussed in Section 9.3, but we disregard this effect for simplicity. Spin-induced ellipticity $\mathcal{E}(t)$ and Faraday rotation $\mathcal{F}(t)$ signals detected by a probe pulse with the carrier frequency ω_{pr} are given (up to a common factor) in accordance with Eq. (6.31) by the expression (Yugova *et al.*, 2009*a*; Glazov *et al.*, 2010):

$$\mathcal{E}(t) + \mathrm{i}\mathcal{F}(t) = \int \mathrm{d}\omega_0^{\mathrm{T}} \mathrm{d}\Omega p(\omega_0^{\mathrm{T}}, \Omega_B) G(\omega_{\mathrm{pr}} - \omega_0^{\mathrm{T}}) \\ \times S_z^+(\omega_0^{\mathrm{T}}, \Omega_B, \omega_{\mathrm{P}}) \cos[\Omega_B t + \varphi] \exp(-t/T_2). \tag{9.18}$$

Here the delay between the pump and probe pulse $t > 0$, function $p(\omega_0^{\mathrm{T}}, \Omega)$ is the joint distribution of optical and Larmor frequencies in quantum dots, which, in absence of nuclear fluctuations, takes a form $p(\omega_0^{\mathrm{T}}, \Omega_B) = \delta[\Omega_B - g(\omega_0^{\mathrm{T}})\mu_B B/\hbar]$, and $G(\Lambda)$ describes spectral sensitivity of the spin-induced signals, see Eq. (6.32). The last two factors in Eq. (9.18) describe the single spin dynamics in the quantum dot, T_2 is the transversal spin relaxation time, and $\varphi = \varphi(\omega_0^{\mathrm{T}}, \Omega_B, \omega_{\mathrm{P}})$ is the initial phase of the spin precession. Functions $S_z^+(\omega_0^{\mathrm{T}}, \Omega_B, \omega_{\mathrm{P}})$ and $\varphi(\omega_0^{\mathrm{T}}, \Omega_B, \omega_{\mathrm{P}})$ can be found from the general solutions of spin dynamics equations as described in Section 6.1 and in more detail in (Yugova *et al.*, 2009*a*).

The detailed analysis and modeling of electron spin dynamics in quantum dots described by Eq. (9.18) is carried out in (Glazov *et al.*, 2010). Here we consider a minimal model, which allows us to obtain simplified analytical description of the situation. Let us take $G(\Lambda)$ in the form

$$G(\Lambda) = (1 + 2\mathrm{i}\Lambda\tau_p) \exp[-(\Lambda\tau_p)^2]. \tag{9.19}$$

At moderate values of detunings, $\Lambda = \omega_{\mathrm{pr}} - \omega_0^{\mathrm{T}}$, between the probe carrier frequency and resonance frequency of the quantum dot, $\Lambda\tau_p \lesssim 1$, function $G(\Lambda)$ given by Eq. (9.19) has a shape similar to that for Rosen–Zener pulse [see Eq. (6.33)]. At $\Lambda\tau_p \gg 1$ the imaginary part of G (that is Faraday rotation signal sensitivity) decays faster than the exact function, which behaves as

$\text{Im}\, G(\Lambda) \sim 1/(\Lambda \tau_p)$. However, it results only in a quantitative difference of Faraday signal behavior calculated in this model and obtained in the exact calculation (Glazov *et al.*, 2010). We assume further that nuclear effects are absent and spin precession frequency is directly connected with the quantum dot resonance frequency $\Omega_B(\omega_0^{\mathrm{T}}) = g(\omega_0^{\mathrm{T}})\mu_B B/\hbar$. Moreover, we choose S_z^+ function in the simplified form

$$S_z^+\left(\omega_0^{\mathrm{T}}, \omega_{\mathrm{P}}\right) = S_0 \exp\left[-\left(\omega_0^{\mathrm{T}} - \omega_{\mathrm{P}}\right)^2 \tau_p^2\right], \tag{9.20}$$

where S_0 is some constant that depends on the pump pulse area and we set $\varphi \equiv 0$, $T_2 \to \infty$. Under these approximations, the integration in Eq. (9.18) gives

$$\mathcal{E}(t) = \sqrt{\frac{\pi}{2\tau_p^2}} \exp\left[\frac{-\Delta^2\tau_p^2/(2\hbar^2) - (\Omega' t)^2}{8\tau_p^2}\right] \cos\left(\tilde{\Omega}_0 t\right), \tag{9.21a}$$

$$\mathcal{F}(t) = \frac{1}{2}\sqrt{\frac{\pi}{2\tau_p^2}} \exp\left[\frac{-\Delta^2\tau_p^2/(2\hbar^2) - (\Omega' t)^2}{8\tau_p^2}\right] \times \left[\frac{2\Delta\tau_p}{\hbar}\cos\left(\tilde{\Omega}_0 t\right) + \frac{\Omega' t}{\tau_p}\sin\left(\tilde{\Omega}_0 t\right)\right]. \tag{9.21b}$$

Here the following notations are introduced: $\Omega' = \mathrm{d}\Omega/\mathrm{d}\omega_0^{\mathrm{T}}$, $\Delta/\hbar = \omega_{\mathrm{P}} - \omega_{\mathrm{pr}}$ is the pump and probe pulses detuning, $\tilde{\Omega}_0 = g(\omega_{\mathrm{pr}})\mu_B B + \Omega'\Delta/(2\hbar)$ is the observed spin precession frequency.

It is seen from Eqs. (9.21) that the temporal dependence of the spin-Faraday rotation and ellipticity signals can be qualitatively different. The ellipticity signal amplitude, Eq. (9.21a), simply decays with time, this decay rate is determined by the spread of Larmor frequencies of excited by the pump pulse electrons. The Faraday signal, Eq. (9.21b), has two contributions [see Eq. (9.21b)]: The first one is similar to the ellipticity, but its amplitude depends sharply on the detuning, $\sim \Delta\tau_p$, while the amplitude of the second contribution ($\propto \sin\tilde{\Omega}_0 t$) contains the linear in time factor. For degenerate pump and probe pulses ($\Delta = 0$) Faraday signal, described by the second term in the brackets of Eq. (9.21b), first grows and then decays, in agreement with experimental data, presented in Figure 9.1.

Such temporal behavior of spin signals is indeed related to the different spectral sensitivity of the spin-Faraday and ellipticity signals, and it is a direct consequence of the correlation between the Larmor frequency and optical transition frequency in a quantum dot, see Eq. (9.17). Figure 9.6 qualitatively illustrates the spin Faraday signal formation for the spectrally degenerate pump and probe pulses. At $t = 0$ the spin distribution is a symmetric function of $\omega_0^{\mathrm{T}} - \omega_{\mathrm{P}}$ and it makes no contribution to the Faraday signal, since it is determined by the convolution of the even with respect to the ω_{P} S_z^+ distribution and the odd function $\text{Im}\, G(\Lambda)$, as shown in Figure 9.6(a). The spin distribution becomes asymmetric with a

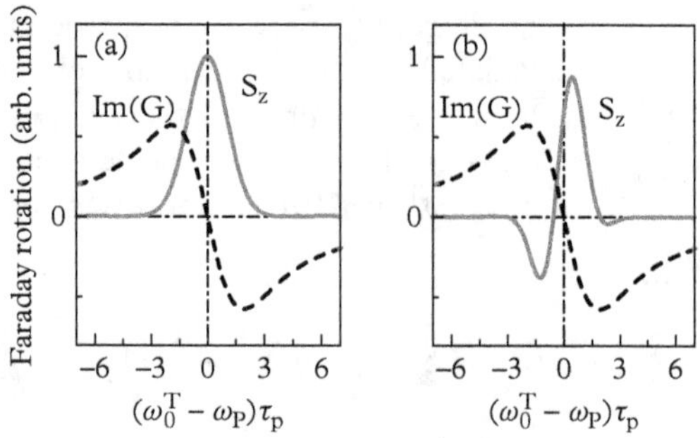

Figure 9.6 *Schematic illustration of Faraday rotation signal formation for spectrally degenerate pump and probe pulses,* $\omega_{pr} = \omega_P$. *Panel (a) corresponds to the zero temporal delay between the pump and probe pulses, panel (b) corresponds to the positive delay, t > 0. The solid curve shows the distribution of the spin z component, while the dashed one shows the spectral sensitivity of Faraday signal,* $\mathrm{Im}\, G(\omega_0^T - \omega_{pr})$. *From (Glazov et al., 2010).*

course of time, since [for $a > 0$, $c > 0$ in Eq. (9.17)] spins in quantum dots with larger optical transition energies, ω_0^T, precess faster than those in dots with smaller transition energies. Therefore, with an increase in the pump and probe pulses temporal separation, the spin distribution function becomes asymmetric with respect to the carrier frequency of the optical pulse, as shown in Figure 9.6(b). Hence, the Faraday rotation signal becomes nonzero at $t > 0$. At large enough delays the electron spin dephasing takes place and spin-Faraday rotation decays.

We note, that the detuning between the probe and pump pulses in the two-color experiment already introduces asymmetry into the spin distribution with respect to the probe pulse carrier frequency, ω_{pr}, and results in the Faraday signal formation at $t = 0$. Therefore, at $\Delta \neq 0$ the decaying with time component of the spin Faraday signal appears, which is described by the first term in the brackets of Eq. (9.21b). The spectral sensitivity of the spin-induced ellipticity, $\mathrm{Re}\, G(\Lambda)$, is even; therefore, the ellipticity signal reflects largely the ensemble average z spin component. Induced ellipticity decays as a function of time due to the spread of the Larmor frequencies. It is in agreement with the experimental data shown in Figure 9.1.

The simplified model presented here qualitatively describes also the differences between the Faraday rotation and ellipticity at negative delays. In the spin precession mode-locking regime described in Section 9.2, the distribution function of spin z component, $S_z^+(\omega_0^T, \Omega, \omega_P)$, has sharp maxima for those quantum dots, where $\Omega(\omega_0^T) T_R = 2\pi N$, see Eq. (9.1). If one allows for the synchronized modes only, the spin signals become even functions of the pump-probe delay, Δt. It implies, in particular, that the spin-Faraday signal at the zero detuning and $\Delta t < 0$ first builds up and afterward decays with an increase of $|\Delta t|$, see Figure 9.1(b). Presence of other spin precession frequencies gives rise to the additional contribution to the signal, which decays at $t > 0$ and is absent at negative delays. Note that the focusing of electron spin precession frequency induced by the electron interaction with the lattice nuclei breaks the correlation between optical transition and spin precession frequencies and weakens the growing part of the spin-Faraday signal.

The model outlined here has a number of limitations. In particular, due to simplified form of the function Eq. (9.19), it does not describe the amplitudes of the Faraday signal at large detunings. For the same reasons, the spectral dependence of g-factor, extracted from the Faraday rotation experiment is different from that predicted by Eq. (9.21b). The complete description of experimental data shown in Figure 9.1, was carried out in (Glazov *et al.*, 2010). Microscopic calculation by using exact function $G(\Lambda)$ and functions $S_z^+(\omega_0^T, \Omega, \omega_P)$, $\varphi(\omega_0^T, \Omega, \omega_P)$, obtained with an allowance for spin precession mode locking, are in good agreement with experiments. Comparison of the theory and experiment is presented in Figure 9.7. The figure shows the spectral dependence of the amplitudes of the induced ellipticity [panel (a)] and the spin-Faraday rotation [panel (b)]. Circles and squares are the experimental data obtained at negative and positive delays, respectively, by means of fitting the data by the equation:

$$\mathcal{F} \propto [\alpha \cos(\Omega \Delta t) + \beta t \sin(\Omega \Delta t)] \exp\left(-\frac{\Delta t^2}{(T_2^*)^2}\right).$$

Curves in Figure 9.7 are the result of the calculation. Figure shows amplitudes of the Faraday and ellipticity signal decaying components. The inset to panel (b) shows amplitudes of the growing-in-time contribution to the Faraday signal.

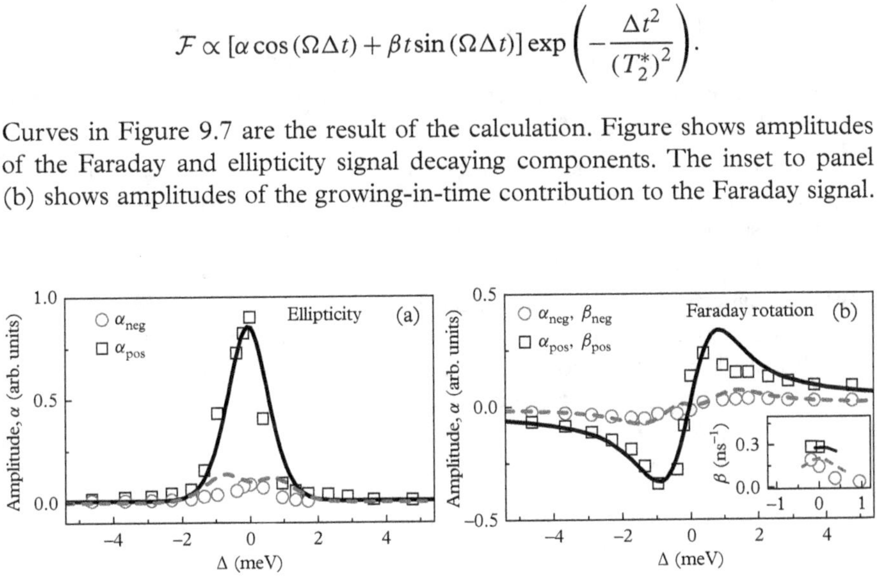

Figure 9.7 *Amplitudes of the spin-induced ellipticity (a) and Faraday rotation (b) signals as functions of the detuning between the pump and probe pulses. Circles show amplitudes of the decaying contributions to spin signals at negative delays, α_{neg}, squares show the amplitudes for positive ones, α_{pos}. The inset to (b) presents the amplitudes of the growing with time component of the Faraday rotation signal: β_{neg} at negative delays (circles) and β_{pos} at positive ones (squares). Solid lines are the modeling results. Measurements were carried out on the structure consisting of 20 layers of InGaAs/GaAs quantum dots with dot density per layer being 10^{10} cm^{-2}, the structure is doped with one electron per dot on average. From (Glazov et al., 2010).*

A good agreement of spectral dependence of signal amplitudes between the experiment and theory is seen from Figure 9.7.

Thus, spin-Faraday and ellipticity signals in inhomogeneous ensembles of quantum dots are formed by different resident electrons. As a result, their behavior as function of the pump and probe pulses delay can be qualitatively different. It is most brightly manifested in the buildup of the Faraday rotation signal as a function of time being a result of the link between the electron g-factor and optical transition energy in a quantum dot.

To summarize, the spin dynamics of charge carriers under periodic excitation demonstrate remarkable phenomena related to the long spin coherence time, T_2, of an individual charge carrier compared with the pump pulse repetition period, T_R. Correspondingly, the spin accumulation of individual electrons or holes takes place.

- The resonant spin amplification takes place if the spin of electrons in an ensemble does not decay during the repetition period, that is, if the spin ensemble is sufficiently homogeneous; for example, for electrons localized on quantum well interface roughnesses.
- The mode locking of electron spin precession takes place in strongly inhomogeneous ensembles, that is in quantum dots, and for rather high excitation powers. In this case the revivals of the spin-Faraday, -Kerr, and ellipticity signals are observed before the next pump pulse arrival.
- The mode locking makes it possible to overcome, to a large extent, the inhomogeneity of the quantum dot ensembles: One can realize single mode spin precession of an electron ensemble at an appropriately selected magnetic field and repetition period.
- The hyperfine interaction of electron and nuclear spins gives rise to the nuclei-induced frequency focusing effect, which makes it possible to make almost all spin precession frequencies in the electron ensemble commensurable with pump-pulse repetition frequency.

10

Fluctuations of Electron and Nuclear Spins

10.1 Electron spin fluctuations

In previous chapters, we analyzed interactions in spin systems in semiconductors and spin dynamics of nuclei and electrons induced by external perturbations: alternating and static electromagnetic fields. By bringing the electron or nuclear spin system out of thermodynamic equilibrium, for example by irradiating a nanosystem with circularly polarized light, one is able to monitor the spin precession in external fields and relaxation via the polarization of the emitted light or by means of the rotation of the polarization plane of a linearly polarized "probe" beam owing to the spin-Faraday effect.

Naturally, the question arises as to whether one can learn anything about the spin dynamics by monitoring the polarization plane rotation of the probe beam in the absence of any excitation. Apart from possible static Faraday (or Kerr) rotation induced by an external magnetic field, if any, or due to natural optical activity of gyrotropic systems, the time-average of the Faraday rotation angle is zero because any spin polarization is absent in the system. However, inevitable fluctuations of the electron spins, $\delta S_z(t)$, Figure 10.1, provide a stochastic contribution to the detected Faraday rotation angle,

$$\delta\vartheta(t) \propto \delta S_z(t).$$

This is in full analogy with the contribution of any spin polarization, induced by the external magnetic field or the optical orientation, to the spin-Faraday or spin-Kerr effect. Here, z is the light propagation axis and, as in Chapter 6, we neglect the delay of light propagating through a sample. By definition, the spin fluctuations are zero on average, $\langle \delta S_\alpha(t)\rangle = 0$, where the angular brackets denote the time averaging, namely,

Electron and Nuclear Spin Dynamics in Semiconductor Nanostructures. M. M. Glazov, Oxford University Press (2018). © M. M. Glazov. DOI: 10.1093/oso/9780198807308.001.0001

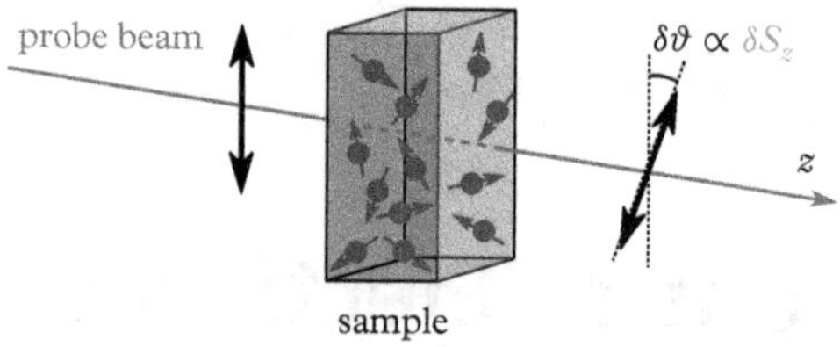

Figure 10.1 *Schematic illustration of spin fluctuation detection: z is the propagation axis of the linearly polarized probe beam, double arrows show the orientation of the linear polarization plane before and after the transmission through a sample with randomly fluctuating spins.*

$$\langle \delta S_\alpha(t) \rangle = \lim_{T\to\infty} \frac{1}{2T} \int_{-T}^{T} \delta S_\alpha(t) dt, \quad \alpha = x, y, z, \tag{10.1}$$

with T being a macroscopically long time that exceeds by far the period of the electron spin precession in magnetic fields, spin relaxation times, and so on. The spin fluctuations, also termed *spin noise*, are characterized by the correlation functions. The *second-order correlation functions* defined in accordance with

$$C_{\alpha\beta}(\tau) = \langle \delta S_\alpha(t+\tau) \delta S_\beta(t) \rangle, \tag{10.2}$$

are the main interest to us. In Eq. (10.2) the averaging takes place over t at a fixed time difference τ. Equations (10.1) and (10.2) are written as if the quantities $\delta S_\alpha(t)$ are classical variables. Such an approach can be justified for systems consisting of many electrons where the total spin of the ensemble can be considered as a classical vector. Taking into account the quantum-mechanical nature of the spin Eq. (10.2) should be written as (Landau and Lifshitz, 2000)

$$C_{\alpha\beta}(\tau) = \langle \{\delta \hat{S}_\alpha(t+\tau) \delta \hat{S}_\beta(t)\}_s \rangle, \tag{10.3}$$

where $\{\hat{A}\hat{B}\}_s = (\hat{A}\hat{B} + \hat{B}\hat{A})/2$ is the symmetrized product of the operators, and the spin fluctuation operator is formally defined as $\delta\hat{S}_\alpha = \hat{S}_\alpha - \langle \hat{S}_\alpha \rangle$, where the angular brackets denote the averaging over the quantum-mechanical state of the system. The definition of the second-order correlation function (10.3) can be applied both to spin ensembles and to single electron spins. In what follows, the curly brackets denoting symmetrization of the operators are, as a rule, omitted. Hereafter, we consider the steady (but not necessary equilibrium) state of the system.

The second-order correlation function of the Faraday rotation angles $\langle \delta\vartheta_F(t+\tau)\vartheta_F(t) \rangle$ is directly proportional to the spin–spin correlation function $C_{zz}(\tau)$, see Eq. (10.3). Hence, by monitoring the noise of the Faraday, Kerr, or ellipticity effects one can measure the electron spin correlation function. The latter contains important information on the spin dynamics. In particular, in equilibrium conditions the correlation functions $C_{\alpha\beta}(\tau)$ are related to the spin susceptibility via the *fluctuation-dissipation theorem* (Landau and Lifshitz, 2000). Therefore, spin-noise studies allow one the determine the spin susceptibility of the system and, correspondingly, to evaluate the important parameters of the spin dynamics such as spin relaxation/decoherence times and spin precession frequencies, see Section 3.1 and Eqs. (3.8) where the response of spins to the static and alternating magnetic fields, that is the spin susceptibility, is addressed.

The detection of the spin noise and, accordingly, the spin susceptibility via the Faraday rotation fluctuations was first performed by E.B. Aleksandrov and V.S. Zapasskii on atomic gas vapors (Aleksandrov and Zapasskii, 1981) at the beginning of 1980s. Much later, these experiments were reproduced (Sørensen *et al.*, 1998; Mitsui, 2000; Crooker *et al.*, 2004) and, what is more important, this method was applied to bulk semiconductors (Oestreich *et al.*, 2005; Crooker *et al.*, 2009; Römer *et al.*, 2010) and semiconductor nanosystems (Müller *et al.*, 2008; Crooker *et al.*, 2010; Li *et al.*, 2012; Dahbashi *et al.*, 2014). As a result, a novel area of the spin dynamics studies *spin-noise spectroscopy*, is formed (Müller *et al.*, 2010; Aleksandrov and Zapasskii, 2011; Zapasskii, 2013; Hübner *et al.*, 2014; Sinitsyn and Pershin, 2016; Dyakonov, 2017). It turned out that spin-noise spectroscopy is particularly useful to study the slow spin dynamics of localized charge carriers because this technique is particularly sensitive to long-time $\gtrsim 1$ ns (i.e., low frequency) spin fluctuations.

Let us start with the presentation of the spin-noise model in the case of electrons or holes localized in quantum dots. Here we describe the basic formalism to calculate the fluctuation spectra, while the detailed discussion of the spin-noise spectra is given in Section 10.2. The spin fluctuation $\delta\boldsymbol{s}(t)$ in a given dot can be described by various techniques among those the Langevin random-force method is quite transparent (Landau and Lifshitz, 2000). Within this approach the fluctuation $\delta\boldsymbol{s}(t)$ obeys the Bloch equation, see Eqs. (2.32) and (7.8),

$$\frac{\partial \delta\boldsymbol{s}(t)}{\partial t} + \frac{\delta\boldsymbol{s}(t)}{\tau_s} + \delta\boldsymbol{s}(t) \times (\boldsymbol{\Omega}_B + \boldsymbol{\Omega}_N) = \boldsymbol{\xi}(t)\,. \tag{10.4}$$

Here, τ_s is the phenomenological electron spin relaxation time unrelated to the hyperfine coupling and caused by, for example, the electron–phonon interaction, see Chapter 8, $\boldsymbol{\Omega}_B$ and $\boldsymbol{\Omega}_N$ are the Larmor precession frequencies related, respectively, to the external magnetic field $\boldsymbol{B}$ and the effective field caused by the hyperfine interaction. Here, for simplicity, the hyperfine interaction of electron and nuclei spins in the quantum dot is described within the semiclassical approach via the effective static magnetic field acting on electron spin, see Section 7.2. The external magnetic field is assumed to be weak enough in order not to affect the nuclear spin fluctuations and the temperature is assumed to be large enough to disregard the thermal polarization of the spins.

The important ingredient in the description of spin fluctuations is the fictitious random (Langevin) force $\boldsymbol{\xi}(t)$ on the right-hand side of Eq. (10.4). It is included in order to support nonzero mean square values of the electron spin components, namely,

$$\langle \delta s_\alpha(t)\delta s_\beta(t)\rangle \equiv \langle \{\delta\hat{s}_\alpha(t)\delta\hat{s}_\beta(t)\}_s\rangle = \frac{\delta_{\alpha\beta}}{4}. \tag{10.5}$$

This equality immediately follows from the definition of the electron spin operator, see Eqs. (2.6) and (2.7), under the assumption of negligible average polarization of the electron. The correlator of the Langevin forces coincides with that for the equilibrium spin decoupled from the magnetic field $\boldsymbol{B}$ and the nuclei, namely,

$$\langle \xi_\alpha(t)\xi_\beta(t+\tau)\rangle = \frac{1}{2\tau_s}\delta_{\alpha\beta}\delta(\tau). \tag{10.6}$$

It is worth stressing that the fictitious random force $\boldsymbol{\xi}(t)$ is not related to any real physical processes and is introduced in Eq. (10.4) in the Langevin approach only in order to provide the proper values of the spin fluctuations in equilibrium (Lax, 1960, 1966; Gantsevich *et al.*, 1979; Landau and Lifshitz, 2000). That is why the correlator of random forces (10.6) is proportional to the δ-function of τ; that is, it has no specific time scale.

In many cases it is instructive to consider the spin noise in the spectral rather than in the temporal domain. To that end, we introduce the Fourier transforms of the fluctuating spin pseudovector as

$$\delta\boldsymbol{s}_\omega = \int_{-\infty}^{+\infty} \delta\boldsymbol{s}(t)\mathrm{e}^{\mathrm{i}\omega t}dt\,, \quad \delta\boldsymbol{s}(t) = \int_{-\infty}^{+\infty} \delta\boldsymbol{s}_\omega \mathrm{e}^{-\mathrm{i}\omega t}\frac{d\omega}{2\pi},$$

and the spin and random-force correlators

$$(\delta s_\alpha \delta s_\beta)_\omega = \int_{-\infty}^{+\infty} \langle \delta s_\alpha(t+\tau)\delta s_\beta(t)\rangle \mathrm{e}^{\mathrm{i}\omega\tau}\,d\tau, \tag{10.7a}$$

$$(\xi_\alpha \xi_\beta)_\omega = \int_{-\infty}^{+\infty} \langle \xi_\alpha(t+\tau)\xi_\beta(t)\rangle \mathrm{e}^{\mathrm{i}\omega\tau}\,d\tau = \frac{1}{2\tau_s}\delta_{\alpha\beta}. \tag{10.7b}$$

According to Eq. (10.6) the correlator of the random forces Fourier transforms is given by

$$\langle \xi_{\alpha,\omega}\xi^*_{\beta,\omega'}\rangle = \delta(\omega-\omega')\delta_{\alpha\beta}\frac{\pi}{\tau_s}\,.$$

Evidently, the Fourier components $\delta\boldsymbol{s}_\omega$ satisfy the vector equation obtained from Eq. (10.4) by the standard replacements $\partial/\partial t \to -\mathrm{i}\omega$, $\delta\boldsymbol{s}(t) \to \delta\boldsymbol{s}_\omega$ and $\boldsymbol{\xi}(t) \to \boldsymbol{\xi}_\omega$.

The solution of Eq. (10.4) for the spin pseudovector reads

$$\delta \boldsymbol{s}_\omega = \tau_{s,\omega} \frac{\boldsymbol{\xi}_\omega + \tau_{s,\omega} \boldsymbol{\Omega} \times \boldsymbol{\xi}_\omega + \tau_{s,\omega}^2 \boldsymbol{\Omega}(\boldsymbol{\xi}_\omega \cdot \boldsymbol{\Omega})}{1 + \Omega^2 \tau_{s,\omega}^2}, \tag{10.8}$$

where

$$\boldsymbol{\Omega} = \boldsymbol{\Omega}_B + \boldsymbol{\Omega}_N, \quad \tau_{s,\omega} = \frac{\tau_s}{1 - \mathrm{i}\omega\tau_s}. \tag{10.9}$$

It is convenient to introduce the linear-response tensor $\boldsymbol{\chi}(\omega)$ defined as

$$\delta s_{\alpha,\omega} = \chi_{\alpha\beta}(\omega) \xi_{\beta,\omega}, \tag{10.10}$$

whose components are explicitly given by

$$\chi_{\alpha\beta}(\omega) = \frac{\tau_{s,\omega}\left(\delta_{\alpha\beta} - \epsilon_{\alpha\beta\gamma}\Omega_\gamma \tau_{s,\omega} + \tau_{s,\omega}^2 \Omega_\alpha \Omega_\beta\right)}{1 + \Omega^2 \tau_{s,\omega}^2}, \tag{10.11}$$

where $\epsilon_{\alpha\beta\gamma}$ is the unit antisymmetric third-rank tensor. By means of the tensor $\boldsymbol{\chi}(\omega)$ we can present the spin fluctuations spectrum in the form

$$(\delta s_\alpha \delta s_\beta)_\omega = \frac{1}{2\tau_s} \sum_\gamma \chi_{\alpha\gamma}(\omega) \chi_{\beta\gamma}^*(\omega) = \frac{1}{2\tau_s} \left[\boldsymbol{\chi}(\omega) \boldsymbol{\chi}^\dagger(\omega)\right]_{\alpha\beta}. \tag{10.12}$$

It is instructive to analyze several limiting cases in order to elucidate the physical meaning of the results, presented in Eqs. (10.11) and (10.12). First, let us assume that both the external and nuclei-induced magnetic fields are absent, $\boldsymbol{\Omega} = 0$. In this situation the response tensor $\boldsymbol{\chi}(\omega)$ reduces to $\chi_{\alpha\beta}(\omega) = \tau_{s,\omega}\delta_{\alpha\beta}$ and we obtain

$$(\delta s_\alpha \delta s_\beta)_\omega = \frac{|\tau_{s,\omega}|^2}{2\tau_s} \delta_{\alpha\beta} = \frac{1}{2} \frac{\tau_s}{1 + (\omega\tau_s)^2} \delta_{\alpha\beta}, \tag{10.13}$$

or, in the time domain,

$$\langle \delta s_\alpha(t+\tau) \delta s_\beta(t) \rangle = \frac{1}{4} \delta_{\alpha\beta} \exp(-|\tau|/\tau_s). \tag{10.14}$$

Equations (10.13) and (10.14) demonstrate that the spin fluctuations decay with the time constant τ_s. Hence, the spin-noise power spectrum is just a Lorentz function with the half-width at the half maximum being τ_s^{-1}.

Second, for an electron spin subjected to a static magnetic field it is convenient to use the Cartesian coordinate frame x_1, y_1, z_1 with the axis $z_1 \parallel \boldsymbol{\Omega}$. In this case

$$\chi(\omega) = \frac{\tau_{s,\omega}}{1+(\Omega\tau_{s,\omega})^2} \times \begin{bmatrix} 1 & -\Omega\tau_{s,\omega} & 0 \\ \Omega\tau_{s,\omega} & 1 & 0 \\ 0 & 0 & 1+(\Omega\tau_{s,\omega})^2 \end{bmatrix}.$$

Substitution of Eq. (10.15) into Eq. (10.12) leads to the following nonzero components of the spin-noise power spectrum tensor $(\delta s_\alpha \delta s_\beta)_\omega$:

$$(\delta s_{z_1}^2)_\omega = \frac{\pi}{2}\Delta_{\tau_s}(\omega), \tag{10.15a}$$

$$(\delta s_{x_1}^2)_\omega = (\delta s_{y_1}^2)_\omega = \frac{\pi}{4}[\Delta_{\tau_s}(\omega-\Omega) + \Delta_{\tau_s}(\omega+\Omega)], \tag{10.15b}$$

$$(\delta s_{y_1}\delta s_{x_1})_\omega = (\delta s_{x_1}\delta s_{y_1})^*_\omega = \frac{2i\omega\Omega\tau_s^2}{1+\tau_s^2(\omega^2+\Omega^2)}(\delta s_{x_1}^2)_\omega, \tag{10.15c}$$

where we have introduced the broadened δ-function as [cf. Eq. (3.13)]

$$\Delta_{\tau_s}(x) = \frac{1}{\pi}\frac{\tau_s}{1+(x\tau_s)^2}.$$

These expressions (10.15) clearly show that the fluctuations of a spin component along the field are the same as in the absence of the field (provided that, of course, $\hbar\Omega \ll k_B T$) and are described by the Lorentzian function corresponding to exponential spin relaxation. The transversal to the field components of the spin fluctuate at the frequency Ω. This is because the fluctuation $\delta\boldsymbol{s}$ precesses around the static magnetic field $\boldsymbol{\Omega}$, see Eq. (7.13). For the same reason, the correlation between the transversal components $(\delta s_{y_1}\delta s_{x_1})_\omega = (\delta s_{x_1}\delta s_{y_1})^*_\omega$ arises, which simply accounts for the conversion of δs_{x_1} to δs_{y_1}, and vice versa, in the course of the spin fluctuation precession.

Let us now analyze certain general properties of the spin correlation functions. By definition, the second-order correlation function $\mathcal{C}_{\alpha\beta}(\tau)$ satisfies the following property

$$\mathcal{C}_{\alpha\beta}(\tau) = \mathcal{C}_{\beta\alpha}(-\tau), \tag{10.16a}$$

as can immediately be seen from Eq. (10.3). It is noteworthy that the time reversal symmetry imposes additional restrictions on the spin correlation functions in equilibrium conditions (Landau and Lifshitz, 2000). In the absence of a magnetic field it does not matter whether δS_α or δS_β is taken at the earlier or later time in Eqs. (10.2) and (10.3). Therefore, at $\boldsymbol{\Omega} = 0$ we have $\mathcal{C}_{\alpha\beta}(\tau) = \mathcal{C}_{\beta\alpha}(\tau) = \mathcal{C}_{\alpha\beta}(-\tau) = \mathcal{C}_{\beta\alpha}(-\tau)$. In the presence of a magnetic field both external and due to nuclear fluctuations one has, in addition to Eq. (10.16a),

$$\mathcal{C}_{\alpha\beta}(\tau;\boldsymbol{\Omega}) = \mathcal{C}_{\beta\alpha}(\tau;-\boldsymbol{\Omega}), \tag{10.16b}$$

because the magnetic field is odd, $\boldsymbol{\Omega} \to -\boldsymbol{\Omega}$, at $t \to -t$ transformation. Similar properties are fulfilled in the frequency domain, in which case $\omega \to -\omega$ replacement corresponds also to the complex conjugation:

$$\tilde{\mathcal{C}}_{\alpha\beta}(\omega) = \tilde{\mathcal{C}}^*_{\beta\alpha}(\omega) = \tilde{\mathcal{C}}_{\beta\alpha}(-\omega), \tag{10.17a}$$

$$\tilde{\mathcal{C}}_{\alpha\beta}(\omega;\boldsymbol{\Omega}) = \tilde{\mathcal{C}}_{\beta\alpha}(\omega;-\boldsymbol{\Omega}), \tag{10.17b}$$

where the Fourier transform $\tilde{\mathcal{C}}_{\alpha\beta}(\omega)$ of the spin correlation function is defined by Eq. (10.7a). These restrictions are known as *Onsager relations* in thermodynamics (Onsager, 1931). Obviously, Eqs. (10.14) and (10.15) derived before satisfy these relations (10.16).

The general expression for the spin-noise spectrum, Eq. (10.12), as well as its limiting forms, Eqs. (10.13) and (10.15) can be derived by different methods. In many practical situations it is convenient to derive and solve the set of kinetic equations for the spin correlators $\mathcal{C}_{\alpha\beta}(\tau)$ (Landau and Lifshitz, 1981). Also, as we already noted, for systems in the thermodynamical equilibrium the fluctuation-dissipation theorem can be used to obtain Eq. (10.12). The quantum-mechanical derivation of the fluctuation-dissipation theorem is presented in many textbooks, see for example (Landau and Lifshitz, 2000; Vasko and Raichev, 2005). However, it is instructive to calculate both the spin susceptibility and spin-noise spectrum within the same approach based on the kinetic (or Bloch) equation for the spin pseudovector and explicitly obtain the relation between the fluctuations and linear response function. To that end we follow (Glazov and Ivchenko, 2012) and introduce the the generalized spin susceptibility $\mu_{\alpha\beta}$ describing the linear response of electron spin to the generalized forces f_α:

$$\delta s_{\alpha,\omega} = \sum_\beta \mu_{\alpha\beta}(\omega) f_{\beta,\omega}, \tag{10.18}$$

where the force $\boldsymbol{f}(t)$ acts as a perturbation to the spin Hamiltonian (Landau and Lifshitz, 2000)

$$\hat{V} = -\sum_\alpha \hat{s}_\alpha f_\alpha, \tag{10.19}$$

In the presence of the static magnetic field characterized by the spin precession frequency $\boldsymbol{\Omega}$, the magnetic susceptibility $\boldsymbol{\mu}(\omega)$ can be found from the kinetic equation for the electron spin $\boldsymbol{s} = \bar{\boldsymbol{s}} + \delta\boldsymbol{s}$, where $\bar{\boldsymbol{s}} \equiv \bar{\boldsymbol{s}}(\boldsymbol{\Omega})$ is the equilibrium (thermal) spin orientation in the static field with the Larmor frequency $\boldsymbol{\Omega}$ and $\delta\boldsymbol{s}$ is the nonequilibrium spin polarization induced by the weak fluctuating force $\boldsymbol{f}$ ($|\boldsymbol{f}| \ll \hbar|\boldsymbol{\Omega}|$), see Eq. (2.32):

$$\frac{\partial \boldsymbol{s}}{\partial t} + \boldsymbol{s} \times (\boldsymbol{\Omega} + \boldsymbol{\Omega}_{\sim}) + \boldsymbol{Q}\{\boldsymbol{s}\} = 0. \tag{10.20}$$

Here, $\boldsymbol{\Omega} + \boldsymbol{\Omega}_{\sim}$ is the total spin precession frequency that includes contributions both caused by the static, $\boldsymbol{\Omega}$, and the alternating, $\boldsymbol{\Omega}_{\sim}$, fields. The collision integral, $\boldsymbol{Q}\{\boldsymbol{s}\}$, describes the relaxation of the nonequilibrium spin polarization toward its equilibrium value and takes the form

$$Q\{\boldsymbol{s}\} = \frac{\boldsymbol{s} - \bar{\boldsymbol{s}}(\Omega + \Omega_{\sim})}{\tau_s}. \tag{10.21}$$

In contrast to the form of Eq. (3.3) used in Chapter 3, here we disregarded the difference between the longitudinal and the transversal relaxation times, but took into account the alternating field, $\boldsymbol{\Omega}_{\sim}$ in the equilibrium value of the spin [cf. Eq. (2.47) for $|g\mu_B B| \ll k_B T$]:

$$\bar{\boldsymbol{s}}(\boldsymbol{\Omega}) = -\frac{\hbar \boldsymbol{\Omega}}{4k_B T}, \; \bar{\boldsymbol{s}}(\boldsymbol{\Omega} + \boldsymbol{\Omega}_{\sim}) = -\frac{\hbar \boldsymbol{\Omega} - \boldsymbol{f}}{4k_B T}, \tag{10.22}$$

assuming that, as before, $\hbar|\boldsymbol{\Omega} + \boldsymbol{\Omega}_{\sim}| \ll k_B T$. As a result, the nonequilibrium spin polarization $\delta \boldsymbol{s}$ satisfies the following linearized equation:

$$-\mathrm{i}\omega \delta \boldsymbol{s} + \delta \boldsymbol{s} \times \boldsymbol{\Omega} + \frac{\delta \boldsymbol{s}}{\tau_s} = \frac{1}{4k_B T}\left(\frac{\boldsymbol{f}}{\tau_s} - \boldsymbol{\Omega} \times \boldsymbol{f}\right), \tag{10.23}$$

which is readily solved by means of Eqs. (10.8) and (10.11) yielding spin susceptibility in the form

$$\mu_{\alpha\beta} = \frac{1}{4k_B T}\left[\frac{\chi_{\alpha\beta}(\omega)}{\tau_s} - \sum_{\mu\nu} \chi_{\alpha\mu}(\omega)\epsilon_{\mu\nu\beta}\Omega_\nu\right]. \tag{10.24}$$

One can readily check that the spin-noise spectra, Eq. (10.12), in agreement with the general theory of (Landau and Lifshitz, 2000), are expressed via the susceptibility $\mu_{\alpha\beta}(\omega)$ as [cf. (Kos *et al.*, 2010)]

$$(\delta s_\alpha \delta s_\beta)_\omega = \frac{\mathrm{i}k_B T}{\omega}\left[\mu^*_{\beta\alpha}(\omega) - \mu_{\alpha\beta}(\omega)\right]. \tag{10.25}$$

This equivalence can be proved by applying the following identity relating bilinear and linear components of the tensor $\boldsymbol{\chi}$:

$$\sum_\gamma \chi_{\alpha\gamma}(\omega)\chi^*_{\beta\gamma}(\omega) = \frac{\mathrm{i}}{2\omega}\Big\{\chi^*_{\beta\alpha}(\omega) - \chi_{\alpha\beta}(\omega) + \sum_{\mu\nu}[\chi_{\alpha\mu}(\omega)\epsilon_{\mu\nu\beta} - \chi^*_{\beta\mu}(\omega)\epsilon_{\mu\nu\alpha}]\Omega_\nu \tau_s\Big\}.$$

Previously, we discussed the relation between the spin-noise spectrum and spin susceptibility of the system. In spin-noise spectroscopy experiments where the electron or nuclear spin fluctuations are detected through the noise of the Faraday or Kerr rotation, a deep connection between the spin-noise and the spin-flip Raman scattering of light arises (Gorbovitskii and Perel, 1983). Indeed, fluctuating spins in the presence of monochromatic probe beam with the frequency $\omega_{\rm pr}$ gives rise to the secondary emission of a semiconductor medium at shifted frequencies, $\omega_{\rm pr} \pm \omega$, where ω is the spin-noise frequency. The difference between the spin-noise spectroscopy and spectroscopy of the spin-flip Raman scattering of light is related to the specifics of the detection: In conventional Raman spectroscopy, the scattered light is first passed through a spectrum analyzer (optical spectrometer) and then is converted into photocurrent, while in spin-noise spectroscopy the scattered light is first converted into a photocurrent, and only after that is the spectral analysis of the photocurrent carried out, see (Glazov and Zapasskii, 2015; Cronenberger and Scalbert, 2016) for detail.

10.2 Spin-noise spectra of localized electrons

Let us now turn to the calculation and analysis of spin-noise spectra in the quantum dot ensembles. A similar approach can be used for any system with localized electrons where the hyperfine interaction controls the electron spin decoherence, see Chapter 7. Within the semiclassical approach presented in Section 7.2 and applied for spin-noise calculations in (Glazov and Ivchenko, 2012), the electron spin-noise spectrum per quantum dot is obtained by averaging Eq. (10.12) over the direction and absolute value of $\mathbf{\Omega}_N$

$$(\delta s_\alpha \delta s_\beta)_\omega = \frac{1}{2\tau_s}\int d\mathbf{\Omega}_N \mathcal{F}(\mathbf{\Omega}_N)\left[\boldsymbol{\chi}(\omega)\boldsymbol{\chi}^\dagger(\omega)\right]_{\alpha\beta}, \tag{10.26}$$

where $\mathcal{F}(\mathbf{\Omega}_N)$ is the distribution function of the nuclear fields acting on electron spins in the quantum dot ensemble, see Eq. (7.9). At zero external magnetic field, the averaging over the directions of $\mathbf{\Omega}_N$ results in the simplification of

$$(\delta s_\alpha \delta s_\beta)_\omega = \delta_{\alpha\beta}(\delta s_\alpha^2)_\omega,$$

with $(\delta s_\alpha^2)_\omega = [(\delta s_1^2)_\omega + (\delta s_2^2)_\omega + (\delta s_3^2)_\omega]/3$. Hence,

$$(\delta s_\alpha^2)_\omega = \frac{\pi}{6}\left\{\Delta(\omega) + \int_0^\infty d\Omega_N F(\Omega_N)\,\left[\Delta(\omega - \Omega_N) + \Delta(\omega + \Omega_N)\right]\right\}, \tag{10.27}$$

where

$$F(\Omega_N) = 4\pi\Omega_N^2 \mathcal{F}(\Omega_N) = \frac{4\Omega_N^2}{\pi^{1/2}\delta_e^3}\exp\left(-\frac{\Omega_N^2}{\delta_e^2}\right), \tag{10.28}$$

is the distribution of absolute values of the hyperfine fields and we recall that δ_e describes the dispersion of nuclear fields. Equation (10.27) demonstrates that the spin-noise spectrum contains two contributions, see the curve marked $\epsilon = 1$ in Figure 10.2. The first one is centered at $\omega = 0$ and stems from fluctuations of the spin component directed along the nuclear field, which is considered as static here. In that way the zero-frequency peak bears information about the single-electron spin relaxation time, τ_s. This peak is related to the long-living tail (nondecaying fraction) of the electron spin relaxation within the central spin model, see Chapter 7. The second contribution to the spin-noise spectrum reflects electron spin precession in random nuclei-induced fields. Provided that $\delta_e \tau_s \gg 1$ this contribution to the spin-noise spectrum reduces, for $\omega > 0$, to $(\pi/6)F(\omega)$ and describes the distribution of the nuclear-induced spin precession fluctuations.

For quantum dots containing resident holes the distribution of the hyperfine fields, $\boldsymbol{\Omega}_N$, is anisotropic and described by the two-parametric Gaussian law [cf. Eq. (7.15)]:

$$\mathcal{F}_h(\boldsymbol{\Omega}) = \frac{1}{\pi^{3/2}\delta_h^3\epsilon^2} \exp\left(-\frac{\Omega_z^2}{\delta_h^2} - \frac{\Omega_\perp^2}{(\epsilon\delta_h)^2}\right), \tag{10.29}$$

with δ_h being the dispersion of the z-component of the hole spin precession frequency and ϵ being the anisotropy parameter. The spin-noise spectrum acquires a more complex form, particularly, for the z-component of spin fluctuation we have (Glazov and Ivchenko, 2012)

$$(\delta s_z^2)_\omega = \frac{\pi}{2} \int d\boldsymbol{\Omega}\, \mathcal{F}_h(\boldsymbol{\Omega}) \times \left\{ \frac{\Omega_z^2}{\Omega^2}\Delta(\omega) + \frac{\Omega_\perp^2}{2\Omega^2}\left[\Delta(\omega - \Omega) + \Delta(\omega + \Omega)\right]\right\}, \tag{10.30}$$

where $\boldsymbol{\Omega}_\perp = (\Omega_x, \Omega_y)$ and $\Omega^2 = \Omega_\perp^2 + \Omega_z^2$.

We note that Eq. (10.27) is just a Fourier transform of Eq. (7.14) [multiplied by $\exp(-t/\tau_s)$], which describes the decoherence of the electron spin in the time domain. The same holds true for the hole spin noise as well, except for the normalization factor. This is because, irrespective of the spin fluctuation origin, the temporal evolution of $\delta \boldsymbol{s}$ is described by the same Bloch equations as the dynamics of average spin polarization. Particularly, the relative spectral weight of the zero-frequency peak is simply related to the nondecaying spin fraction introduced in Chapter 7 and given by Eq. (7.16) for the model of static nuclear fields.

Figure 10.2 shows the hole spin-noise spectra, $(\delta s_z)^2_\omega$, calculated for different values of the anisotropy parameter ϵ in Eq. (10.29). The spin-noise spectrum for electrons corresponds to the curve calculated for $\epsilon = 1$. Each spectrum, in accordance with general expressions (10.27) and (10.30), contains two peaks: one at a zero frequency related to the nondecaying spin polarization fraction

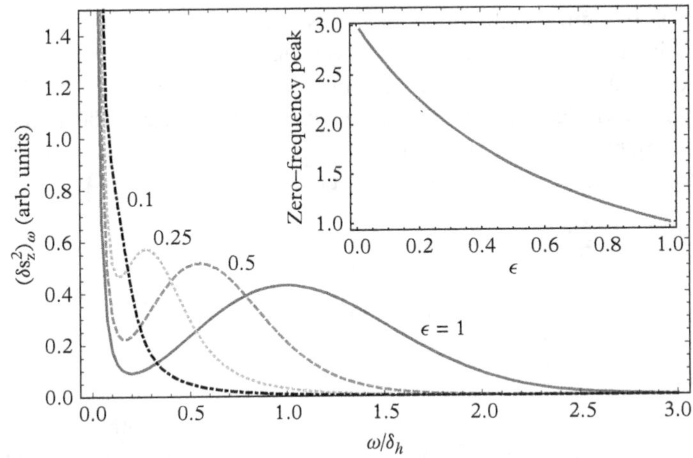

Figure 10.2 *A hole spin-noise spectrum calculated after Eq. (10.30) for different values of anisotropy parameter ϵ in Eq. (10.29) (marked at each curve). Curve for $\epsilon = 1$ corresponds to the isotropic hyperfine interaction also relevant for electrons. Inset shows the amplitude of zero-frequency peak (normalized to its value at $\epsilon = 1$) as a function of the anisotropy parameter calculated for $\delta_h \tau_s = 100$. From (Glazov and Ivchenko, 2012).*

and another, at $\omega \neq 0$, caused by the spin precession around the nuclear spin fluctuations. Note that for $\epsilon < 1$ the height of the zero-frequency peak is enhanced as compared with that for electrons, see the inset in Figure 10.2. In the limit of strong anisotropy, $\epsilon \ll 1$, it is three times higher than in the isotropic case, since, for $\epsilon \to 0$, the effective nuclear field is always directed along the z-axis. The high-frequency peak shifts toward the zero frequency with decreasing ϵ and eventually merges with the zero-frequency peak.

In the presence of an external magnetic field, the charge carrier spin precession frequency $\mathbf{\Omega}$ has, according to Eq. (10.9), two contributions due to (i) interaction with nuclei, $\mathbf{\Omega}_N$, and (ii) interaction with an external field, $\mathbf{\Omega}_B$. Making use of Eq. (10.26) we recast the spectrum of the spin z component fluctuations in the following form [cf. Eq. (10.30)]

$$(\delta s_z^2)_\omega = \frac{\tau_s}{2} \int d\mathbf{\Omega}_N \, \mathcal{F}(\mathbf{\Omega}_N) \times \left\{ \sin^2\theta \frac{1 + (\omega^2 + \Omega^2)\tau_s^2}{[1 + (\omega - \Omega)^2\tau_s^2][1 + (\omega + \Omega)^2\tau_s^2]} + \frac{\cos^2\theta}{1 + \omega^2\tau_s^2} \right\}. \tag{10.31}$$

where θ is the angle between the vector $\mathbf{\Omega} = \mathbf{\Omega}_N + \mathbf{\Omega}_B$ and the z-axis and $\Omega = |\mathbf{\Omega}_B + \mathbf{\Omega}_N|$.

In the transverse magnetic field, $\mathbf{\Omega}_B \perp z$ (Voigt geometry), the noise peak at $\omega \neq 0$ related to the spin precession shifts toward higher frequencies proportional to Ω_B for sufficiently high magnetic fields ($\Omega_B \gg \delta_{e,h}$) and its height slightly increases with an increase in magnetic field (Glazov and Ivchenko, 2012), see also Figures 10.4 and 10.6 discussed later. In fact, for $\Omega_B/\delta_{e,h} \gg 1$, the spin fluctuation dispersion is controlled only by the nuclear spin fluctuations parallel to the external field. In the particular case of negatively charged quantum dots at $\delta_e \tau_s \gg 1$, the spin-noise spectrum can be reduced to (at $\omega > 0$)

$$\langle \delta s_z^2 \rangle_\omega = \frac{\pi}{8} \int_0^\infty \frac{F\left[\sqrt{\Omega_N^2 + (\omega - \Omega_B)^2}\right] \Omega_N d\Omega_N}{\Omega_N^2 + (\omega - \Omega_B)^2} = \frac{\sqrt{\pi}}{4\delta_e} \exp\left(-\frac{(\Omega_B - \omega)^2}{4\delta_e^2}\right). \tag{10.32}$$

This expression is in agreement with its time-domain counterpart, Eq. (7.19), which describes the electron spin beats in transversal magnetic field. With the further increase in magnetic field, the spin-noise spectrum is additionally affected by the inhomogeneous broadening of electron g-factor values similar to the spin dephasing of the localized carriers addressed in Chapter 9, see Eq. (9.11). The transverse magnetic field also suppresses the zero-frequency peak related to the nondecaying spin fraction. As already mentioned in Section 7.2 this is because the higher the field, the smaller the probability of finding the quantum dot with $\boldsymbol{\Omega} = \boldsymbol{\Omega}_N + \boldsymbol{\Omega}_B \parallel z$. In other words, the average value of $\cos^2\theta$ in Eq. (10.31) decreases with an increase in the transverse magnetic field. In contrast, for the magnetic field applied in the Faraday configuration, the zero-frequency peak is enhanced while the precession peak is suppressed.

10.3 Extensions of the semiclassical model

Now we turn to the discussion of various extensions of the semiclassical treatment of the spin fluctuations. First, let us account for the finite electron correlation time, τ_c, at a given localization site by considering an ensemble of quantum dots or donors with bound electrons in the bulk semiconductor. In this approach we assume the nuclear spin fluctuation in each quantum dot or donor to be frozen on the time scale of τ_c, but allow for the electron hopping between the N_{QD} localization sites. This model has been already outlined in Section 7.2, see Eqs. (7.23), (7.24), and (7.25) to study the spin dynamics. Here, to study the spin fluctuations we introduce the correlation functions of the total spin components normalized per electron,

$$\langle \delta S_\alpha(t) \delta S_\beta(t+\tau) \rangle, \quad \delta S_\alpha = \sum_i \delta s_{i,\alpha},$$

where $\alpha, \beta = x, y, z$ enumerate Cartesian components and the subscript $i = 1 \dots N_{QD}$ enumerates the localization sites. Our aim is to calculate the spin-noise power spectrum [cf. Eq. (10.7a)]

$$(\delta S_\alpha \delta S_\beta)_\omega = \int_{-\infty}^{\infty} \langle \delta S_\alpha(t) \delta S_\beta(t+\tau) \rangle \exp(i\omega\tau) d\tau. \tag{10.33}$$

It is instructive to introduce a set of correlation functions

$$C_{\alpha\beta}^{(ij)} \equiv \int_0^{\infty} \langle \delta s_{i,\alpha}(t+\tau) \delta s_{j,\beta}(t) \rangle \exp(i\omega\tau) dt', \tag{10.34}$$

which describe the spin correlations at the same site, $i = j$, or at different sites, $i \neq j$. Here we calculate the spin-noise power spectrum by directly solving the kinetic equations analogous to Eqs. (7.23) for the correlation functions $C_{\alpha\beta}^{(ij)}$ (Landau and Lifshitz, 1981; Glazov and Ivchenko, 2012; Glazov, 2013, 2015):

$$\begin{aligned}\left(-i\omega + \nu_s + \sum_{j'} W_{j'i}\right) C_{\alpha\beta}^{(ij)} + \sum_{\alpha'\beta'} \epsilon_{\alpha\alpha'\beta'} \Omega_{i\beta'} C_{\alpha'\beta}^{(ij)} \\ - \sum_{j'} W_{ij'} C_{\alpha\beta}^{(j'j)} = \frac{1}{4N_{QD}} \delta_{ij} \delta_{\alpha\beta}.\end{aligned} \tag{10.35}$$

Here, $\nu_s = 1/\tau_s$ is the phenomenological spin relaxation time, W_{ij} is the hopping rate from the site i to the site j, see the inset in Figure 10.3, and we recall that $\epsilon_{\alpha\beta\gamma}$ is the Levi–Civita symbol, and $\delta_{\alpha\beta}$ is the Kronecker symbol. The right-hand side of Eq. (10.35) describes the single-time spin fluctuations, see Eq. (10.5), and simultaneous correlations between the different localization sites are absent for noninteracting electrons. Here, as in Section 7.2, we neglect spin-orbit coupling-induced electron spin rotation at the hopping (Kavokin, 2004, 2008; Intronati *et al.*, 2012; Smirnov and Golub, 2017).

Following Section 7.2 we strongly simplify the model and introduce a single correlation time of the electron on the localization site

$$\tau_c = \frac{1}{W_0},$$

assuming that all transition rates W_{ij} are the same; that is, $W_{ij} = W_0/N_{QD}$. We stress that this simplified model applied here to obtain analytical results disregards some important features of the electron hopping transport (Shklovskii and Efros, 1984; Shumilin and Kabanov, 2015; Shumilin *et al.*, 2016).

In the absence of an external magnetic field, the spin-noise spectrum is isotropic $(\delta S_x^2)_\omega = (\delta S_y^2)_\omega = (\delta S_z^2)_\omega$. For definiteness we assume that the probe beam used

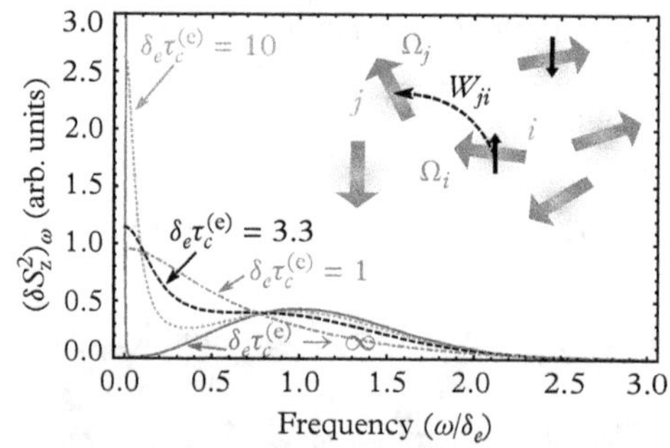

Figure 10.3 *Electron spin-noise power as a function of frequency ω/δ_e and the correlation time $\tau_c\delta_e$ calculated after Eqs. (10.36), (10.37) and (10.39) at $\nu_s/\delta_e = 10^{-4}$. Inset shows a sketch of the ensemble of localized states. Thick arrows denote nuclear spin fluctuations $\boldsymbol{\Omega}_i$, the thin dark arrow denotes electron spin, the curved, dashed arrow shows the electron jump from the occupied site i to the empty site j.*

to measure the spin noise propagates along the z-axis and, correspondingly, calculate $(\delta S_z^2)_\omega = 2\,\mathrm{Re}\{\sum_{ij} C_{zz}^{ij}\}$. Solution of Eqs. (10.35) allows us to obtain the spin-noise spectrum in the form (Glazov, 2015)

$$(\delta S_z^2)_\omega = \frac{\tau_\omega}{4} \frac{\mathcal{A}(\tau_\omega)}{1 - W_0\tau_\omega\mathcal{A}(\tau_\omega)} + \text{c.c.}, \tag{10.36}$$

where c.c. denotes the complex conjugate,

$$\tau_\omega^{-1} = \nu_s + W_0 - \mathrm{i}\omega, \tag{10.37}$$

and

$$\mathcal{A}(\tau_\omega) = \frac{1}{N_{QD}} \sum_i \frac{1 + \Omega_{i,z}^2\tau_\omega^2}{1 + \Omega_i^2\tau_\omega^2}. \tag{10.38}$$

In derivation of Eqs. (10.36) and (10.38) we assumed that the nuclear fields are isotropically distributed. In the case of a macroscopic system where $N_{QD} \to \infty$ and the nuclear fields obey the Gaussian distribution in the form of Eq. (7.9) we obtain

$$\mathcal{A}(\tau_\omega) = \frac{1}{3} + \frac{4}{3(\delta_e\tau_\omega)^2} - \frac{4\sqrt{\pi}\,\mathrm{e}^{1/\delta_e^2\tau_\omega^2}}{3(\delta_e\tau_\omega)^3}\,\mathrm{erfc}\,(1/\delta_e\tau_\omega). \tag{10.39}$$

Here, $\mathrm{erfc}\,(y) = 1 - (2/\sqrt{\pi})\int_0^y \exp(-x^2)\mathrm{d}x$ is the complementary error function. Equations (10.36) and (10.39) describe the spin-noise spectrum for an arbitrary relation between W_0 and δ_e; that is, for an arbitrary value of $\delta_e\tau_c$. For GaAs-based systems with donor-bound electron characteristic values of $\delta_e^{-1} \sim 1 \ldots 10$ ns and electron correlation time at a donor τ_c ranges from ~ 10 ps to ~ 10 ns depending on the donor density, temperature, and so on; see, for example (Kavokin, 2008; Kalevich *et al.*, 2017). The correlation time, τ_c, can be much longer for quantum dot samples in absence of optical excitation.

To analyze the spin-noise spectrum for the arbitrary relation between the electron correlation time and spin precession period in the field of nuclear fluctuation, $\delta_e \tau_c$, it is instructive to address the limits of long and short correlation times, respectively, which have been already briefly discussed in the case of spin dynamics in Chapter 7. We assume that the spin relaxation rate, unrelated to the hyperfine interaction and electron jumps, ν_s, is much smaller than δ_e. In the limit of a long correlation time where $\tau_c \gg 1/\delta_e$ and for $\omega \gg W_0, \nu_s$, one can neglect W_0 and ν_s in the definition of τ_ω, Eq. (10.37), and make use of the fact that for a real y the following relation holds: $\mathrm{Re}\{\mathrm{erfc}\,(iy)\} = 1$. As a result, we arrive, in agreement with Eq. (10.27), at

$$(\delta S_z^2)_\omega = \frac{2\sqrt{\pi}}{3\delta_e^2} \omega^2 \exp\left(-\omega^2/\delta_e\right), \quad \omega, \delta_e \gg W_0, \nu_s. \tag{10.40}$$

It is the precession peak in the spin-noise spectrum whose shape is determined by the distribution function of the nuclear field absolute values $F(\Omega) = 4\pi\Omega^2\mathcal{F}(\Omega)$, Eq. (10.28). At low frequencies, $\omega \ll \delta_e$, it is enough to put $\mathcal{A}(\tau_\omega) = 1/3$ with the result

$$(\delta S_z^2)_\omega = \frac{1}{6} \frac{\Gamma_0}{\Gamma_0^2 + \omega^2}, \quad \omega, W_0, \nu_s \ll \delta_e, \tag{10.41}$$

where $\Gamma_0 = \nu_s + 2W_0/3$. Equation (10.41) describes the zero-frequency peak (also termed the T_1-peak) in the spin-noise spectrum caused by the components conserved during the precession. Effective longitudinal spin relaxation time $T_1 = 1/\Gamma_0$ is limited by the fastest of either the spin-flip process ($\sim \nu_s^{-1}$) or hopping ($\sim W_0^{-1}$). In the latter case, when an electron leaves the site with $\mathbf{\Omega} \parallel z$ its spin starts to precess in the nuclear field fluctuation, unless the field in the new site is also parallel to the z-axis, and stops to contribute to the zero-frequency peak.

In the opposite limit of short correlation time where $W_0 \gg \delta_e$ one needs to develop $\mathcal{A}(\tau_\omega)$ in power series in δ_e with the result $\mathcal{A}(\tau_\omega) = 1 - (\delta_e \tau_\omega)^2$ and

$$(\delta S_z^2)_\omega = \frac{1}{2} \frac{\Gamma_1}{\Gamma_1^2 + \omega^2}, \quad \delta_e \ll \omega \ll W_0, \tag{10.42}$$

where $\Gamma_1 = \delta_e^2/W_0 + \nu_s$. In this limit, the spin-noise spectrum consists of the zero-frequency peak only, whose width decreases with an increase in the hopping rate similarly to the enhancement of the spin lifetime in Eq. (7.25). This is typical for the motional narrowing regime realized for a short correlation time, $\delta_e \tau_c \ll 1$, where the electron spin rotation angle during the correlation time is small at a given site.

Figure 10.3 shows the electron spin-noise power spectrum for different correlation times τ_c. For a the long correlation time where the jumps between the sites are suppressed, the spectrum demonstrates a two-peak structure presented in Figure 10.2 for $\epsilon = 1$ and $\tau_c \to \infty$. An increase in the hopping rate results in the broadening of the zero-frequency peak and suppression of spin precession peaks. If $\tau_c \delta_e \sim 1$ then the zero-frequency and precession peaks merge. A further increase in the hopping rate W_0 (and, correspondingly, decrease in τ_c) results in an increase of the height and the decrease of the width of the spin-noise peak due to the dynamical averaging of the nuclear fields, see Eq. (10.42).

Now we consider the electron spin fluctuations in the presence of an external magnetic field characterized by the electron spin precession frequency $\boldsymbol{\Omega}_B$. As before, we analyze two situations of (i) longitudinal field (parallel to the probe beam propagation axis, i.e., Faraday geometry), $\boldsymbol{\Omega}_B \parallel z$, and (ii) transverse field, $\boldsymbol{\Omega}_B \perp z$; that is, Voigt geometry. The field is assumed to be sufficiently small so as not to affect the hopping rates W_{ij} and not to result in a significant equilibrium electron spin polarization. In the particular case (i), where $\boldsymbol{\Omega}_B \parallel z$ the spin-noise spectrum is, as before, given by Eqs. (10.36)–(10.38) provided that the total spin precession frequency $\boldsymbol{\Omega}_N + \boldsymbol{\Omega}_B$ is taken into account in evaluation of sum in Eq. (10.38). The longitudinal magnetic field thus reduces the effect of nuclear fluctuations on z spin component fluctuations. Hence, in the limit of long correlation time $\tau_c \delta_e \gg 1$ the precession peak gets suppressed and shifted toward the higher frequencies $\omega \sim \Omega_B$, while the zero-frequency peak becomes enhanced. Indeed, provided that $\Omega_B \gg \delta_e, W_0$ the parameter $\mathcal{A}(\tau) \approx 1 - \delta_e^2/\Omega_B^2$ and the spin-noise spectrum has a single-peak form of Eq. (10.42) with $\Gamma_1 = W_0 \delta_e^2/\Omega_B^2 + \nu_s$. Due to rapid Larmor precession of the transverse spin components the nuclear fields dynamically average out and their effect on electron spin noise is reduced, cf. (Dzhioev *et al.*, 2002*b*).

The description of the transverse field effect is more complicated. In the particular case (ii) instead of Eq. (10.36), one has (Glazov, 2015)

$$(S_z^2)_\omega = \frac{\tau_\omega}{4} \frac{\mathcal{A}(\tau_\omega)[1 - W_0 \tau_\omega \mathcal{A}(\tau_\omega)] - (W_0 \tau_\omega)^2 \mathcal{B}^2(\tau_\omega)}{[1 - W_0 \tau_\omega \mathcal{A}(\tau_\omega)]^2 + (W_0 \tau_\omega)^2 \mathcal{B}^2(\tau_\omega)} + \text{c.c.}, \tag{10.43}$$

where $\mathcal{A}(\tau_\omega)$ is given by Eq. (10.38) and $\mathcal{B}(\tau_\omega)$ is defined by

$$\mathcal{B}(\tau_\omega) = \frac{1}{N} \sum_i \frac{\Omega_{x,i} \tau_\omega}{1 + \Omega_i^2 \tau_\omega^2}. \tag{10.44}$$

Here we assumed the magnetic field to be applied along the x-axis, $\boldsymbol{\Omega}_B \parallel x$. An increase of the field Ω_B up to $\sim \delta_e$ results (in the limit of long correlation time

$\delta_e\tau_c \gg 1$) in the suppression of zero-frequency peak and enhancement of the precession peak. Further increase in Ω_B results in the shift of the precession peak to the frequency $\omega \sim \Omega_B$. In the limit of short correlation time the spin precession peak shifts following the Larmor frequency Ω_B. The transformation of the spin-noise spectrum with an increase of the transverse magnetic field is illustrated in Figure 10.4 for intermediate value of the correlation time $\tau_c\delta_e = 3.3$. The figure clearly shows the suppression of the zero-frequency peak and evolution of the spin precession peaks from hyperfine-induced satellites of a zero-frequency peak to the peaks at the Larmor frequency.

To conclude the discussion of the finite correlation time effects on the spin-noise spectra, we note that although Eqs. (10.36) and (10.43) are derived for the particular model of the electron hopping between the localization sites, these expressions can be also applied to the case where the hopping is absent but the nuclear spin dynamics is characterized by a correlation time. This situation may arise if the quadrupole interaction is strong enough and quadrupole axes are randomly distributed in space for different nuclei. In this case, the characteristic τ_c time is related to the magnitude of the quadrupole splittings of nuclear spin states, see Eqs. (3.35) and (7.26).

Here, the spin noise has been calculated within the semiclassical approach to the central spin model where the nuclear spin dynamics was disregarded, see Chapter 7 for details. As we have already noted in Section 7.3, the coupled dynamics of the electron and nuclear spins affects mainly the zero-frequency peak in the spin-noise spectrum related to the nondecaying spin fraction. The analysis performed in Section 7.3 can be directly applied to study the shape of the $\omega = 0$ peak in the spin-noise spectrum that may deviate from the simple Lorentzian one. In particular, the logarithmic asymptotic of the long-time spin polarization tail, $S_z(t) \propto \ln^{-\alpha}(t/T_N)$, Eq. (7.33), which arises in some approaches to the central spin model [see (Khaetskii *et al.*, 2002; Coish and Loss, 2004;

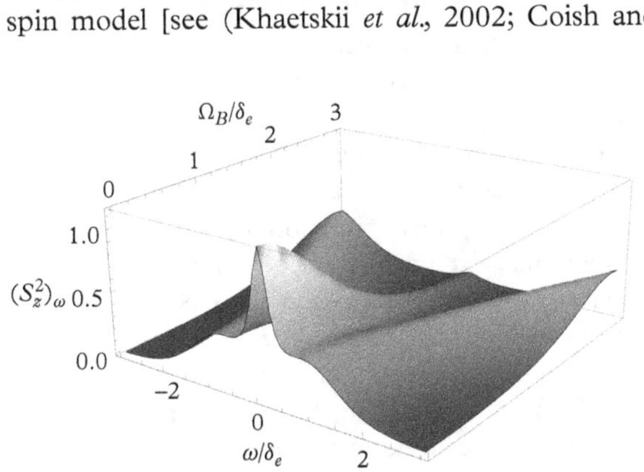

Figure 10.4 *Electron spin-noise power as a function of frequency ω/δ_e and transverse magnetic field Ω_B/δ_e calculated after Eqs. (10.37), (10.43), and (10.44) at $\tau_c\delta_e = 3.3$ and $v_s/\delta_e = 10^{-4}$. From (Glazov, 2015).*

Erlingsson and Nazarov, 2004; Al-Hassanieh *et al.*, 2006; Chen *et al.*, 2007*a*) and Section 7.3 for details] yields low-frequency singularities in the spin-noise spectra $\propto \omega^{-1} L_\alpha[\ln(1/\omega T_N)]$, with $L_\alpha[\ln(1/\omega T_N)]$ being a relatively weak function of the $\ln(1/\omega T_N)$. The detailed theoretical analysis of the spin noise within the central spin model has been performed in (Hackmann and Anders, 2014; Hackmann *et al.*, 2014), where the quantum-mechanical approach has been compared with the semiclassical model. In agreement with the applicability limits of the semiclassical treatment, the precession peak in the spin-noise spectrum calculated disregarding the dynamics of nuclear spins matches well the results of calculations for the frozen nuclear spins. In contrast, the deviations of the low-frequency, $\omega \ll \delta_e$, behavior within the classical and quantum mechanical treatment can be substantial. The results of numerical simulations performed in (Hackmann and Anders, 2014) for the case of n-doped quantum dots can be rather accurately fitted by the $\omega^{-3/4}$ dependence, which implies $t^{-1/4}$ decay of the spin polarization. However, this result cannot rule out the logarithmic long-time spin decay, Eq. (7.33), due to finite time domain studied numerically. Further specifics of the low-frequency spin fluctuations in quantum dots arise due to quadrupole splittings of nuclear spin states that may dominate over the hyperfine interaction with hole or electron spins (Sinitsyn *et al.*, 2012; Hackmann *et al.*, 2015). This point is discussed in more detail in the next section.

10.4 Experiments and modeling

Spin fluctuations of electrons and holes in quantum dot ensembles have been studied in several works (Crooker *et al.*, 2010; Li *et al.*, 2012; Hackmann *et al.*, 2015; Glasenapp *et al.*, 2016), the fluctuations of a single spin of a localized hole have been detected in (Dahbashi *et al.*, 2014), and (Berski *et al.*, 2015) presents the results of experiments on donor-bound electrons in bulk GaAs. Here, we describe the results of (Glasenapp *et al.*, 2016) where the spin noise of electrons and holes in InGaAs quantum dot samples have been studied in detail experimentally and also theoretically in two approaches: the semiclassical one, which assumes frozen nuclear spins, and the quantum-mechanical one based on Chebyshev polynomial expansion of the evolution operator, which accounts for both the hyperfine interaction between the electron and nuclear spins and quadrupole splittings of the nuclear spin states.

Figure 10.5 presents the results of spin-noise spectra measurements for the n- and p-doped InGaAs quantum dots in the left and right columns of the panels, respectively. These data correspond to the power spectra of the Faraday rotation noise in a zero magnetic field, $\boldsymbol{B} = 0$. More precisely, the data in Figure 10.5 show the difference of the $(\delta\vartheta^2)_\omega$ measured at $\boldsymbol{B} = 0$ and at the transversal field $B \sim 10^2$ mT (where the spin noise of charge carriers is beyond the detection range), in which case all contributions to the Faraday rotation fluctuations insensitive to the

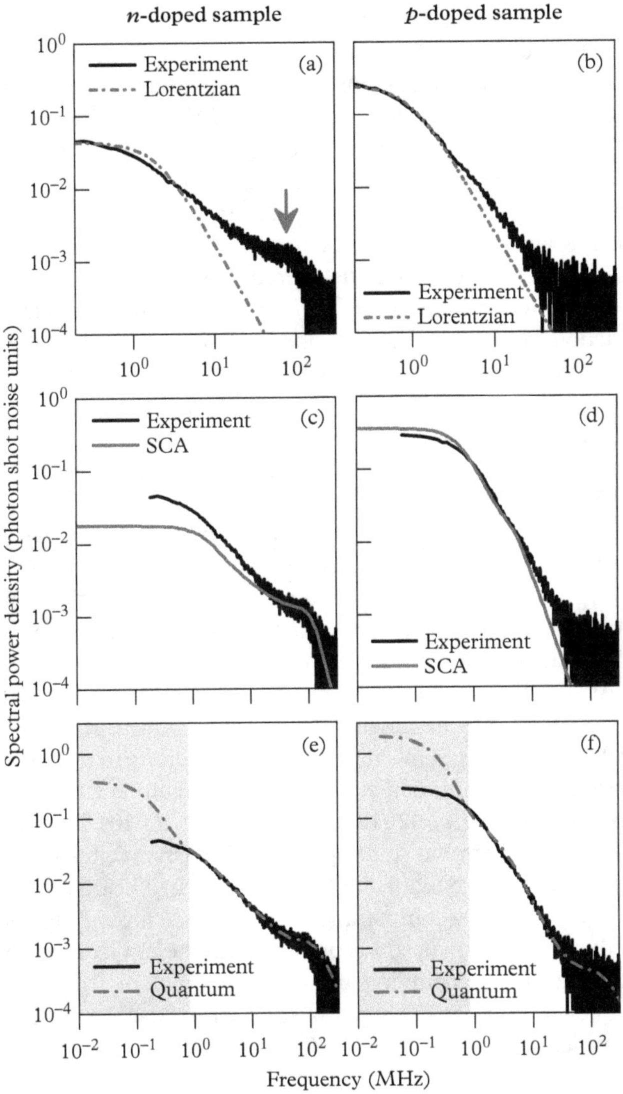

Figure 10.5 *Spin-noise spectra measured on the (a) n-doped and the (b) p-doped quantum dot samples at* $\boldsymbol{B} = 0$. *Panels (c), (d) show also fits (solid curves) to the experimental data with the semiclassical model, Eqs. (10.27) and (10.30), with* $\delta_e = 70$ *MHz,* $\tau_s^e = 5$ *ns,* $\delta_h = 40$ *MHz,* $\epsilon = 0.1$, $\tau_s^h = 80$ *ns (c), and* $\tau_s^h = 240$ *ns (d). The dash-dotted curves in panels (e) and (f) indicate the fits obtained with the quantum approach using* $\lambda_h = 5$, $\delta_e = 110$ *MHz, and* $\delta_h = 16$ *MHz. The electron fraction for the n-doped sample is set to 80%, and to 15% for the p-doped sample. The gray shaded region shows the range where the quantum-mechanical model becomes inaccurate. From (Glasenapp et al., 2016).*

magnetic field are excluded. In the p-doped sample the hole spin noise consists of a single peak centered around zero frequency. The analysis shows that the peak can be rather accurately described by a Lorentzian function. The spin-noise spectrum of the n-doped sample contains, in addition to the zero-frequency peak, a pronounced "shoulder" at a frequency of $\sim$100 MHz, as marked by the arrow in Figure 10.5(a). In agreement with the theory presented in Section 10.2 this peak is related to the electron spin precession caused by the nuclear spin fluctuations. The absence of such a shoulder in the hole spin ensemble signal is a direct indication of the anisotropic nature of the hole-nuclear interaction, where the nuclear-induced effective magnetic field predominantly points along the z-axis. The analysis presented in (Glasenapp *et al.*, 2016) demonstrates the presence of a finite hole fraction in the nominally n-doped sample, which increases the relative weight of the zero-frequency peak. This hole contribution of about 20% inferred from the magnetic-field dependent measurements, see Figure 10.6, has been taken into account in the modelling of the data: The panels (c) and (d) of Figure 10.5 demonstrate the results of the semiclassical calculation and the panels (e) and (f) of Figure 10.5 show the results obtained in the quantum-mechanical approach. Overall, the experimental data are well described by both theoretical approaches. Note that in the range of low frequencies $\lesssim 6$ MHz the quantum-mechanical calculations are not quite accurate due to the small number of nuclei involved in the calculations. Moreover, the parameters used in the quantum-mechanical and semiclassical calculations are somewhat different reflecting inaccuracies in both approaches. As demonstrated in (Glasenapp *et al.*, 2016) the semiclassical model calculations with the same parameters as the quantum-mechanical calculations yields a slightly worse but still in satisfactory agreement with the experiment.

The analysis of the experimental data in (Glasenapp *et al.*, 2016) as well as the results of (Sinitsyn *et al.*, 2012; Hackmann *et al.*, 2015) demonstrate the importance of the quadrupole splittings of nuclear spin states in these strained InGaAs quantum dots, see Section 3.4 for more details. The quadrupole splittings of nuclear spin states may be comparable or even exceed the splitting of the nuclear spin states induced by the hyperfine coupling with the electron. In the simplified approach the quadrupole coupling induced dynamics of the nuclei results in the temporal fluctuations of the nuclear spin, which can be roughly described by a finite correlation time of the hyperfine field, see Eq. (7.26). In accordance with Figure 10.3 and Eq. (10.41) this yields a Lorentzian zero-frequency spin-noise peak. This result is corroborated by sophisticated quantum-mechanical calculations (Hackmann *et al.*, 2015; Glasenapp *et al.*, 2016), see also (Sinitsyn *et al.*, 2012). The information on the slow electron and hole spin dynamics in quantum dots can be conveniently extracted from the spin-noise spectra compared with the pump-probe experiments, see Chapter 9, because, in the spin-noise measurements, limitations related to the finite pulse repetition rate are absent.

Let us now turn to the presentation of the experimental results obtained in the transverse magnetic field, $\boldsymbol{B} \parallel x$. The data for the p-doped quantum dot sample are shown in Figure 10.6(a). At the nonzero magnetic field the spectrum, in agreement

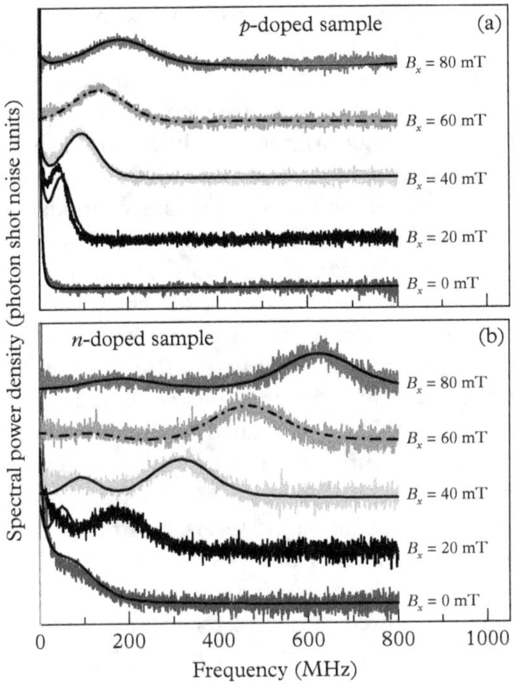

Figure 10.6 *(a) Spin-noise spectra measured on the p-doped quantum dot sample in the transverse magnetic fields indicated at the corresponding curves. Fits with the semiclassical model are shown by the solid lines, using the transversal heavy hole g-factor* $|g_h| = 0.16$ *with a spread of 37%. Other parameters of calculations are the same as for Figure 10.5(d) except for* $\tau_s^h = 16$ *ns at* $B_x \geqslant 20$ *mT. The dashed line at* $B_x = 60$ *mT shows the fit with the quantum model. (b) Spin-noise spectra measured in the n-doped sample at the same transverse magnetic fields, also showing signatures of hole doped quantum dots in the ensemble. Fits with the semiclassical model are depicted by solid lines, with parameters of the hole spin noise being the same as in panel (a). Transversal electron g-factor is* $|g_e| = 0.55$ *with a spread of 7% and the other parameters are the same as for Figure 10.5(c). The fit at* $B_x = 60$ *mT uses the quantum-mechanical approach, with the g-factor being the same as for the semiclassical approach. From (Glasenapp et al., 2016).*

with theoretical model in Section 10.2 consists of two peaks: The first one remains centered at $\omega = 0$, but its amplitude decreases with the increase in magnetic field, and the second one appears at $\omega \neq 0$, shifting with the field toward higher frequencies. The experimental data are adequately described by the semiclassical model, see the solid lines. The results of the quantum-mechanical calculation are given by a dashed line for the exemplary case of $B_x = 60$ mT, which also fits the data quite accurately. The spin fluctuation spectra in the n-doped sample in the transverse magnetic field are shown in Figure 10.6(b). Here, at $B_x \neq 0$ one can see two features in the spin-noise spectra appearing at nonzero frequencies. One of them correlates with the hole Larmor precession, see Figure 10.6(a), and supports assumptions of the presence of *p*-doped quantum dots in the *n*-type sample. Another peak has much higher frequency and can be attributed to the electrons. Again, as in the panel (a), the solid lines demonstrate the calculations within the semiclassical model and the dashed curve for $B_x = 60$ mT is the result of the quantum-mechanical approach. The results of model calculations are in good agreement with experimental data.

The fitting of the spin-noise spectra as a function of magnetic field shown in Figure 10.6 allows one to evaluate accurately the charge carriers Landé factors. It is worth noting that the analysis of the spin correlation functions also provides access to the spread of *g*-factors as illustrated in (Berski *et al.*, 2013).

Another interesting realization of the spin-noise spectroscopy on localized charge carriers in semiconductors is related to donor-bound electrons in bulk GaAs (Berski *et al.*, 2015). Figure 10.7 demonstrates the results of measurements of the spin-noise spectra at a very high purity GaAs layer grown by the molecular beam epitaxy on top of a semi-insulating GaAs substrate. The intentional *n*-type doping by Si donors with a density of $N_D \approx 10^{14}$ cm^{-3} has been realized. The spin-noise power spectrum shown is actually a difference between two Faraday rotation noise spectra measured at a zero magnetic field and finite transversal magnetic field, $B_x = 22$ mT. At the $B_x = 22$ mT the spin precession frequency is about 140 MHz and exceeds by far the frequencies related to the hyperfine interaction. As a result, the positive values of the fluctuation spectrum correspond to the zero-field spin noise, while the negative values describe the spin fluctuations at $\boldsymbol{B} \neq 0$.

As demonstrated by the analysis performed in (Berski *et al.*, 2015) the semiclassical model accounting for the finite correlation time, τ_c, of a donor-bound electron fits experimental data quite well. Both zero-frequency and precession peaks are clearly resolved. The model, Eq. (10.43) and thick curved line in Figure 10.7, fits the data with τ_c and δ_e being the only free parameters and allows extraction of the nuclear field spread $\delta_e \approx 29 \pm 0.3$ MHz. This corresponds to the nuclei-induced fluctuating magnetic field $\delta B_N \approx 4.6$ mT, which is in close agreement with literature data for donor-bound electrons in bulk GaAs, see Eq. (2.60) and (Dzhioev *et al.*, 2002*a*; Colton *et al.*, 2004). The correlation time of $\tau_c \approx 32 \pm 0.6$ ns is also very close to the value reported in (Dzhioev *et al.*, 2002*a*) for a comparable electron density. The spin-noise feature at the frequency of about 140 MHz results from Larmor precession of the electron spin fluctuations in the effective magnetic field given by the sum of the local hyperfine fields and the applied transverse magnetic field. This peak is well described by the theory that includes the spread of the electron *g*-factor. One obtains the average value of the modulus of the electron *g*-factor of 0.46 ± 0.064 and the spread in the order of 10%. The uncertainty of the Landé factor is mainly related to the magnetic field inhomogeneity in the experimental setup.

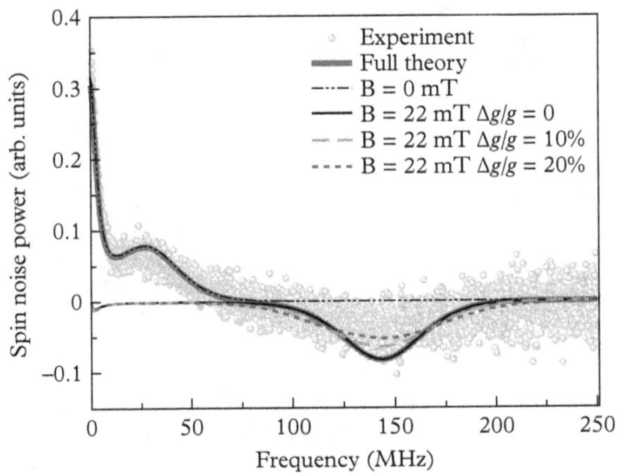

Figure 10.7 *Donor-bound electron spin-noise spectrum. The dots are experimental data showing the difference of the spectra at $B_x = 0$ mT and $B_x = 22$ mT. The curves are the results of theoretical calculations after Eq. (10.43) for different g-factor spreads at $\tau_c = 32$ ns and $\delta_e = 29$ MHz, and electron g-factor $|g| = 0.46$. Data from (Berski et al., 2015).*

10.5 Nonequilibrium spin noise

Spin-noise spectroscopy has emerged as a tool to study spin fluctuations under conditions close to thermal equilibrium, where the excitation of the sample by the probe beam is negligible. In this situation, as shown in Section 10.1, the spin-noise spectrum is directly related to the spin susceptibility of the electron system. This can be realized by using a weak probe beam strongly detuned from the resonance in order to suppress any absorption caused by real transitions in the electronic system. However, in application to semiconductor quantum dot structures, the probe beam is usually tuned to the peak of absorption due to significant inhomogeneity of the dot ensemble (Zapasskii *et al.*, 2013; Glasenapp *et al.*, 2014; Yang *et al.*, 2014; Glasenapp *et al.*, 2016), see also (Huang and Steel, 2011; Glasenapp *et al.*, 2013) where the probe absorption effects on spin noise were observed in bulk semiconductors. Moreover, quantum dots can be placed in optical microcavities, see, for example, (Dahbashi *et al.*, 2012), where, on one hand, the spin-induced Kerr or Faraday rotation is strongly enhanced due to multiple roundtrips of light inside the cavity, but, on the other hand, the field is accumulated in the cavity and, correspondingly, the perturbation of spins by the probe light is enhanced. In atomic systems, nonequilibrium spin fluctuations are also actively studied (Chalupczak and Godun, 2011; Horn *et al.*, 2011) Thus, studies of the spin fluctuations in nonequilibrium conditions has became topical.

By contrast to equilibrium systems where the spin-noise spectrum can be related to a response function by means of the fluctuation-dissipation theorem,

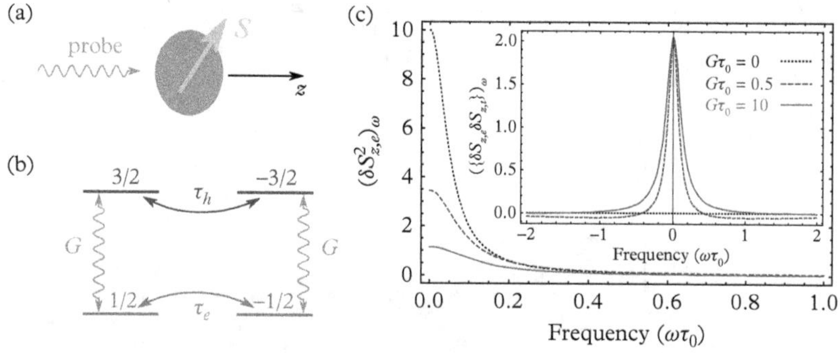

Figure 10.8 *(a) Illustration of a quantum dot with the electron spin subject to the probe beam propagating along the z-axis. (b) Scheme of the states and transitions caused by the absorption of the probe beam and spin relaxation. (c) Electron spin-noise spectra calculated for different trion generation rates $G\tau_0 = 0$, $G\tau_0 = 0.5$ è $G\tau_0 = 10$, $\tau_e/\tau_0 = 20$, $\tau_h/\tau_0 = 5$. The inset shows the spectrum of correlation noise of the electron and trion.*

Eq. (10.25), the general relation between the noise spectrum and any response function is absent for systems out of thermodynamic equilibrium (Kogan and Shul'man, 1969; Gantsevich *et al.*, 1979; Landau and Lifshitz, 1981; Glazov, 2016). This calls for separate analysis of each nonequilibrium realization of the spin noise, see (Glazov, 2016) where various situations have been studied theoretically. The pioneering theoretical study of electron spin fluctuations in the out-of-equilibrium case has been performed in (Ivchenko, 1973) where the effect of circularly polarized radiation, which causes optical orientation of electron spins, on the spin noise has been addressed. According to (Ivchenko, 1973) for circularly polarized light propagating along the z-axis the spectral density of spin z component fluctuations, δS_z, for the ensemble of N electrons can be recast in a simple form

$$(\delta S_z^2)_\omega = \frac{N(1 - r_\omega P_s^2)}{2} \frac{T_s}{1 + (\omega T_s)^2}. \tag{10.45}$$

Here, T_s is the spin lifetime, which accounts for both the spin relaxation processes and the recombination of polarized electrons with unpolarized holes, see Eq. (2.38), P_s is the steady-state spin polarization degree of electrons, Eq. (2.36), and r_ω is a factor taking into account generation-recombination processes. Equation (10.45) demonstrates the suppression of electron spin fluctuations due to spin polarization. An analogous effect takes place in a sufficiently strong static magnetic field $\boldsymbol{B} \parallel z$ as well, where thermal orientation of electron spins, $P_s \propto B_z/k_B T$, is realized. In such a situation, the factor describing spin fluctuations suppression takes a simple form $1 - P_s^2$. This result can be derived from general arguments taking into account that $\int d\omega (\delta S_z^2)_\omega = 2\pi \langle \delta S_z^2(t) \rangle = \pi(1 - P_s^2)/2$.

The absorption of unpolarized probe beam can also affect spin fluctuations of localized electrons due to excitation of trions. To illustrate the effect we consider a quantum dot with a resident electron and employ a four-level model illustrated in Figure 10.8(a,b). Here, the ground state of the system corresponds to the resident electron and it is two-fold spin degenerate, $s_z = \pm 1/2$. The ground state is coupled by the probe beam with the singlet trion with unpaired hole, which is also doubly degenerate in the hole spin component $j_z = \pm 3/2$, This model has been employed in Chapter 6 to describe the optical spin polarization control. The model can be also applied to quantum dots in microcavities operating in the weak coupling regime (Dahbashi *et al.*, 2014; Arnold *et al.*, 2015; Smirnov *et al.*, 2015).

The detailed analysis of the spin fluctuations within such a four level model is presented in (Glazov, 2016; Smirnov *et al.*, 2017). Here we summarize the main results. It follows from the general equations for the single particle density matrix $\rho_{mm'}$, $m, m' = \pm 1/2, \pm 3/2$, that the expressions for the average occupancies of the electron, $f = (\rho_{1/2,1/2} + \rho_{-1/2,-1/2})/2$, and trion, $n = (\rho_{3/2,3/2} + \rho_{-3/2,-3/2})/2$, take the form

$$f = \frac{1}{2}\frac{1+G\tau_0}{1+2G\tau_0}, \quad n = \frac{1}{2} - f. \tag{10.46}$$

Here, we introduced the trion generation rate

$$G = \frac{\gamma |V|^2}{(\omega-\omega_0)^2 + \gamma^2},$$

$V = -dE/(\sqrt{2}\hbar)$, d is the matrix element of a dipole transition between the electron and the trion states, E is the amplitude of the incident field, γ is the damping rate of the off-diagonal elements of the density matrix, $\rho_{\pm 1/2, \pm 3/2}$, and τ_0 is the trion's lifetime. For the linearly polarized probe beam both electron and trion are unpolarized: The spin components $S_{z,e} = (\rho_{1/2,1/2} - \rho_{-1/2,-1/2})/2$, and trion, $S_{z,t} = (\rho_{3/2,3/2} - \rho_{-3/2,-3/2})/2$, are zero in the steady state. Assuming that the spin fluctuations as well as generation processes are slow, that is under conditions that both spin relaxation rates and the generation rate G are by far smaller compared with $[(\omega-\omega_0)^2+\gamma^2]^{1/2}$ [this condition is partially relaxed in (Wiegand *et al.*, 2017)], one can arrange the kinetic equations for the spin correlation functions in the matrix form:

$$\dot{\mathcal{M}} + \mathcal{R}\mathcal{M} = 0. \tag{10.47}$$

Here the matrix elements $\mathcal{M}_{ij}(\tau) = \langle \delta S_{z,i}(t) \delta S_{z,j}(t+\tau)\rangle$, where the subscripts $i, j = e, t$ enumerate the electron and trion states, and the matrix

$$\mathcal{R} = \begin{pmatrix} \tau_e^{-1} + G & -\frac{1}{\tau_0} - G \\ -G & \tau_h^{-1} + \frac{1}{\tau_0} + G \end{pmatrix}, \tag{10.48}$$

describes the relaxation and generation processes, the dot on top denotes the derivative over the time τ. Two more parameters of the theory are introduced in Eq. (10.48): The phenomenological spin relaxation times of the electron τ_e and of the hole-in-trion τ_h, see Figure 10.8(b). This model can also be extended by introducing the correlation functions of all spin components allowing one to account for the electron spin precession in external and in nuclei-induced magnetic fields. In order to calculate the spin-noise spectra we need to supplement Eq. (10.47) with the initial conditions; namely, the values of correlators matrix $\mathcal{M}(\tau = 0)$. The analysis shows that $\langle \delta S_{z,e}^2 \rangle = f/2$, $\langle \delta S_{z,h}^2 \rangle = n/2$, while single-time cross-correlators are zero.

In order to present the solution we introduce an auxiliary 2×2 matrix, $\zeta(\omega)$ according to

$$(-i\omega + \mathcal{R})\zeta = \mathcal{M}(0). \tag{10.49}$$

Hence, the spin-noise spectrum can be recast in agreement with (Landau and Lifshitz, 1981; Glazov, 2016) as

$$(\delta S_{z,i} \delta S_{z,j})_\omega = \frac{1}{2}[\zeta_{ij}(\omega) + \zeta_{ij}^*(-\omega) + \zeta_{ji}(-\omega) + \zeta_{ji}^*(\omega)]. \tag{10.50}$$

To illustrate the results of the model, we present the analytical expression for the electron spin-z component fluctuation power spectrum

$$(\delta S_{z,e}^2)_\omega = \frac{f}{2}\left[\tau_e^{-1} + G - i\omega - \frac{G(G + \tau_0^{-1})}{\tau_h^{-1} + \tau_0^{-1} + G - i\omega}\right]^{-1} + \text{c.c.}, \tag{10.51}$$

which immediately follows from solution of Eqs. (10.47) and (10.51). A similar expression can be derived for the trion spin fluctuations spectrum, $(\delta S_{z,t}^2)_\omega$. The optical transitions between the electron and trion states at $G \neq 0$ result also in a cross-correlation of the electron and trion spins described by the correlation function $\langle \{\delta S_{z,e}(t) \delta S_{z,t}(t+\tau)\} \rangle$; we remind the reader that the curly brackets stand for the symmetrized product $\{AB\} = (AB + BA)/2$.

Calculated electron spin-noise spectra as well as spectra of cross-correlation electron and trion spin fluctuations are shown in Figure 10.8(c). It can be seen that an increase in the probe beam power results in a reduction of the electron spin fluctuation amplitude due to population of the trion state and in a broadening of a peak due to the enhancement of the electron spin relaxation because, for the parameters in question, $\tau_e > \tau_h$. Hence, the pumping electron through the trion results in its depolarization. Under typical conditions where the trion lifetime, τ_0, is considerably shorter compared with the time scales of the electron and hole-in-trion spin relaxation, the trion spin-noise spectrum has a more complex shape and is characterized by the partial contributions of different widths. An inset in Figure 10.8(c) shows the cross-correlation of electron and trion noise spectra, $(\{\delta S_{z,e}\delta S_{z,h}\})_\omega$. We stress that this correlator appears only due to the absorption of the probe beam. It is noteworthy that, since for $\tau = 0$ the cross-correlator $\langle\{\delta S_{z,e}(t)\delta S_{z,t}(t+\tau)\}\rangle$ turns to zero, the frequency integral $\int(\{\delta S_{z,e}\delta S_{z,h}\})_\omega d\omega = 0$, therefore, the cross-correlation fluctuations spectrum changes its sign at a certain frequency. The sign change point depends on the probe power and relaxation times in the system.

The model presented here can, at least semiquantitatively, describe the effects of photoexcitation on the spin fluctuations of single charge carriers in quantum dots reported in (Dahbashi *et al.*, 2014; Glasenapp *et al.*, 2016; Smirnov *et al.*, 2017), see also (Wiegand *et al.*, 2017).

10.6 Manifestations of nuclear spin dynamics in electron spin noise

Another interesting example of spin fluctuations in a nonequilibrium system arises if the nuclear spin system in the crystal is polarized. As discussed in detail in Chapter 5, the absorption of the circularly polarized light in the magnetic field $\boldsymbol{B}$ gives rise to the dynamical nuclear polarization, $\langle \boldsymbol{I} \rangle \neq 0$. In many cases, the nuclear spin system can be described by an effective spin temperature Θ_N, which can be quite different from the temperature T of the electrons and lattice, see Section 5.2 for details. Hence, the system "nuclei+electrons" is in the nonequilibrium state. When the optical pumping is turned off the nuclear temperature relaxes to the lattice temperature with the characteristic longitudinal spin relaxation time T_{1N} [see Eq. (5.21) with $\Theta_0 \to T$ and $T_{1e} \to T_{1N}$] that, as a rule, exceeds by far the electron spin relaxation time τ_s in the system. As a result, the electron spin system remains in equilibrium or close to equilibrium conditions while interacting with nonequilibrium system of host lattice nuclei spins.

It has been shown in (Smirnov, 2015) that such a nonequilibrium nuclear spin polarization and, correspondingly, the Overhauser field exerted by the polarized nuclei on electron, see Eq. (5.42), strongly modifies the spin-noise spectrum of charge carriers. Particularly, if the external field $\boldsymbol{B}$ is applied perpendicularly to the probe light propagation direction z, the electron spin-noise spectrum reads in accordance with Eq. (10.15)

$$(\delta S_z^2)_\omega = \frac{1}{4}\left(\frac{\tau_s}{1+(\omega-\Omega^*)^2\tau_s^2} + \frac{\tau_s}{1+(\omega+\Omega^*)^2\tau_s^2}\right), \qquad (10.52)$$

where $\Omega^* = g\mu_B B_x^*/\hbar$ is the spin precession frequency in the total field $B_x^* = B + B_N$, $B_N \propto 1/\Theta_N$ is the Overhauser field. This expression (10.52) is valid

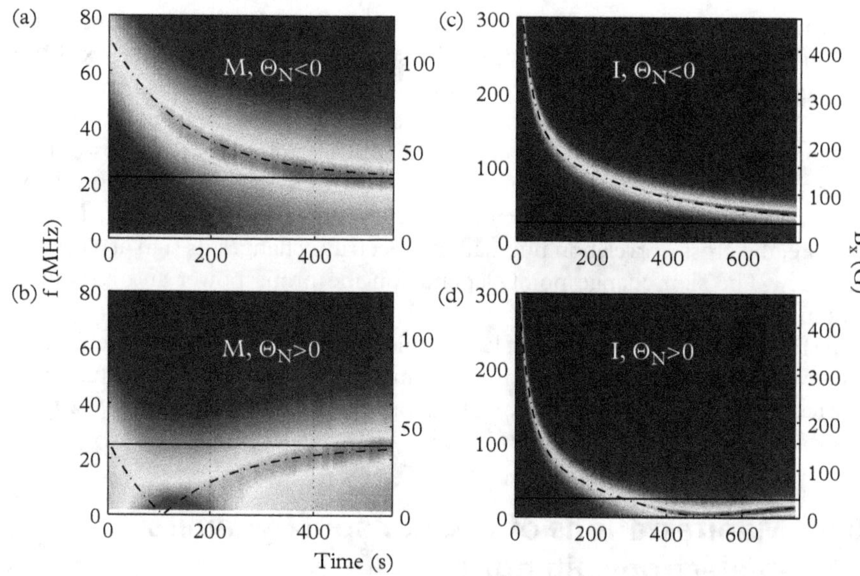

Figure 10.9 *Electron spin-noise spectra measured with time steps in the order of several seconds at the metallic (M, $n \approx 4 \times 10^{16}$ cm^{-3}) [panels (a) and (b)] and insulating (I, $n \approx 2 \times 10^{15}$ cm^{-3}) [(c), (d)] bulk n-type GaAs samples in the external magnetic field $B_{\mathrm{ext}} \approx 4$ mT. Panels (a) and (c) correspond to a negative nuclear temperature ($\Theta_N < 0$), where the nuclear field is parallel to the external one, and panels (b) and (d) correspond to a positive temperature, where the nuclear and external fields are antiparallel. Black lines are the fitting results. White lines show the spin precession frequency of electrons in the external field. From (Ryzhov et al., 2015).*

provided that the quasi-equilibrium electron spin polarization is negligible and, moreover, the effects of nuclear spin fluctuations are disregarded. In Eq. (10.52) the frequency, Ω^*, and, hence, the electron spin-noise spectrum can depend on the laboratory time, t: Over the course of time, the processes of nuclear spin polarization or relaxation take place that result in temporal variations of Θ_N. The possibility to use the quasi-equilibrium form of the electron spin-noise power spectrum, Eq. (10.52), with a time-dependent Ω^* is ensured by the relation, $\tau_s \ll T_{1N}$.

Hence, by measuring the electron spin fluctuation spectra at different moments of time under the conditions of dynamical nuclear polarization one can observe the dynamics of the Overhauser field, B_N, and, correspondingly, of the nuclear spin temperature, Θ_N, as suggested in (Smirnov, 2015). Such experiments have been carried out in (Ryzhov *et al.*, 2015, 2016) on *n*-type bulk GaAs samples. Figure 10.9 presents the results of the electron spin-noise measurements on two samples with different doping levels and, hence, different spin relaxation times of electrons and nuclei. Depending on the conditions of the nuclear spin system preparation, the nuclear spin temperature was either negative or positive, which corresponds to the Overhauser field being, respectively, parallel or antiparallel to the external one. Moreover, under the conditions of experiments in (Ryzhov *et al.*, 2015) the absolute value of B_N exceeded the external field. Therefore, for positive nuclear spin temperature, the frequency, Ω^*, of the electron spin precession goes through zero in the course of nuclear spin relaxation where the Overhauser field exactly compensates the external one, see Figure 10.9(b,d).

In addition to the Overhauser field, an effective "optical" magnetic field was uncovered in (Ryzhov *et al.*, 2016), which acts on the electron spins in the presence of an elliptically polarized electromagnetic wave whose frequency lies in the nominal transparency region of the crystal. This field is caused by the dynamic Zeeman effect induced by circularly polarized light. It results in the splitting of electron Zeeman sublevels $\propto P_c I$, where P_c is the circular degree of radiation and I is the light intensity (Cohen-Tannoudji *et al.*, 2004). In this way, the spin-noise spectroscopy can be used to study not only the effective fields caused by the magnetization of different spin systems, but also the effective fields related to renormalizations of the charge carrier energy dispersion.

10.7 Optically detected nuclear spin noise

Results of (Ryzhov *et al.*, 2015, 2016) summarized in the previous Section 10.6 demonstrate the possibility to study experimentally the dynamics of a nonequilibrium nuclear spin subsystem by means of the spin-noise spectroscopy technique. It is also natural to analyze the possibility to observe nuclear spin fluctuations directly by detecting the Faraday or Kerr rotation noise in the configuration similar to the one depicted in Figure 10.1. In accordance with Section 6.6 nuclear magnetization can be converted to Faraday rotation due to the hyperfine interaction with electron spins; hence, optical manifestations of nuclear spin noise are also expected.

The nuclear spin fluctuations have indeed been observed in the Faraday rotation noise spectrum in bulk GaAs sample with low n-type doping where electrons are bound to donors at low temperatures (Berski *et al.*, 2015). Figure 10.10(a) shows the very low frequency Faraday rotation noise spectrum acquired in the range from 1 to 70 kHz at the same sample where the electron spin noise has been analyzed previously and shown in Figure 10.7; see Section 10.4 for details. In this frequency range the electron contribution to the noise of the Faraday rotation is essentially constant as shown by the dashed horizontal line. At a transverse field of $B = 3.75$ mT, the measured noise spectrum reveals on top of the electron spin noise a clearly resolved fine structure consisting of three additional and very narrow peaks at finite frequencies. These peaks shift linearly for $B \gtrsim 1$ mT as shown in Figure 10.10(b). By the analysis of the peak positions as functions of magnetic fields, their origin can be unambiguously related to the host lattice isotopes ^{75}As, ^{69}Ga, and ^{71}Ga. The ratio of the nuclear to the electron spin noise power is about $\xi = 1.2 \times 10^{-3}$ as found in (Berski *et al.*, 2015) by the corresponding fitting procedure.

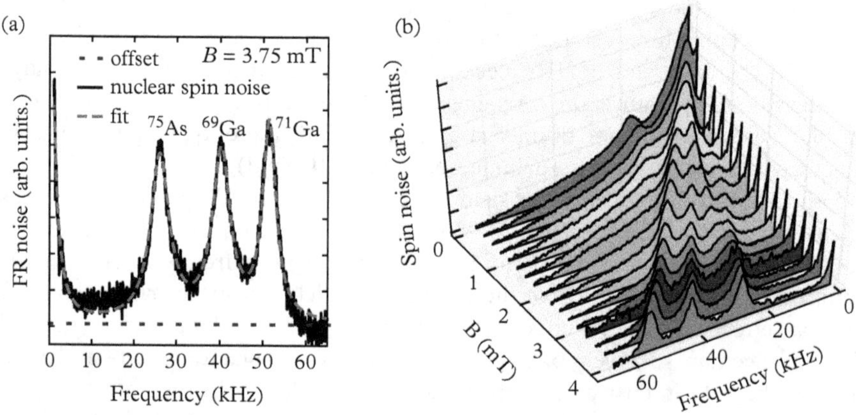

Figure 10.10 *(a) Faraday rotation (FR) noise spectrum (solid line) in the low-frequency range measured at $B = 3.75$ mT on an n-doped bulk GaAs sample. Horizontal dotted line denotes the offset related to $\omega = 0$ contribution to the electron spin noise, see Figure 10.7. Contributions from the individual host lattice isotopes are clearly resolved and labeled accordingly. (b) Nuclear spin-noise spectra as a function of the transverse magnetic field. $T = 4.2$ K. From (Berski et al., 2015).*

The observation of the nuclear spin fluctuations in the Faraday rotation noise spectrum can be readily understood using the same arguments as presented in Section 6.6 where the Faraday rotation induced by the spin polarized host lattice nuclei was discussed. Indeed, the nuclear spin fluctuations $\delta \boldsymbol{I}$ can be largely considered as static from the viewpoint of the electronic spin system. These fluctuations, as already discussed in detail, produce fluctuating effective magnetic field, which leads to the splitting of the electron and hole spin states. As shown in Section 6.6, there are two contributions to nuclei-induced Faraday rotation: (i) the Overhauser field induced splitting of the donor-bound exciton transition line, which is temperature independent, and (ii) a state-filling effect caused by the electron spin polarization in the Overhauser field, which depends on temperature. Due to temporal fluctuations of nuclear spins the splitting slowly varies with time producing the noise of the Faraday, Kerr, or ellipticity effects. The calculations presented in (Berski *et al.*, 2015) demonstrate that the nuclear fluctuation-induced splitting dominates the Faraday rotation noise for quasi-resonant detection on an inhomogeneously broadened donor-bound exciton transition. It is in contrast with the case of strongly detuned probe where the state-filling effect dominates. According to the estimates, the ratio ξ of the nuclei and electron spin-noise contributions is given by $\xi \sim (\Delta_N/\Gamma)^2$, where $\Delta_N \sim 0.1 \ldots 0.3$ μeV is the nuclei-fluctuation induced energetic fluctuation of the bound-exciton line position and and Γ is the homogeneous width of the resonance. Theoretical estimates of $\xi \sim 10^{-3} \ldots 10^{-4}$ (Berski *et al.*, 2015) are in rather good agreement with the experimentally observed value. The splitting-induced origin of the nuclear fluctuations effect on the Faraday rotation noise is also confirmed in (Berski *et al.*, 2015) by temperature-dependent measurements showing that ξ is temperature independent in the studied range between $T = 3.2$ K and 7 K.

In order to calculate the nuclear spin fluctuations microscopically we note that under conditions of the experiment the nuclear spin system remains practically unperturbed. Hence, one can use the fluctuation-dissipation theorem to relate the nuclear spin noise and the nuclear spin susceptibility discussed in Chapter 3. The simplest possible situation is realized in a transverse magnetic field B of several mT, where the absolute value of the nuclear spin precession frequency in the external field exceeds by far the longitudinal, $1/T_{1N}$, and transversal, $1/T_{2N}$, nuclear spin relaxation rates. Therefore, for each isotope ($i = {}^{69}$Ga, ^{71}Ga, and ^{75}As) we introduce the spin-noise spectrum in analogy with Eq. (10.7a)

$$(\delta I^2_{\alpha,i})_\omega = \int_{-\infty}^{\infty} \langle \delta I_{\alpha,i}(t+\tau)\delta I_{\alpha,i}(\tau)\rangle \exp(\mathrm{i}\omega\tau)d\tau. \tag{10.53}$$

The calculation yields for the fluctuating z component of the nuclear spin

$$(\delta I^2_{z,i})_\omega = \frac{5\pi}{4}\left[\Delta_{T_{2N}}(\omega-\omega_{0,i}) + \Delta_{T_{2N}}(\omega+\omega_{0,i})\right], \tag{10.54}$$

where $\omega_{0,i} = -g_n\mu_N B/\hbar = -\gamma_I B$ is the nuclear spin precession frequency in the external field (we recall that g_n and μ_N are the nuclear g-factor and magneton, respectively, γ_I is the nuclear gyromagnetic ratio), and the broadened δ-function $\Delta_{T_2}(\omega)$, Eq. (3.13), describes the simplest possible Lorentzian shape of the nuclear spin-noise spectrum and nuclear spin resonance under condition of $\omega_{0,i}T_2 \gg 1$.

The combination of functions (10.54) for $i = {}^{69}$Ga, ^{71}Ga, and ^{75}As adequately describes the part of the nuclear spin-noise spectrum at positive finite frequencies, see Figure 10.10. However, the model outlined here is insufficient to reproduce: (i) the zero-frequency peak clearly seen in the nuclear spin-noise spectrum, (ii) the complex fine structure of the peaks observed at $B \lesssim 1$ mT, which is clearly visible in Figure 10.10(b), and (iii) the ratio of contributions of different isotopes in the observed signals. These features, which definitely require further studies, are most probably related to the influence of the local fields, that is of the dipole–dipole interaction between the nuclei and of the nuclear quadrupole splittings that may be particularly pronounced for the ^{75}As isotopes in the vicinity of an Si-donor (Berski *et al.*, 2015).

The detection of nuclear spontaneous spin fluctuations by optical spin-noise spectroscopy provides a novel method that enables measurements of nuclear spin dynamics without exciting a nuclear spin polarization, applying strong external magnetic fields to split the nuclear spin sublevels, or using radio frequency pulses as in nuclear magnetic resonance and optically detected magnetic resonance experiments [alternatively, the nuclear spin noise can be detected via electric current fluctuations measurements by superconducting quantum interference device (Sleator *et al.*, 1985; Müller and Jerschow, 2006)]. The suggested optical technique is particularly useful to address nuclear spin dynamics at otherwise inaccessible low magnetic fields.

The spin-noise spectroscopy technique, as a whole, has became now a mature research field where not only the equilibrium but also nonequilibrium spin fluctuations are studied. Due to the correlation character of the measurements, this method provides unique informative abilities (Glazov and Zapasskii, 2015), usually inherent to nonlinear optical methods. More information can be found from studies of higher-order spin correlators (Liu *et al.*, 2010; Starosielec *et al.*, 2010; Liu *et al.*, 2010; Bechtold *et al.*, 2016; Li and Sinitsyn, 2016; Fröhling and Anders, 2017); a field of research to be explored in the future both experimentally and theoretically.

11

Strong Coupling of Electron and Nuclear Spins: Outlook and Prospects

In this concluding chapter, we provide a brief outlook and discuss some prospective systems where novel aspects of the electron and nuclear spin dynamics may appear. In the book we have mainly focused on the systems with localized electrons where the spin–orbit interaction is, to a large extent, quenched and the hyperfine coupling with the host lattice nuclei becomes very important. The stronger the electron localization is, the stronger the effect of the nuclei on electron spin dynamics, and, vice versa, the stronger the nuclei are affected by the electron. It is natural to ask a question about the novel phenomena that may arise in an extreme regime where an electron interacts with a small number of nuclei, ultimately with just one nucleus.

The hyperfine interaction between the spin of electron $\hat{\boldsymbol{s}}$ and that of the single nucleus $\hat{\boldsymbol{I}}$ is described by the Hamiltonian [cf. Eqs. (2.22), (8.39)]

$$\mathcal{H}_{hf} = A(\hat{\boldsymbol{I}} \cdot \hat{\boldsymbol{s}}), \tag{11.1}$$

where A is a constant. It gives rise to the splitting of energy states into two multiplets characterized by a certain value of the total spin $\hat{\boldsymbol{F}} = \hat{\boldsymbol{s}} + \hat{\boldsymbol{I}}, F = I \pm 1/2$, as sketched in Figure 11.1. The energies of multiplets are given by (Landau and Lifshitz, 1977)

$$E_{F,I} = \frac{A}{2}[F(F+1) - I(I+1) - s(s+1)], \tag{11.2}$$

in agreement with Eq. (7.29). The splitting between the multiplets can be large enough to exceed both electron and nuclear spin relaxation rates in the system. As a result, a complex of a strongly coupled electron and nuclear spin is formed, where the hyperfine interaction provides a coherent angular momentum transfer back and forth between the nucleus and electron.

Electron and Nuclear Spin Dynamics in Semiconductor Nanostructures. M. M. Glazov, Oxford University Press (2018). © M. M. Glazov. DOI: 10.1093/oso/9780198807308.001.0001

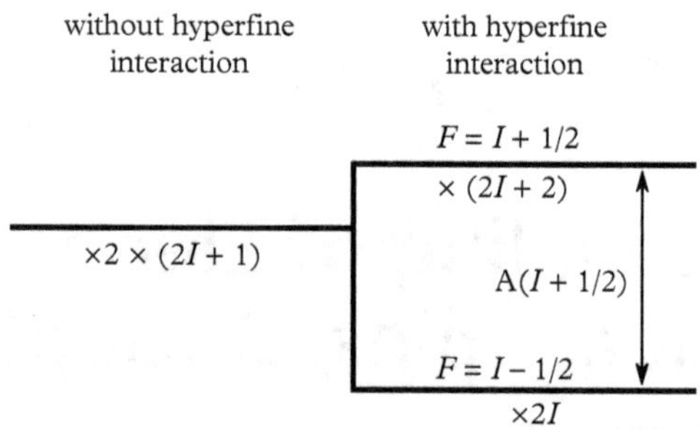

Figure 11.1 *Scheme of energy levels of a complex "electron+nucleus" at a zero magnetic field. Level degeneracies are marked under each state.*

Qualitatively, the strong coupling can be understood considering, as an example, the case of a $I = 1/2$ nucleus. The hyperfine Hamiltonian (11.1) is thus fully analogous to the Hamiltonian describing the exchange interaction of electrons, Eq. (2.18). Denoting the electron spin states with $s_z = +1/2$ and $-1/2$ as $|\uparrow\rangle$ and $|\downarrow\rangle$, respectively, and the corresponding nuclear states as $|\Uparrow\rangle$ and $|\Downarrow\rangle$, one can readily check that the states $|\uparrow\rangle|\Uparrow\rangle$ and $|\downarrow\rangle|\Downarrow\rangle$ are not intermixed by the hyperfine interaction, while the states $|\uparrow\rangle|\Downarrow\rangle$ and $|\downarrow\rangle|\Uparrow\rangle$ are intermixed and form symmetric and antisymmetric combinations split by the energy A, by analogy to the singlet-triplet splitting of electron levels caused by the exchange interaction, see Eqs. (2.16) and (2.17). Quantum beats between the states

$$|\uparrow\rangle|\Downarrow\rangle \quad \longleftrightarrow \quad |\downarrow\rangle|\Uparrow\rangle,$$

give rise to the temporal oscillations of the electron and nuclear spin polarization. In particular, if at $t = 0$ the electron spin was initialized to the spin-up state and the nucleus was unpolarized then

$$S_z(t) = \frac{1}{4}\left[e^{-\gamma_1 t} + e^{-\gamma_2 t}\cos(At/\hbar)\right]. \tag{11.3}$$

Here, γ_1 and γ_2 are the damping rates associated with the spin relaxation and decoherence processes (unrelated to the hyperfine interaction (11.1)), which give rise to the damping of both the beats and mean value of electron spin component. The period of oscillations is determined by the hyperfine coupling constant, which allows direct measurement of the hyperfine coupling strength in the strong coupling regime. In absence of damping related to "external" agents ($\gamma_1 = \gamma_2 = 0$), in contrast to the regime where an electron interacts with a large number of nuclei, here the spin beats do not damp and the time-average value of the spin-z component is $1/4$ rather than $1/6$, see Eq. (7.32) and Chapter 7 for details.

Theoretically, the processes of optical orientation of electron spins and nuclear spin polarization arising in the strongly interacting complex of the "electron+nucleus" have been studied in detail in (Dyakonov and Perel', 1973). This model has been recently extended in (Ivchenko *et al.*, 2015) to account for the specifics of the recombination processes via deep centers in semiconductors, but this regime is still insufficiently studied in semiconductor nanosystems.

In order to realize such a strong coupling regime one needs to have a very efficient localization of electron, that is by deep impurity center, or to ensure isotopic purification of the host lattice, so that under the orbit of localized charge carrier only one nucleus efficiently interacts with the electron. Of course, both approaches can be combined to obtain the best result.

By now, some experimental and theoretical progress has been achieved in spin dynamics studies in such systems. We have already briefly discussed in Section 8.5 a prominent system of bulk isotopically purified Si doped with single (i.e., well separated) donor impurities with a nonzero nuclear spin. Among prominent examples are ^{31}P (Kane, 1998; Schofield *et al.*, 2003; Morley *et al.*, 2008; Morton *et al.*, 2008) and ^{209}Bi (Morley *et al.*, 2010; Bienfait *et al.*, 2016) donors with nuclear spins of $I = 1/2$ and $I = 9/2$, respectively. In such systems, the spin dynamics of the strongly coupled electron–nucleus complex can be addressed by means of a magnetic resonance technique.

Isotopic purification is also possible for II-VI semiconductors, see, for example (Lauck and Schönherr, 1999). The structures with ZnSe doped with fluorine donors have started to attract attention recently due to observation of single photon emission from such structures (Sanaka *et al.*, 2012) and long electron spin relaxation times limited by the hyperfine interaction (Greilich *et al.*, 2012; Heisterkamp *et al.*, 2015*b*). A novel experimental method termed as "spin inertia" has been suggested and realized in (Heisterkamp *et al.*, 2015*b*) to address particularly slow spin dynamics of donor bound electrons in ZnSe:F. Still, there is a long way toward the realization of the strong coupling between the electron and nucleus spins in these materials. The signatures of the strong coupling have been observed in ZnO with ^{115}In acting as a donor (Buß *et al.*, 2016). Here the donor-bound electron spin interacts mainly with $I = 9/2$ nuclear magnetic moment of indium. By means of the time-resolved pump-probe Kerr rotation experiments and resonant spin amplification effect outlined in Chapter 6 and Section 9.2, the authors of (Buß *et al.*, 2016) have observed signatures of the donor-bound electron-nuclear spin beats.

Another interesting system is based on the interstitial Ga^{2+} defects in the dilute nitride GaAsN semiconductor. These defects form relatively deep and well isolated centers where an unpaired electron can be bound (Kalevich *et al.*, 2005; Wang *et al.*, 2009). Interestingly, these centers can be efficiently accessible not only in standard magnetic resonance spectroscopy, but also optically and electrically owing to the giant spin dependent photoconductivity effect discovered in these structures (Kunold *et al.*, 2011). Optical access to the electron and nuclear spin ($I = 3/2$) dynamics of the Ga^{2+} defect is provided through the spin-dependent

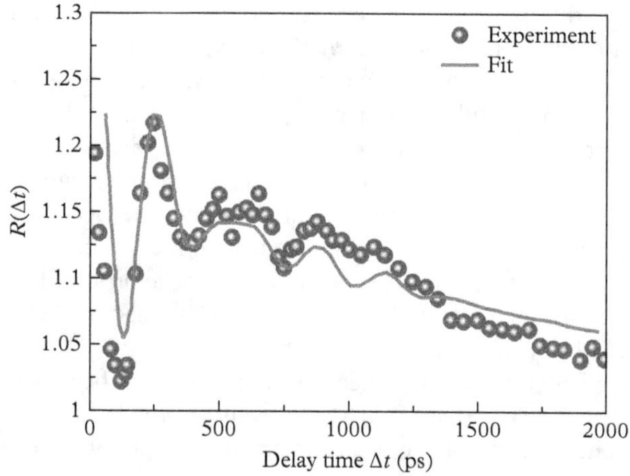

Figure 11.2 *The spin dependent recombination signal signal, that is, the ratio of the probe pulse photoluminescence intensity under a circularly to linearly polarized pump pulse measured as a function of the delay time. The time origin is taken at the arrival of the pump pulse. The solid line is a fit to the data. Data are reproduced from (Azaizia et al., 2017).*

recombination (Weisbuch and Lampel, 1974) via the defect. The pump-probe protocol realized in (Azaizia *et al.*, 2017) has made it possible to observe the electron-nuclear spin beats, Eq. (11.3). The spin dependent recombination signal, $R(\Delta t)$, measured in (Azaizia *et al.*, 2017) as a function of the pump-probe delay time Δt is shown in Figure 11.2 by the points. The solid line shows the fit according to

$$R(\Delta t) = 1 + \alpha \exp(-\gamma_1 \Delta t) + \beta \exp(-\gamma_2 \Delta t)[f_1 \cos(\Omega_1 \Delta t) + f_2 \cos(\Omega_2 \Delta t)],$$

derived in (Azaizia *et al.*, 2017) on the basis of the microscopic model. Apart from the constant background, it includes contributions similar to the ones in Eq. (11.3) due to two stable isotopes ^{69}Ga and ^{71}Ga, f_1 and f_2 are their relative abundances of about 60 and 40%, respectively. The beat frequencies are determined by the hyperfine coupling constants and their ratio $\Omega_2/\Omega_1 = 1.27$ is in very good agreement with the model estimations based on the ratio of corresponding magnetic moments, Eq. (4.8) and Tables 2.1 and 4.1. The authors have been able to extract the hyperfine coupling constants for the defects based on ^{69}Ga and ^{71}Ga placed in different nonequivalent interstitial positions as well as to evaluate the spin relaxation times for the "electron+nucleus" complex.

In this regard, it is also interesting to study the intermediate – mesoscopic – case where an electron interacts with a small number of nuclei, as sketched in Figure 11.3(a), and to address the transition between the simple picture of the spin beats, Eq. (11.3), and the involved process of the spin decoherence within

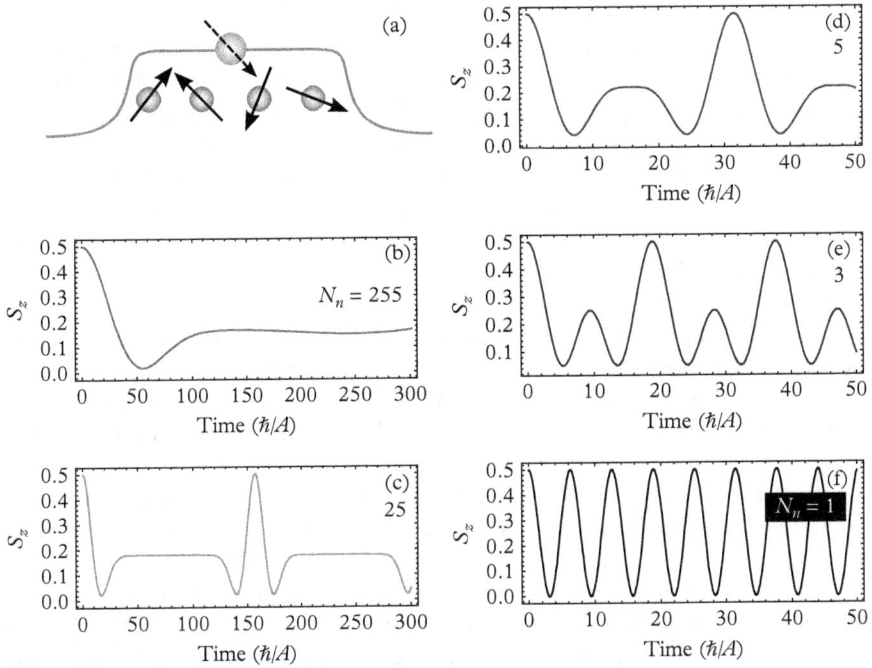

Figure 11.3 *(a) Schematics of the electron interacting with several ($N_n = 4$) nuclei. (b–f) Electron spin dynamics in the complex containing N_n nuclei with spin $I = 1/2$ calculated after Eq. (7.31), Section 7.3.*

the central spin model, see Chapter 7. To illustrate novel features of the spin dynamics in this practically unexplored regime, Figure 11.3(b–f) presents the calculated traces of spin dynamics of an electron interacting with several nuclei. These calculations have been performed in the box model following the approach presented in Section 7.3. It is assumed that the electron is equally distributed between the nucleus, Figure 11.3(a), hence the hyperfine interaction is written in the form

$$\mathcal{H} = \frac{A}{N_n} \sum_{n=1}^{N_n} (\hat{\boldsymbol{I}} \cdot \hat{\boldsymbol{s}}). \tag{11.4}$$

Here, the factor N_n in the denominator accounts for reduction of the hyperfine interaction proportionally to the probability to find electron at a given nucleus, see Eq. (2.22). For a substantial N_n, Figure 11.3(b), the electron spin dynamics closely agrees with the results of the semiclassical approach, Eqs. (7.14), discussed in detail in Chapter 7. The difference becomes visible for a smaller number of nuclei; for example $N_n = 25$ as illustrated in Figure 11.3(c) where the revivals of the electron spin polarization are observed. This is due to the fact that, for a finite number of involved nuclei, the energy spectrum of the system is discrete and all eigenfrequencies are commensurable, see Eq. (7.30). As a result, the electron spin dynamics is periodic with the period of $2\pi\hbar N_n/A$. The decrease in the number of nuclei results in the decrease of the spin dynamics period as illustrated in Figure 11.3(d–f) and transition to the limit of sinusoidal spin beats for $N_n = 1$, Eq. (11.3) ($\gamma_1 = \gamma_2 = 0$), as shown in Figure 11.3(f).

So far, experimentally, only two limits of $N_n \to \infty$, see Section 7.4, and $N_n = 1$ (Buß *et al.*, 2016; Azaizia *et al.*, 2017) have been addressed in some detail. Further experimental studies of the situation where an electron interacts with a single or several nuclei will reveal, most certainly, many peculiarities of the electron spin and nuclear dynamics well beyond simple predictions in Figure 11.3. Accordingly, further theoretical analysis accounting for specifics of realistic systems, such as, for example, quadrupole splittings of the nuclei and finite electron correlation time at a given localization center, is of high importance. These results can be also of importance to study the dynamics and fluctuations of the magnetic moments of paramagnetic ions in diluted magnetic semiconductors (Cronenberger *et al.*, 2015).

Here, it would also be very interesting to address the transition from the regular, that is, time-periodic spin dynamics, as presented in Figure 11.3, to the non- or quasi-periodic dynamics realized with an account for dispersion of the hyperfine interaction constants, quadrupole interactions, or magnetic field. The possibilities to study classical or quantum chaos (Gutzwiller, 1990; Stöckmann, 2006) in spin dynamics in the regime of strong interaction between the electron and nuclear spins could also form an interesting field of further studies where only preliminary steps have been made so far (Georgeot and Shepelyansky, 2000; Mourik *et al.*, 2017). The development and experimental realization of protocols to prepare "entangled" spin states of electron and nuclei – such as,

$$\frac{|\uparrow\rangle|\Downarrow\rangle \pm |\downarrow\rangle|\Uparrow\rangle}{\sqrt{2}}$$

– and manipulate and read out these states remains a challenging task.

While discussing the spin properties of defects in semiconductors it is impossible to avoid mentioning the nitrogen-vacancy center in diamond (NV center), which is formed from the nitrogen atom substituted for a carbon atom and an adjacent vacancy in the host lattice (Fuchs *et al.*, 2009; Doherty *et al.*, 2013; Childress *et al.*, 2014). This system is widely studied theoretically and experimentally due to efficient optical access to the ground $S = 1$ two-electron triplet state of the NV center. Due to mostly spin-less environment of ^{12}C atoms for the NV center, preferable conditions in which to observe the strong interaction between the $S = 1$ two-electron state and a single ^{13}C nucleus spin in the vicinity of the NV center occur. The first experimental steps in this direction have already been made (Jelezko *et al.*, 2004). A similar and very promising system is formed by silicon vacancies in SiC, which exhibit half-integer spin $S = 3/2$ with an intricate fine structure, opening up possibilities for all-optical magnetometry and thermometry applications (Simin *et al.*, 2016; Tarasenko *et al.*, 2017). Theoretical and experimental investigations of the hyperfine interaction effects for such deep defects have also started to attract increased interest today (Falk *et al.*, 2015; Ivády *et al.*, 2015).

Another challenge related to the strong interaction of the electron and nuclear spins is the formation of ordered states of the host lattice nuclei in bulk semiconductors or nanosystems. We have discussed in detail in Chapter 5 the effect of dynamical nuclear polarization related to the transfer of electron spin polarization to the nuclear spin system owing to the hyperfine interaction. It has been demonstrated that one is able to cool the nuclear spin system down to very low effective temperatures (as compared with the lattice temperature), see Section 5.2, such that even a tiny magnetic field can produce substantial orientation of nuclei. The question is whether it is possible to cool down the nuclei in such a way that their spin polarization is produced spontaneously; that is, in the absence of any external magnetic field. It turns out that such a spontaneous magnetization of nuclei due to magnetic (dipole–dipole) interactions between the spins requires such low nuclear spin temperatures far beyond experimental reach. The hyperfine interaction with electron spin allows, in principle, the ordering of the nuclear spins, that is, via the dynamic nuclear self-polarization (D'yakonov and Perel', 1972) already addressed at the end of Section 5.3. In this effect, the nuclei get polarized due to constant spin flux between the nuclear spin system and the electron spin system, which is kept depolarized.

Figure 11.4 illustrates another scenario for the ordering of the nuclear spins due to the hyperfine coupling with electrons, namely, the formation of nuclear spin polaron as predicted by I.A. Merkulov in (Merkulov, 1998). As follows from Eq. (11.4), the energy of the disordered nuclear spin state interacting with the

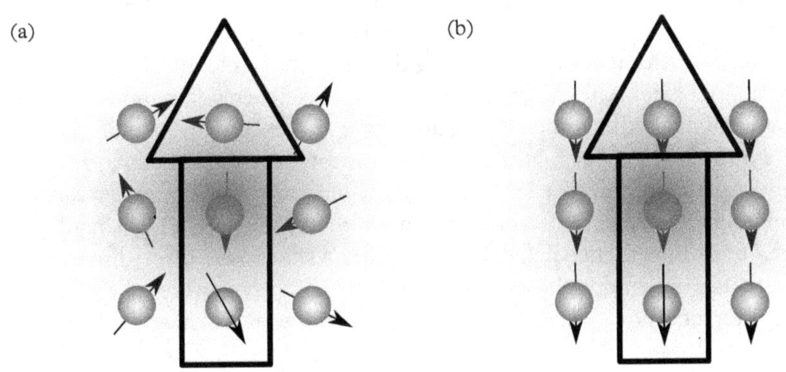

Figure 11.4 *Sketch of the nuclear polaron formation in semiconductor. (a) Spin-polarized electron (large arrow) interacts with disordered nuclear spin system. (b) Nuclei with ordered spins interact with the electron. This state has lower energy.*

spin-polarized electron, Figure 11.4(a), is larger than the energy of the ordered state interacting with the electron, Figure 11.4(b). This is in close analogy with magnetic polaron formation in diluted magnetic semiconductors where the spins of paramagnetic centers align due to the exchange interaction with the spin-polarized charge carrier or exciton (Wolff, 1988; Merkulov *et al.*, 1995). The critical nuclear spin temperature Θ_c where the polaron formation is possible can be easily estimated from the following simple argument: Let us assume that nuclei are polarized in the Knight field exerted by the electron onto the nuclear spins, while the electron is polarized in the Overhauser field created by the same nuclei:

$$\langle I_z \rangle \sim \frac{AS_z}{N_n k_B \Theta_c}, \quad S_z \sim \frac{A\langle I_z \rangle}{k_B T}.$$

Here, T is the lattice temperature, $\langle I_z \rangle$ is the average nuclear spin, and S_z is the average electron spin. This system admits a nontrivial solution $\langle I_z \rangle, S_z \neq 0$ for

$$\Theta_c \sim \frac{A^2}{N_n k_B T}. \tag{11.5}$$

Hence, if nuclei are cooled down below Θ_c their spins become ordered due to polaron formation. For a bulk GaAs semiconductor with donor-bound electrons $\Theta_c \sim 10^{-7}$ K at the lattice temperature in the order of 4 K. Once a polaron is formed, the nuclear spin system behaves like a microscopic magnetic moment, which efficiently interacts with electron spin. Although some indications of polaron formation in quantum dot structures have been demonstrated (Oulton *et al.*, 2007) and a detailed theory of nuclear polaron formation that goes beyond the mean-field approach has been recently developed (Scalbert, 2017), the unambiguous observation of nuclear spin polaron is still absent. Demonstration of nuclear spin polaron may open up a lot of fascinating areas of study of strongly correlated nuclear spin states similar to those realized in magnetic materials.

Concluding this chapter and book, and speaking about prospects and outlook in more general terms, it is worth stressing that modern semiconductor physics is driven, to a large extent, by the progress of material science and nanotechnology. The advancements in these interrelated areas open up exciting possibilities linked with the synthesis of novel materials, formation of nano- and heterostructures with a given geometry, symmetry, and energy spectrum of the charge carriers. New semiconductor systems like bulk and two-dimensional topological insulators, monolayer transition metal dichalcogenides, perovskites and three-dimensional Weyl semimetals are among the examples of material systems where the effects of hyperfine interaction have been scarcely studied and await detailed experimental and theoretical analysis. Novel and completely unexplored avenues for studies of electron and nuclear spin dynamics are offered by hybrid systems where inorganic conventional semiconductors are brought in to contact with organic materials.

Appendix A
Irreducible Representations of Several Relevant Point Groups

In this appendix, we present tables with irreducible representations of the point symmetry groups T_d, D_{2d}, C_{2v}, and C_{3v} relevant for most nanosystems discussed in this book. Further information on the point groups can be found in (Koster *et al.*, 1963) and at the website "32 point groups" (https://goo.gl/q8VZrB).

We use the following notations (general for all discussed point groups): x, y, z are the Cartesian coordinates, $\boldsymbol{r} = (x, y, z)$ is the vector, $\boldsymbol{L} = (L_x, L_y, L_z)$ is the pseudovector, $|M, M_z\rangle$ is the function of angular momentum M with the z-component equal to M_z, see Eq. (2.56). A particular coordinate system is specified for each studied point group.

A.1 T_d point group

The T_d point symmetry group is relevant for bulk zinc-blende lattice semiconductors such as GaAs or CdTe. Axes $x \parallel [100]$, $y \parallel [010]$, $z \parallel [001]$ are the cubic axes. We denote as $\boldsymbol{\sigma} = (\sigma_x, \sigma_y, \sigma_z)$ the pseudovector composed of Pauli matrices acting in the electron (spin-$1/2$, Γ_6) spin space, and $\boldsymbol{J} = (\mathcal{J}_x, \mathcal{J}_y, \mathcal{J}_z)$ the pseudovector composed of angular momentum $3/2$ basic matrices acting in the space of four Γ_8 basic functions. Table A.1 contains information on the irreducible representations of the point group.

A.2 D_{2d} point group

The D_{2d} point symmetry group is relevant for quantum well structures with symmetric heteropotential grown along a $z \parallel [001]$ cubic axis. Such point symmetry can also be realized for quantum dots grown on the (001) plane with symmetric confinement potential and a square or cylindrical base. Here, in addition to the cubic axes x, y, z, we also use another frame $x_1 \parallel [1\bar{1}0]$, $y_1 \parallel [110]$ and $z_1 \parallel [001]$. Table A.2 contains information on the irreducible representations of the point group.

A.3 C_{2v} point group

The C_{2v} point symmetry group is relevant for quantum well structures with asymmetric heteropotential grown along the $z \parallel [001]$ cubic axis as well as for asymmetric quantum dot structures grown on the (001) plane. We use the frame of main axes in C_{2v} point group:

Table A.1 *Irreducible representations of the T_d point group and examples of basic functions.*

Representation	Exemplary basic functions
A_1 or Γ_1	$1; x^2+y^2+z^2; xyz$
(scalar)	
A_2 or Γ_2	$x^4(y^2-z^2)+y^4(z^2-x^2)+z^4(x^2-y^2);$
(pseudoscalar)	$L_xL_yL_z$
E or Γ_{12}	$[\sqrt{3}(x^2-y^2), 2z^2-x^2-y^2]$
F_1 or Γ_{25}	$[x(y^2-z^2), y(z^2-x^2), z(x^2-y^2)];$
(pseudovector)	$(L_x, L_y, L_z); (\sigma_x, \sigma_x, \sigma_z)$
F_2 or Γ_{15}	$(x, y, z); (yz, zx, xy)$
(vector)	
Γ_6	$(\vert 1/2, -1/2\rangle,\ \vert 1/2, 1/2\rangle)$
$\Gamma_7 = \Gamma_2 \otimes \Gamma_6$	$(L_xL_yL_z\vert 1/2, -1/2\rangle,\ L_xL_yL_z\vert 1/2, 1/2\rangle)$
Γ_8	$(\vert 3/2, -3/2\rangle, \vert 3/2, -1/2\rangle, \vert 3/2, 1/2\rangle, \vert 3/2, 3/2\rangle)$

Table A.2 *Irreducible representations of the D_{2d} point group and examples of basic functions.*

Representation	Exemplary basic functions
A_1 or Γ_1	$1; x^2+y^2; z^2;$
(scalar)	$xyz = -\frac{1}{2}(x_1^2 - y_1^2)z$
A_2 or Γ_2 (z-pseudovector)	$z(x^2-y^2) = 2x_1y_1z_1; \sigma_z$
B_1 or Γ_3	$x^2-y^2 = 2x_1y_1$
B_2 (z-vector) or Γ_4	$z; xy = -\frac{1}{2}(x_1^2 - y_1^2)$
E or Γ_5	$(x, y); (\sigma_x, -\sigma_y); (yz, zx);$
	$(x_1, y_1); (\sigma_{y_1}, \sigma_{x_1}); (-x_1z, y_1z);$
	$(L_x, -L_y)$
Γ_6	$(\vert 1/2, -1/2\rangle,\ \vert 1/2, 1/2\rangle)$
$\Gamma_7 = \Gamma_3 \otimes \Gamma_6$	$[(x^2-y^2)\vert 1/2, -1/2\rangle,\ (x^2-y^2)\vert 1/2, 1/2\rangle]$

Table A.3 *Irreducible representations of the C_{2v} point group and examples of basic functions.*

Representation	Exemplary basic functions
A^+ or Γ_1	$1; x_1^2; y_1^2; z_1$
A^- or Γ_3	$x_1y_1; L_{z_1}; \sigma_{z_1}$
B^+ or Γ_4	$y_1; L_{x_1}; \sigma_{x_1}$
B^- or Γ_2	$x_1; L_{y_1}; \sigma_{y_1}$
Γ_5	$(\vert 1/2,-1/2\rangle, \vert 1/2,1/2\rangle)$

Table A.4 *Irreducible representations of the C_{3v} point group and examples of basic functions. Here, the momentum quantization axis is z_2, that is, the states $\vert M,M_{z'}\rangle$ possess definite angular momentum projection onto the [111] axis.*

Representation	Exemplary basic functions
A_1 or Γ_1	$1; x_2^2+y_2^2; z_2^2; z_2; \sigma_{y_2}^{(h)}$
A_2 or Γ_2	$\sigma_{z_2}^{(e)}; \sigma_{z_2}^{(h)}; \sigma_{x_2}^{(h)}$
E or Γ_3	$(x_2,y_2); (x_2z_2,y_2z_2)$;
	$(L_{y_2},-L_{x_2}); (\sigma_{y_2}^{(e)},-\sigma_{x_2}^{(e)})$
Γ_4	$(\vert 1/2,-1/2\rangle, \vert 1/2,1/2\rangle)$
Γ_5	$2^{-1/2}(\vert 3/2,-3/2\rangle - \mathrm{i}\vert 3/2,3/2\rangle)$
Γ_6	$2^{-1/2}(\vert 3/2,3/2\rangle - \mathrm{i}\vert 3/2,-3/2\rangle)$

$x_1 \parallel [1\bar{1}0]$, $y_1 \parallel [110]$ and $z_1 \parallel [001]$. Table A.3 contains information on the irreducible representations of the point group.

A.4 C_{3v} point group

The C_{3v} point group is relevant for quantum well and quantum dot structures with the growth axis $z_2 \parallel [111]$. We use the frame of axes $x_2 \parallel [11\bar{2}]$, $y_2 \parallel [\bar{1}10]$ and $z_2 \parallel [111]$. Note that one of the vertical reflection planes σ_v contains the axes x_2 and z_2. We denote $\boldsymbol{\sigma}^{(e)} = \left(\sigma_{x_2}^{(e)}, \sigma_{y_2}^{(e)}, \sigma_{z_2}^{(e)}\right)$ as the pseudovector composed of Pauli matrices acting in the electron (spin-1/2, Γ_4) basis and as $\boldsymbol{\sigma}^{(h)} = \left(\sigma_{x_2}^{(h)}, \sigma_{y_2}^{(h)}, \sigma_{z_2}^{(h)}\right)$ the set of Pauli matrices acting in the basis of two states $\vert 3/2,3/2\rangle$ and $\vert 3/2,-3/2\rangle$ being the heavy-hole Bloch basis functions (reducible $\Gamma_5 \oplus \Gamma_6$ representation), see Table A.4.

Appendix B
Outline of the $k \cdot p$ Method

The $\boldsymbol{k} \cdot \boldsymbol{p}$-perturbation theory is one of most efficient methods of description of the band structure in semiconductors and semiconductor nanosystems. It allows one to account for the effects of external fields as well. Here, we outline the derivation of the $\boldsymbol{k} \cdot \boldsymbol{p}$ method in the simplest possible case where the electron states in the vicinity of the Γ-point (Brillouin zone center, $\boldsymbol{k} = 0$) are of interest.

We consider a bulk semiconductor and disregard for a moment the spin-orbit interaction. The single-electron states are described within the framework of the Schrödinger equation with the Hamiltonian [cf. (2.2)]

$$\mathcal{H}_0 = \frac{\hat{\boldsymbol{p}}^2}{2m_0} + V(\boldsymbol{r}), \tag{B.1}$$

where $\hat{\boldsymbol{p}} = -\mathrm{i}\hbar\partial/\partial\boldsymbol{r}$ is the electron momentum operator, m_0 is the free electron mass and $V(\boldsymbol{r}) = V(\boldsymbol{r}+\boldsymbol{a})$ is the periodic potential with $\boldsymbol{a}$ being a translation vector. In accordance with the Bloch theorem the electron wavefunction can be sought in the form of Eq. (2.1):

$$\Psi_{\boldsymbol{k},n}(\boldsymbol{r}) = \exp(\mathrm{i}\boldsymbol{k}\cdot\boldsymbol{r})\,U_{\boldsymbol{k},n}(\boldsymbol{r}), \tag{B.2}$$

where $U_{\boldsymbol{k},n}(\boldsymbol{r}) = U_{\boldsymbol{k},n}(\boldsymbol{r}+\boldsymbol{a})$ is the periodic function (Bloch amplitude) and $\boldsymbol{k}$ is the electron quasi-wavevector ($\hbar\boldsymbol{k}$ is the quasi-momentum of the electron), the subscript n enumerates energy bands and includes the spin index as well. Substituting the wavefunction in the form of Eq. (B.2) into Eq. (B.1) we obtain the following equation for the periodic amplitude $U_{\boldsymbol{k},n}(\boldsymbol{r})$:

$$\left[\mathcal{H}_0 + \frac{\hbar}{m_0}(\boldsymbol{k}\cdot\hat{\boldsymbol{p}}) + \frac{\hbar^2 k^2}{2m_0}\right] U_{\boldsymbol{k},n}(\boldsymbol{r}) = E_{\boldsymbol{k},n} U_{\boldsymbol{k},n}(\boldsymbol{r}), \tag{B.3}$$

where $E_{\boldsymbol{k},n}$ is the electron dispersion in the band n. To describe the electron states in the vicinity of the Γ-point the term

$$\frac{\hbar}{m_0}(\boldsymbol{k}\cdot\hat{\boldsymbol{p}}) = -\frac{\mathrm{i}\hbar}{m_0}(\boldsymbol{k}\cdot\nabla)$$

in Eq. (B.3) can be considered a small perturbation.

To that end, we note that the set of the Bloch amplitudes $U_{0,n}$, which correspond to the Γ-point, forms a complete set of the basis functions periodic with the lattice period. As a result, the Bloch amplitude at $\boldsymbol{k} \neq \boldsymbol{0}$ can be decomposed as

$$U_{\boldsymbol{k},n} = \sum_{n'} C_{n'}(\boldsymbol{k})\,U_{\boldsymbol{0},n'}, \tag{B.4}$$

where $C_{n'}(\boldsymbol{k})$ are the coefficients. We have the following set of equations for these expansion coefficients

$$\left(E_{0,n} + \frac{\hbar^2 k^2}{2m_0}\right) C_n(\boldsymbol{k}) + \frac{\hbar}{m_0} \sum_{n'} (\boldsymbol{k} \cdot \boldsymbol{p}_{nn'}) C_{n'}(\boldsymbol{k}) = E C_n(\boldsymbol{k}), \tag{B.5}$$

where $\boldsymbol{p}_{nn'} = \int d\boldsymbol{r} U^*_{0,n}(\boldsymbol{r}) \hat{\boldsymbol{p}} U_{0,n'}(\boldsymbol{r})$ are the interband momentum operator matrix elements. Equations (B.5) can be arranged in the matrix form, see Eq. (2.3) of the main text. Solution of Eqs. (B.5) provide the expansion coefficients $C_n(\boldsymbol{k})$ and energy dispersions $E = E_{\boldsymbol{k},n}$ for the allowed bands.

Usually, within the $\boldsymbol{k} \cdot \boldsymbol{p}$ approach, several nearest energy bands are treated exactly, while the contribution of remote bands is accounted for in the framework of the perturbation theory. Enumerating by the subscript r the remote bands (considered perturbatively), by the subscript n the nearest bands (treated exactly), and expressing the coefficients $C_r(\boldsymbol{k})$ via $C_n(\boldsymbol{k})$ in the first-order perturbation theory, we arrive at the following form of the effective Hamiltonian acting in the space of the nearest bands:

$$\begin{aligned} \mathcal{H}_{nn'}(\boldsymbol{k}) = &\left(E_{0,n} + \frac{\hbar^2 k^2}{2m_0}\right) \delta_{nn'} + \frac{\hbar}{m_0} (\boldsymbol{k} \cdot \boldsymbol{p}_{nn'}) \\ &+ \frac{\hbar^2}{2m_0^2} \sum_r (\boldsymbol{k} \cdot \boldsymbol{p}_{nr})(\boldsymbol{k} \cdot \boldsymbol{p}_{rn'}) \left(\frac{1}{E_{0,n} - E_{0,r}} + \frac{1}{E_{0,n'} - E_{0,r}} \right). \end{aligned} \tag{B.6}$$

This effective Hamiltonian is valid up to the second order in the $\boldsymbol{k} \cdot \boldsymbol{p}$ interaction. The regular approach to derive higher-order terms in $\boldsymbol{k} \cdot \boldsymbol{p}$ mixing is presented in (Bir and Pikus, 1974).

If all bands but the only band of interest (n) are accounted for in the second order, the effective Hamiltonian takes a particularly simple form

$$\mathcal{H}(\boldsymbol{k}) = \frac{\hbar^2}{2} \sum_{\alpha\beta} m^{-1}_{\alpha\beta} k_\alpha k_\beta,$$

with the tensor of inverse effective masses:

$$\frac{1}{m_{\alpha\beta}} = \frac{\delta_{\alpha\beta}}{m_0} + \frac{1}{m_0^2} \sum_{n'} \frac{p_{\alpha,nn'} p_{\beta,n'n} + p_{\beta,nn'} p_{\alpha,n'n}}{E_{0,n} - E_{0,n'}}. \tag{B.7}$$

In order to account for the spin–orbit interaction the Hamiltonian, $\mathcal{H}_0$, should be supplemented with the spin–orbit term

$$\mathcal{H}_{so} = \frac{\hbar^2}{4m_0^2 c^2} \left(\boldsymbol{\sigma} \cdot \left[\frac{\partial V(\boldsymbol{r})}{\partial \boldsymbol{r}} \times \hat{\boldsymbol{p}} \right] \right). \tag{B.8}$$

Within the $\boldsymbol{k} \cdot \boldsymbol{p}$ approach the perturbation now takes the form (Bir and Pikus, 1974; Abakumov *et al.*, 1991)

$$\frac{\hbar}{m_0}(\boldsymbol{k}\cdot\hat{\boldsymbol{\pi}}),$$

where

$$\hat{\boldsymbol{\pi}} = \hat{\boldsymbol{p}} + \frac{\hbar}{4m_0c^2}[\sigma \times \nabla V].$$

Of course, the basic functions have to be chosen with account for the spin–orbit interaction, Eq. (B.8). For example, for the Γ_8 and Γ_7 bands these functions should be taken in the form of Eqs. (2.12) and (2.13). As a rule, the difference between $\hat{\boldsymbol{\pi}}$ and $\hat{\boldsymbol{p}}$ is minor and can be neglected.

The $\boldsymbol{k}\cdot\boldsymbol{p}$ approach allows one to include the effect of external fields and elastic strain, which vary smoothly on the scale of the lattice constant. To that end the potential energy associated with an external field is added to the Hamiltonian (B.6) and the wavevector $\boldsymbol{k}$ should be replaced by the operator $\boldsymbol{k} - (e/\hbar c)\boldsymbol{A}$, where $\boldsymbol{A}$ is the vector potential of the external electromagnetic field. Additionally, the "bare" Zeeman effect of the external magnetic field described by the Hamiltonian $\mathcal{H}_B = \mu_B(\hat{\boldsymbol{\sigma}}\cdot\boldsymbol{B})$ [cf. Eq. (2.8)] should be included. This effective Hamiltonian method allows, in particular, to evaluate the electron and hole Zeeman splittings. For example, in GaAs-type semiconductor the conduction band electron g-factor at the Γ-point can be recast as

$$g = 2 - \frac{4}{3}\frac{|p_{cv}|^2}{m_0}\frac{\Delta}{E_g(E_g+\Delta)} + \Delta g. \tag{B.9}$$

Here, p_{cv} is the momentum matrix element between the conduction and the valence band states, E_g is the band gap, Δ is the spin–orbit splitting of the valence band, and Δg is the small contribution related to remote bands (Roth *et al.*, 1959).

References

Abakumov, V. N., Perel, V. I., and Yassievich, I. N. (1991). *Nonradiative recombination in semiconductors*. North Holland, Amsterdam.

Abalmassov, V. A. (2013). On the possibility of the dynamic self-polarization of nuclear spins in a quantum dot. *JETP Lett.*, **98**, 266.

Abalmassov, V. A. (2014). *Theoretical study of behavior of electron and nuclear spins in a quantum dot*. Ph.D. thesis, Novosibirsk.

Abragam, A. (2002). *Principles of nuclear magnetism*. Oxford Science Publications.

Abragam, A. and Bleaney, B. (1970). *Electron paramagnetic resonance of transition ions*. International Series of Monographs on Physics. Clarendon Press, Oxford.

Abragam, A. and Goldman, M. (1978). Principles of dynamic nuclear polarisation. *Reports on Progress in Physics*, **41**, 395.

Abragam, A. and Goldman, M. (1982). *Nuclear magnetism: Order and disorder*. Clarendon Press, Oxford.

Abragam, A. and Proctor, W. G. (1958). Spin temperature. *Phys. Rev.*, **109**, 1441.

Abrahams, E. (1957). Donor electron spin relaxation in silicon. *Phys. Rev.*, **107**, 491.

Adachi, S. (2005). *Properties of group-IV, III-V and II-VI semiconductors*. John Wiley & Sons, Ltd, Chichester.

Adachi, S. (2009). *Properties of semiconductor alloys*. John Wiley & Sons, Ltd, Chichester.

Adams, F. D., Look, D. C., Brown, L. C., and Locker, D. R. (1971). Nuclear-magnetic-resonance studies of the semiconductor-to-metal transition in chlorine-doped cadmium sulfide. *Phys. Rev. B*, **4**, 2115.

Al-Hassanieh, K. A., Dobrovitski, V. V., Dagotto, E., and Harmon, B. N. (2006). Numerical modeling of the central spin problem using the spin-coherent-state p representation. *Phys. Rev. Lett.*, **97**, 037204.

Aleksandrov, E. B. and Zapasskii, V. S. (1981). Magnetic resonance in the Faraday-rotation noise spectrum. *JETP*, **54**, 64.

Aleksandrov, E. B. and Zapasskii, V. S. (2011). Spin noise spectroscopy. *Journal of Physics: Conference Series*, **324**, 012002.

Amasha, S., MacLean, K., Radu, I. P., Zumbühl, D. M., Kastner, M. A., Hanson, M. P., and Gossard, A. C. (2008). Electrical control of spin relaxation in a quantum dot. *Phys. Rev. Lett.*, **100**, 046803.

Androvitsaneas, P., Young, A. B., Schneider, C., Maier, S., Kamp, M., Höfling, S., Knauer, S., Harbord, E., Hu, C. Y., Rarity, J. G., and Oulton, R. (2016). Charged quantum dot micropillar system for deterministic light-matter interactions. *Phys. Rev. B*, **93**, 241409.

Arnold, C., Demory, J., Loo, V., Lemaître, A., Sagnes, I., Glazov, M., Krebs, O., Voisin, P., Senellart, P., and Lanco, L. (2015). Macroscopic rotation of photon polarization induced by a single spin. *Nat Commun*, **6**, 6236.

Aronov, A. G. and Ivchenko, E. L. (1973). Dichroism and optical anisotropy of media with oriented spins of free electrons. *Sov. Phys. Solid State*, **15**, 160.

Aronov, A. G. and Lyanda-Geller, Y. B. (1989). Nuclear electric resonance and orientation of carrier spins by an electric field. *JETP Lett.*, **50**, 431.

Artemova, E. S. and Merkulov, I. A. (1985). Nuclear field and Faraday effect in semiconductors. *Sov. Phys. Solid State*, **27**, 941.

Astakhov, G. V., Glazov, M. M., Yakovlev, D. R., Zhukov, E. A., Ossau, W., Molenkamp, L. W., and Bayer, M. (2008). Time-resolved and continuous-wave optical spin pumping of semiconductor quantum wells. *Semiconductor Science and Technology*, **23**, 114001.

Astakhov, G. V., Kochereshko, V. P., Yakovlev, D. R., Ossau, W., Nurnberger, J., Faschinger, W., and Landwehr, G. (2000). Oscillator strength of trion states in ZnSe-based quantum wells. *Phys. Rev. B*, **62**, 10345.

Astakhov, G. V., Kochereshko, V. P., Yakovlev, D. R., Ossau, W., Nurnberger, J., Faschinger, W., Landwehr, G., Wojtowicz, T., Karczewski, G., and Kossut, J. (2002). Optical method for the determination of carrier density in modulation-doped quantum wells. *Phys. Rev. B*, **65**, 115310.

Atature, M., Dreiser, J., Badolato, A., Hogele, A., Karrai, K., and Imamoglu, A. (2006). Quantum-dot spin-state preparation with near-unity fidelity. *Science*, **312**, 551.

Atature, M., Dreiser, J., Badolato, A., and Imamoglu, A. (2007). Observation of Faraday rotation from a single confined spin. *Nat. Phys.*, **3**, 101.

Averkiev, N. S. and Golub, L. E. (1999). Giant spin relaxation anisotropy in zinc-blende heterostructures. *Phys. Rev. B*, **60**, 15582.

Averkiev, N. S., Golub, L. E., Gurevich, A. S., Evtikhiev, V. P., Kochereshko, V. P., Platonov, A. V., Shkolnik, A. S., and Efimov, Yu. P. (2006). Spin-relaxation anisotropy in asymmetrical (001) $Al_xGa_{1-x}As$ quantum wells from Hanle-effect measurements: Relative strengths of Rashba and Dresselhaus spin-orbit coupling. *Phys. Rev. B*, **74**, 033305.

Averkiev, N. S., Golub, L. E., and Willander, M. (2002). Spin relaxation anisotropy in two-dimensional semiconductor systems. *J. Phys.: Condens. Matter*, **14**, R271.

Awschalom, D. D., Halbout, J. M., von Molnar, S., Siegrist, T., and Holtzberg, F. (1985). Dynamic spin organization in dilute magnetic systems. *Phys. Rev. Lett.*, **55**, 1128.

Azaizia, S., Carrere, H., Sandoval-Santana, J. C., Ibarra-Sierra, V. G., Kalevich, V. K., Ivchenko, E. L., Bakaleinikov, L. A., Marie, X., Amand,T.,Kunold, A., and Balocchi, A. (2018). Electron-nuclear coherent spin oscillations probed by spin-dependent recombination. *Phys. Rev. B* **97**, 155201.

Badescu, S. C., Lyanda-Geller, Y. B., and Reinecke, T. L. (2005). Asymmetric exchange between electron spins in coupled semiconductor quantum dots. *Phys. Rev. B*, **72**, 161304.

Baranov, P. G., von Bardeleben, H. J., Jelezko, F., and Wrachtrup, J. (2017). *Magnetic resonance of semiconductors and their nanostructures: Basic and advanced applications*. Springer Series in Materials Science. Springer, Heielberg.

Barnes, E., Cywiński, Ł., and Das Sarma, S. (2011). Master equation approach to the central spin decoherence problem: Uniform coupling model and role of projection operators. *Phys. Rev. B*, **84**, 155315.

Bastard, G. (1988). *Wave mechanics applied to semiconductor heterostructures*. Monographies de physique. Les Éditions de Physique.

Bechtold, A., Li, F., Müller, K., Simmet, T., Ardelt, P.-L., Finley, J. J., and Sinitsyn, N. A. (2016). Quantum effects in higher-order correlators of a quantum-dot spin qubit. *Phys. Rev. Lett.*, **117**, 027402.

Bechtold, A., Rauch, D., Li, F., Simmet, T., Ardelt, P.-L., Regler, A., Muller, K., Sinitsyn, N. A., and Finley, J. J. (2015). Three-stage decoherence dynamics of an electron spin qubit in an optically active quantum dot. *Nat Phys*, **11**, 1005.

Belhadj, T., Kuroda, T., Simon, C.-M., Amand, T., Mano, T., Sakoda, K., Koguchi, N., Marie, X., and Urbaszek, B. (2008). Optically monitored nuclear spin dynamics in individual gaas quantum dots grown by droplet epitaxy. *Phys. Rev. B*, **78**, 205325.

Belousov, Y. M., Gorelkin, V. N., Mikaélyan, A. L., Miloserdin, V. Y., and Smilga, V. P. (1979). Study of metals by means of positive muons. *Soviet Physics Uspekhi*, **22**, 679.

Belykh, V. V., Evers, E., Yakovlev, D. R., Fobbe, F., Greilich, A., and Bayer, M. (2016). Extended pump-probe Faraday rotation spectroscopy of the submicrosecond electron spin dynamics in *n*-type GaAs. *Phys. Rev. B*, **94**, 241202.

Berestetskii, V. B., Pitaevskii, L. P., and Lifshitz, E. M. (1999). *Quantum electrodynamics, Second Edition (Vol. 4)*. Butterworth-Heinemann, Oxford.

Berezovsky, J., Mikkelsen, M. H., Stoltz, N. G., Coldren, L. A., and Awschalom, D. D. (2008). Picosecond coherent optical manipulation of a single electron spin in a quantum dot. *Science*, **320**, 349.

Berger, J. D., Lyngnes, O., Gibbs, H. M., Khitrova, G., Nelson, T. R., Lindmark, E. K., Kavokin, A. V., Kaliteevski, M. A., and Zapasskii, V. V. (1996). Magnetic-field enhancement of the exciton-polariton splitting in a semiconductor quantum-well microcavity: The strong coupling threshold. *Phys. Rev. B*, **54**, 1975.

Berkovits, V. L. and Safarov, V. I. (1977). Optical observation of multiphoton NMR of absorption in semiconducting crystals. *JETP Lett.*, **26**, 256.

Berski, F., Hübner, J., Oestreich, M., Ludwig, A., Wieck, A. D., and Glazov, M. (2015). Interplay of electron and nuclear spin noise in *n*-type GaAs. *Phys. Rev. Lett.*, **115**, 176601.

Berski, F., Kuhn, H., Lonnemann, J. G., Hübner, J., and Oestreich, M. (2013). Ultrahigh bandwidth spin noise spectroscopy: Detection of large *g*-factor fluctuations in highly-*n*-doped GaAs. *Phys. Rev. Lett.*, **111**, 186602.

Beschoten, B. (2005). Spin coherence in semiconductors. In *Magnetism goes Nano, 36th Spring School 2005, Schriften des Forschungzentrums Julich, Matter and Materials, Vol. 26* (ed. T. B. S. Blugel and C. Schneider), pp. E7.1–7.27.

Bester, G. and Zunger, A. (2005). Cylindrically shaped zinc-blende semiconductor quantum dots do not have cylindrical symmetry: Atomistic symmetry, atomic relaxation, and piezoelectric effects. *Phys. Rev. B*, **71**, 45318.

Beugeling, W., Uhrig, G. S., and Anders, F. B. (2016). Quantum model for mode locking in pulsed semiconductor quantum dots. *Phys. Rev. B*, **94**, 245308.

Beugeling, W., Uhrig, G. S., and Anders, F. B. (2017). Influence of the nuclear Zeeman effect on mode locking in pulsed semiconductor quantum dots. *ArXiv e-prints*, 1704.01468.

Bienfait, A., Pla, J. J., Kubo, Y., Zhou, X., Stern, M., Lo, C. C., Weis, C. D., Schenkel, T., Vion, D., Esteve, D., Morton, J. J. L., and Bertet, P. (2016). Controlling spin relaxation with a cavity. *Nature*, **531**, 74.

Bimberg, D., Grundmann, M., and Ledenstov, N. N. (1999). *Quantum dot heterostructures*. Wiley-VCH Verlag, Weinheim.

Bir, G. L., Aronov, A. G., and Pikus, G. E. (1975). Spin relaxation of electrons due to scattering by holes. *Sov. Phys. JETP*, **42**, 705.

Bir, G. L. and Pikus, G. E. (1974). *Symmetry and strain-induced effects in semiconductors*. Wiley/Halsted Press.

Bloch, F. (1946). Nuclear induction. *Phys. Rev.*, **70**, 460.

Bloembergen, N. (1949). On the interaction of nuclear spins in a crystalline lattice. *Physica*, **15**, 386.

Bloembergen, N. (1954). Nuclear magnetic relaxation in semiconductors. *Physica*, **20**, 1130.

Bluhm, H., Foletti, S., Neder, I., Rudner, M., Mahalu, D., Umansky, V., and Yacoby, A. (2011). Dephasing time of GaAs electron-spin qubits coupled to a nuclear bath exceeding 200 μs. *Nat. Phys.*, **7**, 109.

Boguslawski, P. and Gorczyca, I. (1994). Atomic-orbital interpretation of electronic structure of III-V semiconductors: GaAs versus AlAs. *Semicond. Sci. Techn.*, **9**, 2169.

Bolef, D. I. and Menes, M. (1959). Nuclear magnetic resonance acoustic absorption in KI and KBr. *Phys. Rev.*, **114**, 1441.

Bonadeo, N. H., Erland, J., Gammon, D., Park, D., Katzer, D. S., and Steel, D. G. (1998). Coherent optical control of the quantum state of a single quantum dot. *Science*, **282**, 1473.

Bortz, M. and Stolze, J. (2007*a*). Exact dynamics in the inhomogeneous central-spin model. *Phys. Rev. B*, **76**, 014304.

Bortz, M. and Stolze, J. (2007*b*). Spin and entanglement dynamics in the central-spin model with homogeneous couplings. *J. Stat. Mech.: Theory Exp.*, **2007**, P06018.

Bracker, A. S., Gammon, D., and Korenev, V. L. (2008). Fine structure and optical pumping of spins in individual semiconductor quantum dots. *Semiconductor Sci. Techn.*, **23**, 114004.

Brand, M. A., Malinowski, A., Karimov, O. Z., Marsden, P. A., Harley, R. T., Shields, A. J., Sanvitto, D., Ritchie, D. A., and Simmons, M. Y. (2002). Precession and motional slowing of spin evolution in a high mobility two-dimensional electron gas. *Phys. Rev. Lett.*, **89**, 236601.

Braun, P.-F., Eble, B., Lombez, L., Urbaszek, B., Marie, X., Amand, T., Renucci, P., Krebs, O., Lemaître, A., Voisin, P., Kalevich, V. K., and Kavokin, K. V. (2006*a*). Spin relaxation of positive trions in InAs/GaAs quantum dots: the role of hyperfine interaction. *Physica Status Solidi (B)*, **243**, 3917.

Braun, P.-F., Marie, X., Lombez, L., Urbaszek, B., Amand, T., Renucci, P., Kalevich, V. K., Kavokin, K. V., Krebs, O., Voisin, P., and Masumoto, Y. (2005). Direct observation of the electron spin relaxation induced by nuclei in quantum dots. *Phys. Rev. Lett.*, **94**, 116601.

Braun, P.-F., Urbaszek, B., Amand, T., Marie, X., Krebs, O., Eble, B., Lemaitre, A., and Voisin, P. (2006*b*). Bistability of the nuclear polarization created through optical pumping in $In_{1-x}Ga_xAs$ quantum dots. *Phys. Rev. B*, **74**, 245306.

Brossel, J., Kastler, A., and Winter, J. (1952). Gréation optique d'une inégalité de population entre les sous-niveaux zeeman de l'état fondamental des atomes. *J. Phys. Radium*, **13**, 668.

Brown, S. W., Kennedy, T. A., and Gammon, D. (1998). Optical NMR from single quantum dots. *Solid State NMR*, **11**, 49.

Brown, S. W., Kennedy, T. A., Gammon, D., and Snow, E. S. (1996). Spectrally resolved Overhauser shifts in single $GaAs/Al_xGa_{1-x}As$ as quantum dots. *Phys. Rev. B*, **54**, R17339.

Brunner, K., Abstreiter, G., Böhm, G., Tränkle, G., and Weimann, G. (1994). Sharp-line photoluminescence and two-photon absorption of zero-dimensional biexcitons in a GaAs/AlGaAs structure. *Phys. Rev. Lett.*, **73**, 1138.

Bulutay, C. (2012). Quadrupolar spectra of nuclear spins in strained $In_xGa_{1-x}As$ quantum dots. *Phys. Rev. B*, **85**, 115313.

Bulutay, C., Chekhovich, E. A., and Tartakovskii, A. I. (2014). Nuclear magnetic resonance inverse spectra of InGaAs quantum dots: Atomistic level structural information. *Phys. Rev. B*, **90**, 205425.

Buratto, S. K., Hwang, J. Y., Kurur, N. D., Shykind, D. N., and Weitekamp, D. P. (1993). Fourier-transform time-sequenced optical nuclear magnetic resonance of gallium arsenide. *Bull. Magn. Reson.*, **15**, 190.

Buß, J. H., Rudolph, J., Wassner, T. A., Eickhoff, M., and Hägele, D. (2016). Optical manipulation of a multilevel nuclear spin in ZnO: Master equation and experiment. *Phys. Rev. B*, **93**, 155204.

Bychkov, Y. A. and Rashba, E. I. (1984). Oscillatory effects and the magnetic susceptibility of carriers in inversion layers. *J. Phys. C: Solid State*, **17**, 6039.

Camenzind, L. C., Yu, L., Stano, P., Zimmerman, J., Gossard, A. C., Loss, D., and Zumbühl, D. M. (2017). Hyperfine-phonon spin relaxation in a single-electron GaAs quantum dot. *ArXiv e-prints*, 1711.01474.

Cardona, M., Christensen, N. E., and Fasol, G. (1988). Relativistic band structure and spin-orbit splitting of zinc-blende-type semiconductors. *Phys. Rev. B*, **38**, 1806.

Cardona, M., Shaklee, K. L., and Pollak, F. H. (1967). Electroreflectance at a semiconductor-electrolyte interface. *Phys. Rev.*, **154**, 696.

Carter, S. G., Chen, Z., and Cundiff, S. T. (2007). Ultrafast below-resonance Raman rotation of electron spins in GaAs quantum wells. *Phys. Rev. B*, **76**, 201308.

Carter, S. G., Shabaev, A., Economou, S. E., Kennedy, T. A., Bracker, A. S., and Reinecke, T. L. (2009). Directing nuclear spin flips in InAs quantum dots using detuned optical pulse trains. *Phys. Rev. Lett.*, **102**, 167403.

Chadi, D. J. (1976). Angular momentum decomposition of $k = 0$ Bloch functions in group IV and zincblende crystals. *Solid State Communications*, **20**, 361.

Chalupczak, W. and Godun, R. M. (2011). Near-resonance spin-noise spectroscopy. *Phys. Rev. A*, **83**, 032512.

Chekhovich, E. A., Glazov, M. M., Krysa, A. B., Hopkinson, M., Senellart, P., Lemaitre, A., Skolnick, M. S., and Tartakovskii, A. I. (2013). Element-sensitive measurement of the hole-nuclear spin interaction in quantum dots. *Nat Phys*, **9**, 74.

Chekhovich, E. A., Kavokin, K. V., Puebla, J., Krysa, A. B., Hopkinson, M., Andreev, A. D., Sanchez, A. M., Beanland, R., Skolnick, M. S., and Tartakovskii, A. I. (2012). Structural analysis of strained quantum dots using nuclear magnetic resonance. *Nat Nano*, **7**, 646.

Chekhovich, E. A., Krysa, A. B., Skolnick, M. S., and Tartakovskii, A. I. (2011). Direct measurement of the hole-nuclear spin interaction in single InP/GaInP quantum dots using photoluminescence spectroscopy. *Phys. Rev. Lett.*, **106**, 027402.

Chekhovich, E. A., Makhonin, M. N., Kavokin, K. V., Krysa, A. B., Skolnick, M. S., and Tartakovskii, A. I. (2010). Pumping of nuclear spins by optical excitation of spin-forbidden transitions in a quantum dot. *Phys. Rev. Lett.*, **104**, 066804.

Chekhovich, E. A., Ulhaq, A., Zallo, E., Ding, F., Schmidt, O. G., and Skolnick, M. S. (2017). Measurement of the spin temperature of optically cooled nuclei and GaAs hyperfine constants in GaAs/AlGaAs quantum dots. *Nat. Mater.*, **16**, 982.

Chen, G., Bergman, D. L., and Balents, L. (2007*a*). Semiclassical dynamics and long-time asymptotics of the central-spin problem in a quantum dot. *Phys. Rev. B*, **76**, 045312.

Chen, Y. S., Huang, J., Reuter, D., Ludwig, A., Wieck, A. D., and Bacher, G. (2011*a*). Optically detected nuclear magnetic resonance in *n*-GaAs using an on-chip microcoil. *Appl. Phys. Lett.*, **98**, 081911.

Chen, Y. S., Reuter, D., Wieck, A. D., and Bacher, G. (2011*b*). Dynamic nuclear spin resonance in *n*-GaAs. *Phys. Rev. Lett.*, **107**, 167601.

Chen, Z., Bratschitsch, R., Carter, S. G., Cundiff, S. T., Yakovlev, D. R., Karczewski, G., Wojtowicz, T., and Kossut, J. (2007*b*). Electron spin polarization through interactions between excitons, trions, and the two-dimensional electron gas. *Phys. Rev. B*, **75**, 115320.

Cherbunin, R. V., Auer, T., Greilich, A., Ignatiev, I. V., Oulton, R., Bayer, M., Yakovlev, D. R., Kozlov, G. G., Reuter, D., and Wieck, A. D. (2008). Manifestations of the hyperfine interaction between electron and nuclear spins in singly-charged (In,Ga)As/GaAs quantum dots. *ArXiv e-prints*, 0805.3247.

Cherbunin, R. V., Vladimirova, M., Kavokin, K. V., Mikhailov, A. V., Kopteva, N. E., Lagoudakis, P. G., and Kavokin, A. V. (2015). Significant photoinduced Kerr rotation achieved in semiconductor microcavities. *Phys. Rev. B*, **91**, 205308.

Childress, L., Walsworth, R., and Lukin, M. (2014). Atom-like crystal defects: From quantum computers to biological sensors. *Physics Today*, **67**, 38.

Clementi, E. and Raimondi, D. L. (1963). Atomic screening constants from scf functions. *J. Chem. Phys.*, **38**, 2686.

Cohen-Tannoudji, C., Dupont-Roc, J., and Grynberg, G. (2004). *Atom-photon interactions. Basic processes and applications*. John Wiley & Sons, Inc.

Coish, W. A. and Loss, Daniel (2004). Hyperfine interaction in a quantum dot: Non-Markovian electron spin dynamics. *Phys. Rev. B*, **70**, 195340.

Coish, W. A., Loss, Daniel, Yuzbashyan, E. A., and Altshuler, B. L. (2007). Quantum versus classical hyperfine-induced dynamics in a quantum dot. *J. Appl. Phys.*, **101**, 081715.

Colton, J. S., Kennedy, T. A., Bracker, A. S., Miller, J. B., and Gammon, D. (2004). Dependence of optically oriented and detected electron spin resonance on donor concentration in *n*-GaAs. *Solid State Comm.*, **132**, 613.

Crankshaw, S., Sedgwick, F. G., Moewe, M., Chang-Hasnain, C., Wang, H., and Chuang, S.-L. (2009). Electron spin polarization induced by linearly polarized light in a (110) GaAs quantum-well waveguide. *Phys. Rev. Lett.*, **102**, 206604.

Cronenberger, S. and Scalbert, D. (2016). Quantum limited heterodyne detection of spin noise. *Rev. Sci. Instr.*, **87**, 093111.

Cronenberger, S., Scalbert, D., Ferrand, D., Boukari, H., and Cibert, J. (2015). Atomic-like spin noise in solid-state demonstrated with manganese in cadmium telluride. *Nat. Commun.*, **6**, 8121.

Crooker, S. A., Brandt, J., Sandfort, C., Greilich, A., Yakovlev, D. R., Reuter, D., Wieck, A. D., and Bayer, M. (2010). Spin noise of electrons and holes in self-assembled quantum dots. *Phys. Rev. Lett.*, **104**, 036601.

Crooker, S. A., Cheng, L., and Smith, D. L. (2009). Spin noise of conduction electrons in *n* -type bulk GaAs. *Phys. Rev. B*, **79**, 035208.

Crooker, S. A., Rickel, D. G., Balatsky, A. V., and Smith, D. L. (2004). Spectroscopy of spontaneous spin noise as a probe of spin dynamics and magnetic resonance. *Nature*, **431**, 49.

Crooker, S. A. and Smith, D. L. (2005). Imaging spin flows in semiconductors subject to electric, magnetic, and strain fields. *Phys. Rev. Lett.*, **94**, 236601.

Cywiński, Ł., Dobrovitski, V. V., and Das Sarma, S. (2010). Spin echo decay at low magnetic fields in a nuclear spin bath. *Phys. Rev. B*, **82**, 035315.

Dahbashi, R., Hübner, J., Berski, F., Pierz, K., and Oestreich, M. (2014). Optical spin noise of a single hole spin localized in an (InGa)As quantum dot. *Phys. Rev. Lett.*, **112**, 156601.

Dahbashi, R., Hübner, J., Berski, F., Wiegand, J., Marie, X., Pierz, K., Schumacher, H. W., and Oestreich, M. (2012). Measurement of heavy-hole spin dephasing in (InGa)As quantum dots. *Appl. Phys. Lett.*, **100**, 031906.

Davies, J. H. (1998). *The physics of low-dimensional semiconductors*. Cambridge University Press.

De Greve, K., McMahon, P. L., Press, D., Ladd, T. D., Bisping, D., Schneider, C., Kamp, M., Worschech, L., Hofling, S., Forchel, A., and Yamamoto, Y. (2011). Ultrafast coherent control and suppressed nuclear feedback of a single quantum dot hole qubit. *Nat Phys*, **7**, 872.

Doherty, M. W., Manson, N. B., Delaney, P., Jelezko, F., Wrachtrup, J., and Hollenberg, L. C. L. (2013). The nitrogen-vacancy colour centre in diamond. *Phys. Rep.*, **528**, 1.

Döhrmann, S., Hagele, D., Rudolph, J., Bichler, M., Schuh, D., and Oestreich, M. (2004). Anomalous spin dephasing in (110) GaAs quantum wells: Anisotropy and intersubband effects. *Phys. Rev. Lett.*, **93**, 147405.

Dousse, A., Suffczynski, J., Beveratos, A., Krebs, O., Lemaitre, A., Sagnes, I., Bloch, J., Voisin, P., and Senellart, P. (2010). Ultrabright source of entangled photon pairs. *Nature*, **466**, 217.

Dresselhaus, G. (1955). Spin-orbit coupling effects in zinc blende structures. *Phys. Rev.*, **100**, 580.

Dupertuis, M. A., Karlsson, K. F., Oberli, D. Y., Pelucchi, E., Rudra, A., Holtz, P. O., and Kapon, E. (2011). Symmetries and the polarized optical spectra of exciton complexes in quantum dots. *Phys. Rev. Lett.*, **107**, 127403.

Durnev, M. V., Glazov, M. M., and Ivchenko, E. L. (2014). Spin-orbit splitting of valence subbands in semiconductor nanostructures. *Phys. Rev. B*, **89**, 075430.

Durnev, M. V., Glazov, M. M., Ivchenko, E. L., Jo, M., Mano, T., Kuroda, T., Sakoda, K., Kunz, S., Sallen, G., Bouet, L., Marie, X., Lagarde, D., Amand, T., and Urbaszek, B. (2013). Magnetic field induced valence band mixing in [111] grown semiconductor quantum dots. *Phys. Rev. B*, **87**, 085315.

Dutt, M. V. Gurudev, C., Jun, L., Bo, X., Xiaodong, L., X., Berman, P. R., Steel, D. G., Bracker, A. S., Gammon, D., Economou, S. E., Liu, R.-B., and Sham, L. J. (2005). Stimulated and spontaneous optical generation of electron spin coherence in charged GaAs quantum dots. *Phys. Rev. Lett.*, **94**, 227403.

Dyakonov, M. I. and Kachorovskii, V. Y. (1986). Spin relaxation of two-dimensional electrons in noncentrosymmetric semiconductors. *Sov. Phys. Semicond.*, **20**, 110.

D'yakonov, M. I., Marushchak, V. A., Perel', V. I., and Titkov, A. N. (1986). The effect of strain on the spin relaxation of conduction electrons in Ill-V semiconductors. *JETP*, **63**, 655.

D'yakonov, M. I., Merkulov, I. A., and Perel', V. I. (1979). Optical-orientation anisotropy produced in semiconductors by quadrupole splitting of the spin levels of the lattice nuclei. *JETP*, **49**, 160.

D'yakonov, M. I., Merkulov, I. A., and Perel', V. I. (1980). Instabilities in the spin system of optically oriented electrons and nuclei in semiconductors. *JETP*, **51**, 175.

Dyakonov, M. I. and Perel, V. I. (1971). Spin orientation of electrons associated with interband absorption of light in semiconductors. *Sov. Phys. JETP*, **33**, 1053.

Dyakonov, M. I. and Perel', V. I (1972). Spin relaxation of conduction electrons in noncentrosymmetric semiconductors. *Sov. Phys. Solid State*, **13**, 3023.

D'yakonov, M. I., Perel', V. I., Berkovits, V. L., and Safarov, V. I. (1975). Optical effects due to polarization of nuclei in semiconductors. *JETP*, **40**, 950.

Dyakonov, M. I. (1975). Optical recording of dynamic self-polarization of nuclei in semiconductors. *JETP*, **40**, 746.

Dyakonov, M. I. (ed.) (2017). *Spin physics in semiconductors, Second Edition*, Springer Series in Solid-State Sciences, Vol. 157. Springer International Publishing.

D'yakonov, M. I. and Perel', V. I. (1972). Dynamic self-polarization of nuclei in solids. *JETP Lett.*, **16**, 398.

Dyakonov, M. I. and Perel', V. I. (1973). Hyperfine interaction in optical orientation of electrons in semiconductors. *JETP*, **36**, 995.

Dyakonov, M. I. and Perel', V. I. (1974). Optical orientation in a system of electrons and lattice nuclei in semiconductors. Theory. *JETP*, **38**, 177.

D'yakonov, M. I. and Perel', V. I. (1975). Cooling of a system of nuclear spins following optical orientation of electrons in semiconductors. *JETP*, **41**, 759.

Dyakonov, M. I. and Perel, V. I. (1984). Theory of optical spin orientation of electrons and nuclei in semiconductors. In *Optical orientation* (ed. F. Meier and B. P. Zakharchenya), North-Holland, Amsterdam-Oxford-New-York-Tokyo p. 11.

Dzhioev, R. I., Kavokin, K.V., Korenev, V.L., Lazarev, M.V., Meltser, B. Ya., Stepanova, M. N., Zakharchenya, B. P., Gammon, D., and Katzer, D. S. (2002*a*). Low-temperature spin relaxation in *n*-type GaAs. *Phys. Rev. B*, **66**, 245204.

Dzhioev, R. I. and Korenev, V. L. (2007). Stabilization of the electron-nuclear spin orientation in quantum dots by the nuclear quadrupole interaction. *Phys. Rev. Lett.*, **99**, 037401.

Dzhioev, R. I., Korenev, V. L., Merkulov, I. A., Zakharchenya, B. P., Gammon, D., Efros, A. L., and Katzer, D. S. (2002*b*). Manipulation of the spin memory of electrons in *n*-GaAs. *Phys. Rev. Lett.*, **88**, 256801.

Dzhioev, R. I., Zakharchenya, B. P., Korenev, V. L., and Lazarev, M. V. (1999). Interaction between the exciton and nuclear spin systems in a self-organized ensemble of InP/InGaP size-quantized islands. *Phys. of Solid State*, **41**, 2014.

Dzhioev, R. I., Zakharchenya, B. P., Korenev, V. L., Pak, P. E., Tkachuk, M. N., Vinokurov, D. A., and Tarasov, I. S. (1998). Dynamic polarization of nuclei in a self-organized ensemble of quantum-size *n*-InP/InGaP islands. *JETP Lett.*, **68**, 745.

Dzhioev, R. I., Zakharchenya, B. P., Korenev, V. L., and Stepanova, M. N. (1997). Spin diffusion of optically oriented electrons and photon entrainment in *n*-gallium arsenide. *Physics of the Solid State*, **39**, 1765.

Eble, B., Desfonds, P., Fras, F., Bernardot, F., Testelin, C., Chamarro, M., Miard, A., and Lemaître, A. (2010). Hole and trion spin dynamics in quantum dots under excitation by a train of circularly polarized pulses. *Phys. Rev. B*, **81**, 045322.

Eble, B., Krebs, O., Lemaître, A., Kowalik, K., Kudelski, A., Voisin, P., Urbaszek, B., Marie, X., and Amand, T. (2006). Dynamic nuclear polarization of a single charge-tunable InAs/GaAs quantum dot. *Phys. Rev. B*, **74**, 081306.

Economou, S. E. and Barnes, E. (2014). Theory of dynamic nuclear polarization and feedback in quantum dots. *Phys. Rev. B*, **89**, 165301.

Economou, S. E., Sham, L. J., Wu, Y., and Steel, D. G. (2006). Proposal for optical U(1) rotations of electron spin trapped in a quantum dot. *Phys. Rev. B*, **74**, 205415.

Edmonds, A. R. (1974). *Angular momentum in quantum mechanics*. Princeton University Press, Princeton, New Jersey.

Efros, Al. L. and Rosen, M. (2000). The electronic structure of semiconductor nanocrystals. *Ann. Rev. Mat. Sci.*, **30**, 475.

Ekimov, A. I. and Safarov, V. I. (1972). Optical electron-nuclear resonance in semiconductors. *JETP Lett.*, **15**, 319.

Eles, P. T. and Michal, C. A. (2010). Two-photon excitation in nuclear magnetic and quadrupole resonance. *Prog. NMR Spec.*, **56**, 232.

Elliott, R. J. (1954). Theory of the effect of spin-orbit coupling on magnetic resonance in some semiconductors. *Phys. Rev.*, **96**, 266.

Elzerman, J. M., Hanson, R., Willems van Beveren, L. H., Witkamp, B., Vandersypen, L. M. K., and Kouwenhoven, L. P. (2004). Single-shot read-out of an individual electron spin in a quantum dot. *Nature*, **430**, 431.

Erlingsson, S. I. and Nazarov, Y. V. (2004). Evolution of localized electron spin in a nuclear spin environment. *Phys. Rev. B*, **70**, 205327.

Erlingsson, S. I., Nazarov, Y. V., and Fal'ko, V. I. (2001). Nucleus-mediated spin-flip transitions in GaAs quantum dots. *Phys. Rev. B*, **64**, 195306.

Falk, A. L., Klimov, P. V., Ivády, V., Szász, K., Christle, D. J., Koehl, W. F., Gali, Á., and Awschalom, D. D. (2015). Optical polarization of nuclear spins in silicon carbide. *Phys. Rev. Lett.*, **114**, 247603.

Fal'ko, V. I., Altshuler, B. L., and Tsyplyatyev, O. (2005). Anisotropy of spin splitting and spin relaxation in lateral quantum dots. *Phys. Rev. Lett.*, **95**, 076603.

Fallahi, P., Yılmaz, S. T., and Imamoğlu, A. (2010). Measurement of a heavy-hole hyperfine interaction in ingaas quantum dots using resonance fluorescence. *Phys. Rev. Lett.*, **105**, 257402.

Faribault, A. and Schuricht, D. (2013*a*). Integrability-based analysis of the hyperfine-interaction-induced decoherence in quantum dots. *Phys. Rev. Lett.*, **110**, 040405.

Faribault, A. and Schuricht, D. (2013*b*). Spin decoherence due to a randomly fluctuating spin bath. *Phys. Rev. B*, **88**, 085323.

Feller, W. (1968). *An introduction to probability theory and its applications, Vol. 1*. Wiley.

Fermi, E. (1930). Uber die magnetischen momente der atomkerne. *Zeitschrift fur Physik*, **60**, 320.

Fischer, J., Coish, W. A., Bulaev, D. V., and Loss, D. (2008). Spin decoherence of a heavy hole coupled to nuclear spins in a quantum dot. *Phys. Rev. B*, **78**, 155329.

Fokina, L. V., Yugova, I. A., Yakovlev, D. R., Glazov, M. M., Akimov, I. A., Greilich, A., Reuter, D., Wieck, A. D., and Bayer, M. (2010). Spin dynamics of electrons and holes in InGaAs/GaAs quantum wells at millikelvin temperatures. *Phys. Rev. B*, **81**, 195304.

Fras, F., Eble, B., Siarry, B., Bernardot, F., Miard, A., Lemaître, A., Testelin, C., and Chamarro, M. (2012). Hole spin mode locking and coherent dynamics in a largely inhomogeneous ensemble of *p*-doped InAs quantum dots. *Phys. Rev. B*, **86**, 161303.

Fröhling, N. and Anders, F. B. (2017). Long-time coherence in fourth-order spin correlation functions. *Phys. Rev. B*, **96**, 045441.

Fu, K.-M. C., Santori, C., Stanley, C., Holland, M. C., and Yamamoto, Y. (2005). Coherent population trapping of electron spins in a high-purity *n*-type GaAs semiconductor. *Phys. Rev. Lett.*, **95**, 187405.

Fuchs, G. D., Dobrovitski, V. V., Toyli, D. M., Heremans, F. J., and Awschalom, D. D. (2009). Gigahertz dynamics of a strongly driven single quantum spin. *Science*, **326**, 1520.

Fuller, G. H. (1976). Nuclear spins and moments. *J. Phys. Chem. Ref. Data*, **5**, 835.

Gammon, D., Brown, S. W., Snow, E. S., Kennedy, T. A., Katzer, D. S., and Park, D. (1997). Nuclear spectroscopy in single quantum dots: nanoscopic Raman scattering and nuclear magnetic resonance. *Science*, **277**, 85.

Gammon, D., Efros, Al. L., Kennedy, T. A., Rosen, M., Katzer, D. S., Park, D., Brown, S. W., Korenev, V. L., and Merkulov, I. A. (2001). Electron and nuclear spin interactions in the optical spectra of single GaAs quantum dots. *Phys. Rev. Lett.*, **86**, 5176.

Gammon, D., Snow, E. S., Shanabrook, B. V., Katzer, D. S., and Park, D. (1996). Fine structure splitting in the optical spectra of single GaAs quantum dots. *Phys. Rev. Lett.*, **76**, 3005.

Gangadharaiah, S., Sun, J., and Starykh, O. A. (2008). Spin-orbit-mediated anisotropic spin interaction in interacting electron systems. *Phys. Rev. Lett.*, **100**, 156402.

Gantmakher, V. F. and Levinson, Y. B. (1987). *Carrier scattering in metals and semiconductors*. North-Holland Publishing Company.

Gantsevich, S., Gurevich, V., and Katilius, R. (1979). Theory of fluctuations in nonequilibrium electron gas. *La Rivista del Nuovo Cimento (1978-1999)*, **2**, 1.

Gaudin, M. (1976). Diagonalisation d'une classe d'hamiltoniens de spin. *J. Phys. France*, **37**, 1087.

Gavini, A. and Cardona, M. (1970). Modulated piezoreflectance in semiconductors. *Phys. Rev. B*, **1**, 672.

Georgeot, B. and Shepelyansky, D. L. (2000). Quantum chaos border for quantum computing. *Phys. Rev. E*, **62**, 3504.

Giri, R., Cronenberger, S., Glazov, M. M., Kavokin, K. V., Lemaître, A., Bloch, J., Vladimirova, M., and Scalbert, D. (2013). Nondestructive measurement of nuclear magnetization by off-resonant Faraday rotation. *Phys. Rev. Lett.*, **111**, 087603.

Giri, R., Cronenberger, S., Vladimirova, M., Scalbert, D., Kavokin, K. V., Glazov, M. M., Nawrocki, M., Lemaître, A., and Bloch, J. (2012). Giant photoinduced Faraday rotation due to the spin-polarized electron gas in an n-GaAs microcavity. *Phys. Rev. B*, **85**, 195313.

Glasenapp, P., Greilich, A., Ryzhov, I. I., Zapasskii, V. S., Yakovlev, D. R., Kozlov, G. G., and Bayer, M. (2013). Resources of polarimetric sensitivity in spin noise spectroscopy. *Phys. Rev. B*, **88**, 165314.

Glasenapp, P., Sinitsyn, N. A., Yang, Luyi, Rickel, D. G., Roy, D., Greilich, A., Bayer, M., and Crooker, S. A. (2014). Spin noise spectroscopy beyond thermal equilibrium and linear response. *Phys. Rev. Lett.*, **113**, 156601.

Glasenapp, P., Smirnov, D. S., Greilich, A., Hackmann, J., Glazov, M. M., Anders, F. B., and Bayer, M. (2016). Spin noise of electrons and holes in (In,Ga)As quantum dots: Experiment and theory. *Phys. Rev. B*, **93**, 205429.

Glazov, M. M. (2012). Coherent spin dynamics of electrons and excitons in nanostructures (a review). *Phys. of the Solid State*, **54**, 1.

Glazov, M. M. (2013). Coherent spin dynamics in semiconductor quantum dots. *J. Appl. Phys.*, **113**, 136503.

Glazov, M. M. (2015). Spin noise of localized electrons: Interplay of hopping and hyperfine interaction. *Phys. Rev. B*, **91**, 195301.

Glazov, M. M. (2016). Spin fluctuations of nonequilibrium electrons and excitons in semiconductors. *JETP*, **122**, 472.

Glazov, M. M. and Ivchenko, E. L. (2008). Resonant spin amplification in nanostructures with anisotropic spin relaxation and spread of the electronic g-factor. *Semiconductors*, **42**, 951.

Glazov, M. M. and Ivchenko, E. L. (2012). Spin noise in quantum dot ensembles. *Phys. Rev. B*, **86**, 115308.

Glazov, M. M. and Kulakovskii, V. D. (2009). Spin-orbit effect on electron-electron interaction and the fine structure of electron complexes in quantum dots. *Phys. Rev. B*, **79**, 195305.

Glazov, M. M., Yugova, I. A., and Efros, A. L. (2012). Electron spin synchronization induced by optical nuclear magnetic resonance feedback. *Phys. Rev. B*, **85**, 041303(R).

Glazov, M. M., Yugova, I. A., Spatzek, S., Schwan, A., Varwig, S., Yakovlev, D. R., Reuter, D., Wieck, A. D., and Bayer, M. (2010). Effect of pump-probe detuning on the Faraday rotation and ellipticity signals of mode-locked spins in (In,Ga)As/GaAs quantum dots. *Phys. Rev. B*, **82**, 155325.

Glazov, M. M. and Zapasskii, V. S. (2015). Linear optics, raman scattering, and spin noise spectroscopy. *Opt. Express*, **23**, 11713.

Godden, T. M., Quilter, J. H., Ramsay, A. J., Wu, Yanwen, Brereton, P., Boyle, S. J., Luxmoore, I. J., Puebla-Nunez, J., Fox, A. M., and Skolnick, M. S. (2012). Coherent optical control of the spin of a single hole in an InAs/GaAs quantum dot. *Phys. Rev. Lett.*, **108**, 017402.

Goldman, M. (2001). Formal theory of spin-lattice relaxation. *J. Magn. Reson.*, **149**, 160.

Golub, L.E. and Ivchenko, E.L. (2004). Spin splitting in symmetrical SiGe quantum wells. *Phys. Rev. B*, **69**, 115333.

Gorbovitskii, B. M. and Perel, V. I. (1983). Aleksandrov and Zapasskii experiment and the Raman effect. *Opt. Spectrosc.*, **54**, 229.

Gotschy, B., Denninger, G., Obloh, H., Wilkening, W., and Schneider, J. (1989). Overhauser shift and dynamic nuclear polarization in InP. *Solid State Comm.*, **71**, 629.

Greilich, A., Carter, S. G., Kim, D., Bracker, A. S., and Gammon, D. (2011). Optical control of one and two hole spins in interacting quantum dots. *Nature Photon.*, **5**, 702.

Greilich, A., Economou, Sophia E., Spatzek, S., Yakovlev, D. R., Reuter, D., Wieck, A. D., Reinecke, T. L., and Bayer, M. (2009*a*, Apr). Ultrafast optical rotations of electron spins in quantum dots. *Nature Physi.*, **5**, 262.

Greilich, A., Oulton, R., Zhukov, E. A., Yugova, I. A., Yakovlev, D. R., Bayer, M., Shabaev, A., Efros, Al. L., Merkulov, I. A., Stavarache, V., Reuter, D., and Wieck, A. (2006*a*). Optical control of spin coherence in singly charged (In,Ga)As/GaAs quantum dots. *Phys. Rev. Lett.*, **96**, 227401.

Greilich, A., Pawlis, A., Liu, F., Yugov, O. A., Yakovlev, D. R., Lischka, K., Yamamoto, Y., and Bayer, M. (2012). Spin dephasing of fluorine-bound electrons in ZnSe. *Phys. Rev. B*, **85**, 121303.

Greilich, A., Shabaev, A., Yakovlev, D. R., Efros, Al. L., Yugova, I. A., Reuter, D., Wieck, A. D., and Bayer, M. (2007). Nuclei-induced frequency focusing of electron spin coherence. *Science*, **317**, 1896.

Greilich, A., Spatzek, S., Yugova, I. A., Akimov, I. A., Yakovlev, D. R., Efros, Al. L., Reuter, D., Wieck, A. D., and Bayer, M. (2009*b*). Collective single-mode precession of electron spins in an ensemble of singly charged (In,Ga)As/GaAs quantum dots. *Phys. Rev. B*, **79**, 201305.

Greilich, A., Yakovlev, D. R., Shabaev, A., Efros, Al. L., Yugova, I. A., Oulton, R., Stavarache, V., Reuter, D., Wieck, A., and Bayer, M. (2006*b*). Mode locking of electron spin coherences in singly charged quantum dots. *Science*, **313**, 341.

Griesbeck, M., Glazov, M. M., Korn, T., Sherman, E. Y., Waller, D., Reichl, C., Schuh, D., Wegscheider, W., and Schüller, C. (2009). Cyclotron effect on coherent spin precession of two-dimensional electrons. *Phys. Rev. B*, **80**, 241314.

Griesbeck, M., Glazov, M. M., Sherman, E. Y., Schuh, D., Wegscheider, W., Schüller, C., and Korn, T. (2012). Strongly anisotropic spin relaxation revealed by resonant spin amplification in (110) GaAs quantum wells. *Phys. Rev. B*, **85**, 085313.

Gryncharova, E. I. and Perel', V. I. (1977). Relaxation of nuclear spins interacting with holes in semiconductors. *Sov. Phys. Semicond.*, **11**, 997.

Gueron, M. (1964). Density of the conduction electrons at the nuclei in indium antimonide. *Phys. Rev.*, **135**, A200.

Gutzwiller, M. C. (1990). *Chaos in classical and quantum mechanics*. Springer, New York.

Gywat, O., Krenner, H. J., and Berezovsky, J. (2010). *Spins in optically active quantum dots: Concepts and methods*. John Wiley & Sons, Ltd, Chichester.

Hackmann, J. and Anders, F. B. (2014). Spin noise in the anisotropic central spin model. *Phys. Rev. B*, **89**, 045317.

Hackmann, J., Glasenapp, P., Greilich, A., Bayer, M., and Anders, F. B. (2015). Influence of the nuclear electric quadrupolar interaction on the coherence time of hole and electron spins confined in semiconductor quantum dots. *Phys. Rev. Lett.*, **115**, 207401.

Hackmann, J., Smirnov, D. S., Glazov, M. M., and Anders, F. B. (2014). Spin noise in a quantum dot ensemble: From a quantum mechanical to a semi-classical description. *Physica Status Solidi (B)*, **251**, 1270.

Hanson, R., Kouwenhoven, L. P., Petta, J. R., Tarucha, S., and Vandersypen, L. M. K. (2007). Spins in few-electron quantum dots. *Rev. Mod. Phys.*, **79**, 1217.

Harmon, N. J. and Flatté, M. E. (2013). Distinguishing spin relaxation mechanisms in organic semiconductors. *Phys. Rev. Lett.*, **110**, 176602.

Harrison, R. J. and Sagalyn, P. L. (1962). Trace relations for tensors relating electric fields and elastic strains to nuclear quadrupole effects. *Phys. Rev.*, **128**, 1630.

Hayes, R. R., Kiselev, A. A., Borselli, M. G., Bui, S. S., Croke, III, E. T., Deelman, P. W., Maune, B. M., Milosavljevic, I., Moon, J.-S., Ross, R. S., Schmitz, A. E., Gyure, M. F., and Hunter, A. T. (2009). Lifetime measurements (T1) of electron spins in Si/SiGe quantum dots. *ArXiv e-prints*, 0908.0173.

Heiss, D., Kroutvar, M., Finley, J. J., and Abstreiter, G. (2005). Progress towards single spin optoelectronics using quantum dot nanostructures. *Solid State Commun.*, **135**, 591.

Heisterkamp, F., Greilich, A., Zhukov, E. A., Kirstein, E., Kazimierczuk, T., Korenev, V. L., Yugova, I. A., Yakovlev, D. R., Pawlis, A., and Bayer, M. (2015*a*). Inhomogeneous nuclear spin polarization induced by helicity-modulated optical excitation of fluorine-bound electron spins in ZnSe. *Phys. Rev. B*, **92**, 245441.

Heisterkamp, F., Zhukov, E. A., Greilich, A., Yakovlev, D. R., Korenev, V. L., Pawlis, A., and Bayer, M. (2015*b*). Longitudinal and transverse spin dynamics of donor-bound electrons in fluorine-doped ZnSe: Spin inertia versus Hanle effect. *Phys. Rev. B*, **91**, 235432.

Hess, H. F., Betzig, E., Harris, T. D., Pfeiffer, L. N., and West, K. W. (1994). Near-field spectroscopy of the quantum constituents of a luminescent system. *Science*, **264**, 1740.

Högele, A., Kroner, M., Latta, C., Claassen, M., Carusotto, I., Bulutay, C., and Imamoglu, A. (2012). Dynamic nuclear spin polarization in the resonant laser excitation of an InGaAs quantum dot. *Phys. Rev. Lett.*, **108**, 197403.

Horn, H., Müller, G. M., Rasel, E. M., Santos, L., Hübner, J., and Oestreich, M. (2011). Spin-noise spectroscopy under resonant optical probing conditions: Coherent and nonlinear effects. *Phys. Rev. A*, **84**, 043851.

Hu, C. Y., Munro, W. J., and Rarity, J. G. (2008*a*). Deterministic photon entangler using a charged quantum dot inside a microcavity. *Phys. Rev. B*, **78**, 125318.

Hu, C. Y., Young, A., O'Brien, J. L., Munro, W. J., and Rarity, J. G. (2008*b*). Giant optical Faraday rotation induced by a single-electron spin in a quantum dot: Applications to entangling remote spins via a single photon. *Phys. Rev. B*, **78**, 085307.

Huang, C.-W. and Hu, X. (2010). Theoretical study of nuclear spin polarization and depolarization in self-assembled quantum dots. *Phys. Rev. B*, **81**, 205304.

Huang, Q. and Steel, D. S. (2011). Optical excitation effects on spin-noise spectroscopy in semiconductors. *Phys. Rev. B*, **83**, 155204.

Hübner, J., Berski, F., Dahbashi, R., and Oestreich, M. (2014). The rise of spin noise spectroscopy in semiconductors: From acoustic to GHz frequencies. *Physica Status Solidi (B)*, **251**, 1824.

Ibarra-Sierra, V. G., Sandoval-Santana, J. C., Azaizia, S., Carrère, H., Bakaleinikov, L. A., Kalevich, V. K., Ivchenko, E. L., Marie, X., Amand, T., Balocchi, A., and Kunold, A. (2017). Electron-nuclear spin dynamics of Ga^{2+} paramagnetic centers probed by spin-dependent recombination: A master equation approach. *Phys. Rev. B*, **95**, 195204.

Intronati, G. A., Tamborenea, P. I., Weinmann, D., and Jalabert, R. A. (2012). Spin relaxation near the metal-insulator transition: Dominance of the dresselhaus spin-orbit coupling. *Phys. Rev. Lett.*, **108**, 016601.

Ivády, V., Szász, K., Falk, A. L., Klimov, P. V., Christle, D. J., Janzén, E., Abrikosov, I. A., Awschalom, D. D., and Gali, A. (2015). Theoretical model of dynamic spin polarization of nuclei coupled to paramagnetic point defects in diamond and silicon carbide. *Phys. Rev. B*, **92**, 115206.

Ivchenko, E. L., Kaminski, A. Y., and Roessler, U. (1996). Heavy-light hole mixing at zinc-blende (001) interfaces under normal incidence. *Phys. Rev. B*, **54**, 5852.

Ivchenko, E., Kochereshko, V., Platonov, A., Yakovlev, D., Waag, A., Ossau, W., and Landwehr, G. (1997). Resonance optical spectroscopy of long-period quantum-well structures. *Phys. Solid State*, **39**, 1852.

Ivchenko, E. L. (1973). Fluctuations of spin polarization of free carriers in semiconductors. *Sov. Phys. Solid State*, **7**, 998.

Ivchenko, E. L. (2005). *Optical spectroscopy of semiconductor nanostructures*. Alpha Science, Harrow, UK.

Ivchenko, E. L., Bakaleinikov, L. A., and Kalevich, V. K. (2015). Spin-dependent recombination and hyperfine interaction at deep defects. *Phys. Rev. B*, **91**, 205202.

Ivchenko, E. L. and Kiselev, A. A. (1992). Electron g factor of quantum wells and superlattices. *Sov. Phys. Semicond.*, **26**, 827.

Ivchenko, E. L. and Pikus, G. E. (1997). *Superlattices and other heterostructures*. Springer.

Jackson, J. D. (1962). *Classical electrodynamics*. John Wiley & Sons, Inc.

Jancu, J.-M., Scholz, R., Beltram, F., and Bassani, F. (1998). Empirical *spds∗* tight-binding calculation for cubic semiconductors: General method and material parameters. *Phys. Rev. B*, **57**, 6493.

Jancu, J.-M., Scholz, R., de Andrada e Silva, E. A., and Rocca, G. C. La (2005). Atomistic spin-orbit coupling and $\boldsymbol{k}\cdot\boldsymbol{p}$ parameters in III-V semiconductors. *Phys. Rev. B*, **72**, 193201.

Jäschke, N., Fischer, A., Evers, E., Belykh, V. V., Greilich, A., Bayer, M., and Anders, F. B. (2017). Nonequilibrium nuclear spin distribution function in quantum dots subject to periodic pulses. *Phys. Rev. B*, **96**, 205419.

Jaynes, E. T. and Cummings, F. W. (1963). Comparison of quantum and semiclassical radiation theories with application to the beam maser. *Proceedings of the IEEE*, **51**, 89.

Jelezko, F., Gaebel, T., Popa, I., Domhan, M., Gruber, A., and Wrachtrup, J. (2004). Observation of coherent oscillation of a single nuclear spin and realization of a two-qubit conditional quantum gate. *Phys. Rev. Lett.*, **93**, 130501.

Johnson, A. C., Petta, J. R., Taylor, J. M., Yacoby, A., Lukin, M. D., Marcus, C. M., Hanson, M. P., and Gossard, A. C. (2005). Triplet-singlet spin relaxation via nuclei in a double quantum dot. *Nature*, **435**, 925.

Kagan, Y. (1999). Theory of coherent phenomena and fundamentals in nuclear resonant scattering. *Hyperfine Interactions*, **123**, 83.

Kaji, R., Adachi, S., Sasakura, H., and Muto, S. (2008). Hysteretic response of the electron-nuclear spin system in single $In_{0.75}Al_{0.25}As$ quantum dots: Dependences on excitation power and polarization. *Phys. Rev. B*, **77**, 115345.

Kalevich, V. K., Zakharchenya, B. P., Kavokin, K. V., Petrov, A. V., Le Jeune, P., Marie, X., Robart, D., Amand, T., Barrau, J., and Brousseau, M. (1997). Determination of the sign of the conduction-electron g factor in semiconductor quantum wells by means of the Hanle effect and spin-quantum-beat techniques. *Phys. Solid State*, **39**, 681.

Kalevich, V. K. (1986). Optically induced nuclear magnetic resonance in semiconductors. *Sov. Phys. Solid State*, **28**, 1947.

Kalevich, V. K., Ivchenko, E. L., Afanasiev, M. M., Shiryaev, A. Yu., Egorov, A. Yu., Ustinov, V. M., Pal, B., and Masumoto, Y. (2005). Spin-dependent recombination in GaAsN solid solutions. *JETP Lett.*, **82**, 455.

Kalevich, V. K., Kavokin, K. V., Merkulov, Igor, and Vladimirova, M. R. (2017). Dynamic nuclear polarization and nuclear fields. In *Spin physics in semiconductors, Second Edition*, (ed. M. I. Dyakonov), Springer Series in Solid-State Sciences, Vol. 157, pp. 309. Springer International Publishing.

Kalevich, V. K. and Korenev, V. L. (1993). Electron g-factor anisotropy in asymmetric GaAs/AlGaAs quantum well. *JETP Lett.*, **57**, 571.

Kalevich, V. K., Korenev, V. L., and Fedorova, O. M. (1990). Optical polarization of nuclei in GaAs/AIGaAs quantum-well structures. *JETP Lett.*, **52**, 349.

Kalevich, V. K., Kul'kov, V. D., Merkulov, I. A., and Fleisher, V. G. (1982). Two-spin resonance transitions of optically oriented nuclei in a semiconductor lattice. *Sov. Phys. Solid State*, **24**, 1195.

Kaminski, A., Nazarov, Y. V., and Glazman, L. I. (2000). Universality of the Kondo effect in a quantum dot out of equilibrium. *Phys. Rev. B*, **62**, 8154.

Kane, B. E. (1998). A silicon-based nuclear spin quantum computer. *Nature*, **393**, 133.

Karlsson, K. F., Dupertuis, M. A., Oberli, D. Y., Pelucchi, E., Rudra, A., Holtz, P. O., and Kapon, E. (2010). Fine structure of exciton complexes in high-symmetry quantum dots: Effects of symmetry breaking and symmetry elevation. *Phys. Rev. B*, **81**, 161307.

Kasprzak, J., Reitzenstein, S., Muljarov, E. A., Kistner, C., Schneider, C., Strauss, M., Hofling, S., Forchel, A., and Langbein, W. (2010). Up on the Jaynes–Cummings ladder of a quantum-dot/microcavity system. *Nature Mater.*, **9**, 304.

Kato, Y. K., Myers, R. C., Gossard, A. C., and Awschalom, D. D. (2004). Observation of the spin Hall effect in semiconductors. *Science*, **306**, 1910.

Kavokin, A., Baumberg, J. J., Malpuech, G., and Laussy, F. P. (2011). *Microcavities*. Oxford University Press, Oxford.

Kavokin, K. V. (2004). Symmetry of anisotropic exchange interactions in semiconductor nanostructures. *Phys. Rev. B*, **69**, 075302.

Kavokin, K. V. (2008). Spin relaxation of localized electrons in n-type semiconductors. *Semicond. Sci. Technol.*, **23**, 114009.

Kavokin, K. V. and Koudinov, A. V. (2013). Dynamical polarization of nuclear spins by acceptor-bound holes in a zinc-blende semiconductor. *Phys. Rev. B*, **88**, 235202.

Kehr, K. W., Honig, G., and Richter, D. (1978). Stochastic theory of spin depolarization of muons diffusing in the presence of traps. *Zeitschrift für Physik B Conden. Matt.*, **32**, 49.

Kennedy, T. A., Shabaev, A., Scheibner, M., Efros, Al. L., Bracker, A. S., and Gammon, D. (2006). Optical initialization and dynamics of spin in a remotely doped quantum well. *Phys. Rev. B*, **73**, 045307.

Khaetskii, A. V. (2001). Spin relaxation in semiconductor mesoscopic systems. *Physica E*, **10**, 27.

Khaetskii, A. V., Loss, D., and Glazman, L. (2002). Electron spin decoherence in quantum dots due to interaction with nuclei. *Phys. Rev. Lett.*, **88**, 186802.

Khaetskii, A. V. and Nazarov, Y. V. (2001). Spin-flip transitions between Zeeman sublevels in semiconductor quantum dots. *Phys. Rev. B*, **64**, 125316.

Khutsishvili, G. R. (1966). Spin diffusion. *Soviet Physics Uspekhi*, **8**, 743.

Kikkawa, J. M. and Awschalom, D. D. (1998). Resonant spin amplification in *n*-type GaAs. *Phys. Rev. Lett.*, **80**, 4313.

Kikkawa, J. M. and Awschalom, D. D. (1999). Lateral drag of spin coherence in gallium arsenide. *Nature*, **397**, 139.

Kikkawa, J. M. and Awschalom, D. D. (2000). All-optical magnetic resonance in semiconductors. *Science*, **287**, 473.

Kikkawa, J. M., Smorchkova, I. P., Samarth, N., and Awschalom, D. D. (1997). Room-temperature spin memory in two-dimensional electron gases. *Science*, **277**, 1284.

Kim, D., Carter, S. G., Greilich, A., Bracker, A. S., and Gammon, D. (2011). Ultrafast optical control of entanglement between two quantum-dot spins. *Nat. Phys.*, **7**, 223.

Kiselev, A. A. (1993). Catastrophes in an electron-nuclear spin system with an anisotropic electron *g* factor. *Phys. Solid State*, **35**, 114.

Knipp, P. A. and Reinecke, T. L. (1995). Coupling between electrons and acoustic phonons in semiconductor nanostructures. *Phys. Rev. B*, **52**, 5923.

Kogan, S. M. and Shul'man, A. Y. (1969). Theory of fluctuations in a noneqilibrium electron gas. *JETP*, **29**, 467.

Koh, A. K. and Miller, D. J. (1985). Hyperfine coupling constants and atomic parameters for electron paramagnetic resonance data. *Atomic Data and Nuclear Data Tables*, **33**, 235.

Korenev, V. L. (1999). Dynamic self-polarization of nuclei in low-dimensional systems. *JETP Lett.*, **70**, 129.

Korenev, V. L. (2007). Nuclear spin nanomagnet in an optically excited quantum dot. *Phys. Rev. Lett.*, **99**, 256405.

Korenev, V. L. (2011). Multiple stable states of a periodically driven electron spin in a quantum dot using circularly polarized light. *Phys. Rev. B*, **83**, 235429.

Korn, T. (2010). Time-resolved studies of electron and hole spin dynamics in modulation-doped GaAs/AlGaAs quantum wells. *Phys. Reports*, **494**, 415.

Korn, T., Kugler, M., Griesbeck, M., Schulz, R., Wagner, A., Hirmer, M., Gerl, C., Schuh, D., Wegscheider, W., and Schueller, C. (2010). Engineering ultralong spin coherence in two-dimensional hole systems at low temperatures. *New J. Phys.*, **12**, 043003.

Kos, S. S., Balatsky, A. V., Littlewood, P. B., and Smith, D. L. (2010). Spin noise of itinerant fermions. *Phys. Rev. B*, **81**, 064407.

Kosaka, H., Inagaki, T., Rikitake, Y., Imamura, H., Mitsumori, Y., and Edamatsu, K. (2009). Spin state tomography of optically injected electrons in a semiconductor. *Nature*, **457**, 702.

Koster, G. F., Wheeler, R. G., Dimmock, J. O., and Statz, H. (1963). *Properties of the thirty-two point groups*. MIT Press.

Kotur, M., Dzhioev, R. I., Vladimirova, M., Jouault, B., Korenev, V. L., and Kavokin, K. V. (2016). Nuclear spin warm up in bulk *n*-GaAs. *Phys. Rev. B*, **94**, 081201.

Kozlov, G. (2007). Exactly solvable spin dynamics of an electron coupled to a large number of nuclei; the electron-nuclear spin echo in a quantum dot. *JETP*, **105**, 803.

Kozlov, G. G. (2008). The limiting nuclear polarization in a quantum dot under optical electron-spin orientation and applicability of the box-model of the electron-nuclear dynamics. *ArXiv e-prints*, 0801.1391.

Krebs, O., Eble, B., Lemaitre, A., Voisin, P., Urbaszek, B., Amand, T., and Marie, X. (2008). Hyperfine interaction in InAs/GaAs self-assembled quantum dots: dynamical nuclear polarization versus spin relaxation. *Comptes Rendus Physique*, **9**, 874.

Krebs, O., Maletinsky, P., Amand, T., Urbaszek, B., Lemaître, A., Voisin, P., Marie, X., and Imamoglu, A. (2010). Anomalous Hanle effect due to optically created transverse Overhauser field in single InAs/GaAs quantum dots. *Phys. Rev. Lett.*, **104**, 056603.

Kroutvar, M., Ducommun, Y., Heiss, D., Bichler, M., Schuh, D., Abstreiter, G., and Finley, J. J. (2004). Optically programmable electron spin memory using semiconductor quantum dots. *Nature*, **432**, 81.

Kubo, R. and Toyabe, T. (1967). A stochastic model for low field resonance and relaxation. In *Magnetic and electric resonance and relaxation* (ed. R. Blinc), p. 810. North Holland, Amsterdam.

Kugler, M., Korzekwa, K., Machnikowski, P., Gradl, C., Furthmeier, S., Griesbeck, M., Hirmer, M., Schuh, D., Wegscheider, W., Kuhn, T., Schüller, C., and Korn, T. (2011). Decoherence-assisted initialization of a resident hole spin polarization in a *p*-doped semiconductor quantum well. *Phys. Rev. B*, **84**, 085327.

Kunold, A., Balocchi, A., Zhao, F., Amand, T., Abdallah, N. Ben, Harmand, J. C., and Marie, X. (2011). Giant spin-dependent photo-conductivity in GaAsN dilute nitride semiconductor. *Phys. Rev. B*, **83**, 165202.

Kusrayev, Y. G., Koudinov, A. V., Aksyanov, I. G., Zakharchenya, B. P., Wojtowicz, T., Karczewski, G., and Kossut, J. (1999). Extreme in-plane anisotropy of the heavy-hole *g* factor in (001)-CdTe/CdMnTe quantum wells. *Phys. Rev. Lett.*, **82**, 3176.

Kuznetsova, M. S., Flisinski, K., Gerlovin, I. Ya., Petrov, M. Y., Ignatiev, I. V., Verbin, S. Y., Yakovlev, D. R., Reuter, D., Wieck, A. D., and Bayer, M. (2014). Nuclear magnetic resonances in (In,Ga)As/GaAs quantum dots studied by resonant optical pumping. *Phys. Rev. B*, **89**, 125304.

Lai, C. W., Maletinsky, P., Badolato, A., and Imamoglu, A. (2006). Knight-field-enabled nuclear spin polarization in single quantum dots. *Phys. Rev. Lett.*, **96**, 167403.

Landau, L. D. and Lifshitz, E. M. (1975). *The classical theory of fields*. Butterworth-Heinemann, Oxford.

Landau, L. D. and Lifshitz, E. M. (1976). *Mechanics*. Butterworth-Heinemann, Oxford.

Landau, L. D. and Lifshitz, E. M. (1977). *Quantum mechanics: Non-relativistic theory*. Butterworth-Heinemann, Oxford.

Landau, L. D. and Lifshitz, E. M. (1981). *Physical kinetics*. Butterworth-Heinemann, Oxford.

Landau, L. D. and Lifshitz, E. M. (2000). *Statistical physics, Part 1*. Butterworth-Heinemann, Oxford.

Landau, L. D. and Lifshitz, E. M. (2004). *Electrodynamics of continuous media (vol. 8)*. Butterworth-Heinemann, Oxford.

Landau, L. D., Pitaevskii, L. P., Kosevich, A. M., and Lifshitz, E. M. (1999). *Theory of elasticity*. Butterworth-Heinemann, Oxford.

Landau, L. and Smorodinsky, Y. (2014). *Lectures on nuclear theory*. Springer.

Latta, C., Hogele, A., Zhao, Y., Vamivakas, A. N., Maletinsky, P., Kroner, M., Dreiser, J., Carusotto, I., Badolato, A., Schuh, D., Wegscheider, W., Atature, M., and Imamoglu, A. (2009). Confluence of resonant laser excitation and bidirectional quantum-dot nuclear-spin polarization. *Nat. Phys.*, **5**, 758.

Latta, C., Srivastava, A., and Imamoğlu, A. (2011). Hyperfine interaction-dominated dynamics of nuclear spins in self-assembled InGaAs quantum dots. *Phys. Rev. Lett.*, **107**, 167401.

Lauck, R. and Schönherr, E. (1999). Isotopically pure ZnSe crystals grown from the vapor. *J. Cryst. Growth*, **197**, 513.

Lax, M. (1960). Fluctuations from the nonequilibrium steady state. *Rev. Mod. Phys.*, **32**, 25.

Lax, M. (1966). Classical noise IV: Langevin methods. *Rev. Mod. Phys.*, **38**, 541.

Leyland, W. J. H., John, G. H., Harley, R. T., Glazov, M. M., Ivchenko, E. L., Ritchie, D. A., Farrer, I., Shields, A. J., and Henini, M. (2007). Enhanced spin-relaxation time due to electron-electron scattering in semiconductor quantum wells. *Phys. Rev. B*, **75**, 165309.

Li, F. and Sinitsyn, N. A. (2016). Universality in higher order spin noise spectroscopy. *Phys. Rev. Lett.*, **116**, 026601.

Li, Y., Sinitsyn, N., Smith, D. L., Reuter, D., Wieck, A. D., Yakovlev, D. R., Bayer, M., and Crooker, S. A. (2012). Intrinsic spin fluctuations reveal the dynamical response function of holes coupled to nuclear spin baths in (In,Ga)As quantum dots. *Phys. Rev. Lett.*, **108**, 186603.

Ling, P. A. (2005). *Quantum dots: Research developments*. Horizons in world physics. Nova Science Publishers.

Linpeng, X., Karin, T., Durnev, M. V., Barbour, R., Glazov, M. M., Sherman, E. Y., Watkins, S. P., Seto, S., and Fu, K.-M. C. (2016). Longitudinal spin relaxation of donor-bound electrons in direct band-gap semiconductors. *Phys. Rev. B*, **94**, 125401.

Liu, R.-B., Fung, S.-H., Fung, H.-K., Korotkov, A. N., and Sham, L. J. (2010). Dynamics revealed by correlations of time-distributed weak measurements of a single spin. *New J. Phys.*, **12**, 013018.

Liu, R.-B., Yao, W., and Sham, L. J. (2007). Control of electron spin decoherence caused by electron–nuclear spin dynamics in a quantum dot. *New J. Phys.*, **9**, 226.

Loo, V., Lanco, L., Krebs, O., Senellart, P., and Voisin, P. (2011). Single-shot initialization of electron spin in a quantum dot using a short optical pulse. *Phys. Rev. B*, **83**, 033301.

Look, D. C. and Moore, D. L. (1972). Nuclear-magnetic-resonance measurement of the conduction-electron g factor in CdTe. *Phys. Rev. B*, **5**, 3406.

Lu, C.-Y., Zhao, Y., Vamivakas, A. N., Matthiesen, C., Fält, S., Badolato, A., and Atatüre, M. (2010). Direct measurement of spin dynamics in InAs/GaAs quantum dots using time-resolved resonance fluorescence. *Phys. Rev. B*, **81**, 035332.

Machnikowski, P. and Kuhn, T. (2010). Theory of the time-resolved Kerr rotation in ensembles of trapped holes in semiconductor nanostructures. *Phys. Rev. B*, **81**, 115306.

Madelung, O. (2004). *Semiconductors: Data handbook, Third Edition*, Springer.

Madelung, O., Rössler, U., and Schulz, M. (ed.) (1999). *II–VI and I–VII compounds; Semimagnetic compounds*. Landolt-Börnstein – Group III Condensed Matter, Volume 41B. Springer-Verlag.

Maialle, M. Z., de Andrada e Silva, E. A., and Sham, L. J. (1993). Exciton spin dynamics in quantum wells. *Phys. Rev. B*, **47**, 15776.

Maletinsky, P., Kroner, M., and Imamoglu, A. (2009). Breakdown of the nuclear-spin-temperature approach in quantum-dot demagnetization experiments. *Nat. Phys.*, **5**, 407.

Maletinsky, P., Lai, C. W., Badolato, A., and Imamoglu, A. (2007). Nonlinear dynamics of quantum dot nuclear spins. *Phys. Rev. B*, **75**, 035409.

Mano, T., Abbarchi, M., Kuroda, T., McSkimming, B., Ohtake, A., Mitsuishi, K., and Sakoda, K. (2010). Self-assembly of symmetric GaAs quantum dots on (111)A substrates: Suppression of fine-structure splitting. *Appl. Phys. Expr.*, **3**, 065203.

Marie, X., Amand, T., Le Jeune, P., Paillard, M., Renucci, P., Golub, L. E., Dymnikov, V. D., and Ivchenko, E. L. (1999). Hole spin quantum beats in quantum-well structures. *Phys. Rev. B*, **60**, 5811.

Marple, D. T. F. and Ehrenreich, H. (1962). Dielectric constant behavior near band edges in CdTe and Ge. *Phys. Rev. Lett.*, **8**, 87.

Marquardt, F. and Abalmassov, V. A. (2005). Spin relaxation in a quantum dot due to Nyquist noise. *Phys. Rev. B*, **71**, 165325.

Masumoto, Y. and Takagahara, T. (2002). *Semiconductor quantum dots: Physics, spectroscopy and applications.* Lecture Notes in Computer Science. Springer.

Meier, F. and Zakharchenya, B. (eds) (1984). *Optical orientation.* North-Holland, Amsterdam.

Merkulov, I. A. (1998). Formation of a nuclear spin polaron under optical orientation in GaAs-type semiconductors. *Phys. Solid State*, **40**, 930.

Merkulov, I. A. (2002). Spin systems of quantum dots. *Physics-Uspekhi*, **45**, 1293.

Merkulov, I. A., Alvarez, G., Yakovlev, D. R., and Schulthess, T. C. (2010). Long-term dynamics of the electron-nuclear spin system of a semiconductor quantum dot. *Phys. Rev. B*, **81**, 115107.

Merkulov, I. A., Efros, Al. L., and Rosen, M. (2002). Electron spin relaxation by nuclei in semiconductor quantum dots. *Phys. Rev. B*, **65**, 205309.

Merkulov, I. A., Yakovlev, D. R., Kavokin, K. V., Mackh, G., Ossau, W., Waag, A., and Landwehr, G. (1995). Hierarchy of relaxation times in the formation of an excitonic magnetic polaron in (CdMn)Te. *JETP Lett.*, **62**, 335.

Mikkelsen, M. H., Berezovsky, J., Stoltz, N. G., Coldren, L. A., and Awschalom, D. D. (2007). Optically detected coherent spin dynamics of a single electron in a quantum dot. *Nature Physics*, **3**, 770.

Mishra, S. and Bray, R. (1977). Surface-ripple mechanism for Brillouin scattering of reflected light from bulk acoustic waves. *Phys. Rev. Lett.*, **39**, 222.

Mitsui, T. (2000). Spontaneous noise spectroscopy of an atomic magnetic resonance. *Phys. Rev. Lett.*, **84**, 5292.

Mkhitaryan, V. V. and Dobrovitski, V. V. (2015). Hyperfine-induced spin relaxation of a diffusively moving carrier in low dimensions: Implications for spin transport in organic semiconductors. *Phys. Rev. B*, **92**, 054204.

Mocek, R. W., Korenev, V. L., Bayer, M., Kotur, M., Dzhioev, R. I., Tolmachev, D. O., Cascio, G., Kavokin, K. V., and Suter, D. (2017). High-efficiency optical pumping of nuclear polarization in a GaAs quantum well. *Phys. Rev. B*, **96**, 201303.

Morley, G. W., McCamey, D. R., Seipel, H. A., Brunel, L.-C., van Tol, J., and Boehme, C. (2008). Long-lived spin coherence in silicon with an electrical spin trap readout. *Phys. Rev. Lett.*, **101**, 207602.

Morley, G. W., Warner, M., Stoneham, A. M., Greenland, P. T., van Tol, J., Kay, C. W. M., and Aeppli, G. (2010). The initialization and manipulation of quantum information stored in silicon by bismuth dopants. *Nat Mater*, **9**, 725.

Morton, J. R. and Preston, K. F. (1978). Atomic parameters for paramagnetic resonance data. *J. Magn. Reson. (1969)*, **30**, 577.

Morton, J. J. L., Tyryshkin, A. M., Brown, R. M., Shankar, S., Lovett, B. W., Ardavan, A., Schenkel, T., Haller, E. E., Ager, J. W., and Lyon, S. A. (2008). Solid-state quantum memory using the ^{31}P nuclear spin. *Nature*, **455**, 1085.

Mourik, V., Asaad, S., Firgau, H., Pla, J. J., Holmes, C., Milburn, G. J., McCallum, J. C., and Morello, A. (2017). An experimentally realizable single-atom chaotic driven top in silicon. *ArXiv e-prints*, 1703.04852.

Muhonen, J. T., Dehollain, J. P., Laucht, A., Hudson, F. E., Kalra, R., Sekiguchi, T., Itoh, K. M., Jamieson, D. N., McCallum, J. C., Dzurak, A. S., and Morello, A. (2014). Storing quantum information for 30 seconds in a nanoelectronic device. *Nat. Nano.*, **9**, 986.

Müller, G. M., Oestreich, M., Roemer, M., and Huebner, J. (2010). Semiconductor spin noise spectroscopy: Fundamentals, accomplishments, and challenges. *Physica E*, **43**, 569.

Müller, G. M., Römer, M., Schuh, D., Wegscheider, W., Hübner, J., and Oestreich, M. (2008). Spin noise spectroscopy in GaAs (110) quantum wells: Access to intrinsic spin lifetimes and equilibrium electron dynamics. *Phys. Rev. Lett.*, **101**, 206601.

Müller, N. and Jerschow, A. (2006). Nuclear spin noise imaging. *Proc. Nat. Acad. Sci.*, **103**, 6790.

Nakamura, A., Paget, D., Hermann, C., Weisbuch, C., Lampel, G., and Cavenett, B. C (1979). Optical detection of electron spin resonance in CDTE. *Solid State Comm.*, **30**, 411.

Nestoklon, M. O., Golub, L. E., and Ivchenko, E. L. (2006). Spin and valley-orbit splittings in SiGe/Si heterostructures. *Phys. Rev. B*, **73**, 235334.

Nestoklon, M. O., Tarasenko, S. A., Benchamekh, R., and Voisin, P. (2016). Spin splitting of electron states in lattice-mismatched (110)-oriented quantum wells. *Phys. Rev. B*, **94**, 115310.

Nikolaenko, A. E., Chekhovich, E. A., Makhonin, M. N., Drouzas, I. W., Van'kov, A. B., Skiba-Szymanska, J., Skolnick, M. S., Senellart, P., Martrou, D., Lemaître, A., and Tartakovskii, A. I. (2009). Suppression of nuclear spin diffusion at a GaAs/Al_xGa_{1-x}As interface measured with a single quantum-dot nanoprobe. *Phys. Rev. B*, **79**, 081303.

Novikov, V. A. and Fleisher, V. G. (1976). Cooling of the spin system of semiconductor lattice nuclei in the field of electrons oriented by light. *JETP*, **44**, 410.

Obata, Y. (1963). Nuclear magnetic relaxation in transition metals. *J. Phys. Soc. Jpn*, **18**, 1020.

Oestreich, M., Römer, M., Haug, R. J., and Hägele, D. (2005). Spin noise spectroscopy in GaAs. *Phys. Rev. Lett.*, **95**, 216603.

Ohkawa, F. J. and Uemura, Y. (1974). Quantized surface states of a narrow-gap semiconductor. *J. Phys. Soc. Jpn*, **37**, 1325.

O'Leary, S. and Wang, H. (2008). Manipulating nonlinear optical responses from spin-polarized electrons in a two-dimensional electron gas via exciton injection. *Phys. Rev. B*, **77**, 165309.

Onsager, L. (1931). Reciprocal relations in irreversible processes. I. *Phys. Rev.*, **37**, 405.

Oulton, R., Greilich, A., Verbin, S. Yu., Cherbunin, R. V., Auer, T., Yakovlev, D. R., Bayer, M., Merkulov, I. A., Stavarache, V., Reuter, D., and Wieck, A. D. (2007). Subsecond spin relaxation times in quantum dots at zero applied magnetic field due to a strong electron-nuclear interaction. *Phys. Rev. Lett.*, **98**, 107401.

Overhauser, A. W. (1953). Polarization of nuclei in metals. *Phys. Rev.*, **92**, 411.

Paget, D. (1982). Optical detection of NMR in high-purity GaAs: Direct study of the relaxation of nuclei close to shallow donors. *Phys. Rev. B*, **25**, 4444.

Paget, D., Amand, T., and Korb, J.-P. (2008, Jun). Light-induced nuclear quadrupolar relaxation in semiconductors. *Phys. Rev. B*, **77**, 245201.

Paget, D., Lampel, G., Sapoval, B., and Safarov, V. I. (1977). Low field electron-nuclear spin coupling in gallium arsenide under optical pumping conditions. *Phys. Rev. B*, **15**, 5780.

Pavlov, S. T. (1966). Spin-lattice relaxation of conduction electrons in semiconductors of $n-$InSb type. *Sov. Phys. Solid State*, **8**, 719.

Pavlov, S. T. and Firsov, Y. A. (1966). Spin flip interaction of electrons with optical phonons in semiconductors. *Sov. Phys. Solid State*, **7**, 2131.

Pavlov, S. T. and Firsov, Y. A. (1967). On spin-phonon interaction of electrons and oscillations of longitudinal magnetoresistance in semiconductors. *Sov. Phys. Solid State*, **9**, 1394.

Pershan, P. S., van der Ziel, J. P., and Malmstrom, L. D. (1966). Theoretical discussion of the inverse Faraday effect, Raman scattering, and related phenomena. *Phys. Rev.*, **143**, 574.

Persson, C. and Zunger, A. (2003). *s–d* coupling in zinc-blende semiconductors. *Phys. Rev. B*, **68**, 073205.

Petrov, M. and Yakovlev, S. (2012). Comparison of quantum-mechanical and semiclassical approaches for an analysis of spin dynamics in quantum dots. *JETP*, **115**, 326.

Petrov, M. Y., Ignatiev, I. V., Poltavtsev, S. V., Greilich, A., Bauschulte, A., Yakovlev, D. R., and Bayer, M. (2008). Effect of thermal annealing on the hyperfine interaction in InAs/GaAs quantum dots. *Phys. Rev. B*, **78**, 045315.

Petrov, M. Y., Kozlov, G. G., Ignatiev, I. V., Cherbunin, R. V., Yakovlev, D. R., and Bayer, M. (2009). Coupled electron-nuclear spin dynamics in quantum dots: A graded box model approach. *Phys. Rev. B*, **80**, 125318.

Petrovićfi, M. D. and Vukmirović, N. (2012). Spin relaxation in CdTe quantum dots with a single Mn atom. *Phys. Rev. B*, **85**, 195311.

Petta, J. R., Johnson, A. C., Taylor, J. M., Laird, E. A., Yacoby, A., Lukin, M. D., Marcus, C. M., Hanson, M. P., and Gossard, A. C. (2005). Coherent manipulation of coupled electron spins in semiconductor quantum dots. *Science*, **309**, 2180.

Phelps, C., Sweeney, T., Cox, R. T., and Wang, H. (2009). Ultrafast coherent electron spin flip in a modulation-doped CdTe quantum well. *Phys. Rev. Lett.*, **102**, 237402.

Pikus, G. E., Maruschak, V. A., and Titkov, A. N. (1988). Spin splitting of energy-bands and spin relaxation of carriers in cubic III-V crystals. *Sov. Phys. Semicond.*, **22**, 115.

Pines, D., Bardeen, J., and Slichter, C. P. (1957). Nuclear polarization and impurity-state spin relaxation processes in silicon. *Phys. Rev.*, **106**, 489.

Pitaevskii, L. P. (1961). Electric forces in a transparent dispersive medium. *JETP*, **12**, 1008.

Poggio, M. and Awschalom, D. D. (2005). High-field optically detected nuclear magnetic resonance in GaAs. *Appl. Phys. Lett.*, **86**, 182103.

Press, D., De Greve, K., McMahon, P. L., Ladd, T. D., Friess, B., Schneider, C., Kamp, M., Hofling, S., Forchel, A., and Yamamoto, Y. (2010). Ultrafast optical spin echo in a single quantum dot. *Nat. Photon.*, **4**, 367.

Provotorov, B. N. (1962). Magnetic resonance saturation in crystals. *JETP*, **14**, 1126.

Purcell, E. M. (1946). Spontaneous emission probabilities at radio frequencies. *Phys. Rev.*, **69**, 681.

Purcell, E. M. and Pound, R. V. (1951). A nuclear spin system at negative temperature. *Phys. Rev.*, **81**, 279.

Rabi, I. I., Zacharias, J. R., Millman, S., and Kusch, P. (1938). A new method of measuring nuclear magnetic moment. *Phys. Rev.*, **53**, 318.

Ramsay, A. J. (2010). A review of the coherent optical control of the exciton and spin states of semiconductor quantum dots. *Semicond. Science Technol.*, **25**, 103001.

Rashba, E. I. and Sheka, V. I. (1959). Symmetry of energy bands in crystals of wurtzite type. II. Symmetry of bands with spin-orbit interaction included. *Fiz. Tverd. Tela: Collected Papers*, **2**, 162.

Römer, M., Bernien, H., Müller, G., Schuh, D., Hübner, J., and Oestreich, M. (2010). Electron-spin relaxation in bulk GaAs for doping densities close to the metal-to-insulator transition. *Phys. Rev. B*, **81**, 075216.

Rosen, N. and Zener, C. (1932). Double Stern–Gerlach experiment and related collision phenomena. *Phys. Rev.*, **40**, 502.

Roth, L. M., Lax, B., and Zwerdling, S. (1959). Theory of optical magneto-absorption effects in semiconductors. *Phys. Rev.*, **114**, 90.

Rottner, K., Helbig, R., and Müller, G. (1993). Piezoelectric constant of InP. *Appl. Phys. Lett.*, **62**, 352.

Roundy, R. C. and Raikh, M. E. (2014). Spin relaxation of a diffusively moving carrier in a random hyperfine field. *Phys. Rev. B*, **90**, 201203.

Ryabchenko, S. M. and Semenov, Y. G. (1983). Spin-correlation effects for a large-radius electron center in a magnetically mixed semiconductor. *JETP*, **57**, 825.

Ryzhov, I. I., Kozlov, G. G., Smirnov, D. S., Glazov, M. M., Efimov, Y. P., Eliseev, S. A., Lovtcius, V. A., Petrov, V. V., Kavokin, K. V., Kavokin, A. V., and Zapasskii, V. S. (2016). Spin noise explores local magnetic fields in a semiconductor. *Sci. Rep.*, **6**, 21062.

Ryzhov, I. I., Poltavtsev, S. V., Kavokin, K. V., Glazov, M. M., Kozlov, G. G., Vladimirova, M., Scalbert, D., Cronenberger, S., Kavokin, A. V., Lemaître, A., Bloch, J., and Zapasskii, V. S. (2015). Measurements of nuclear spin dynamics by spin-noise spectroscopy. *Appl. Phys. Lett.*, **106**, 242405.

Sallen, G., Kunz, S., Amand, T., Bouet, L., Kuroda, T., Mano, T., Paget, D., Krebs, O., Marie, X., Sakoda, K., and Urbaszek, B. (2014). Nuclear magnetization in gallium arsenide quantum dots at zero magnetic field. *Nat. Communs.*, **5**, 3268.

Sallen, G., Urbaszek, B., Glazov, M. M., Ivchenko, E. L., Kuroda, T., Mano, T., Kunz, S., Abbarchi, M., Sakoda, K., Lagarde, D., Balocchi, A., Marie, X., and Amand, T. (2011). Dark-bright mixing of interband transitions in symmetric semiconductor quantum dots. *Phys. Rev. Lett.*, **107**, 166604.

Sanaka, K., Pawlis, A., Ladd, T. D., Sleiter, D. J., Lischka, K., and Yamamoto, Y. (2012). Entangling single photons from independently tuned semiconductor nanoemitters. *Nano Lett.*, **12**, 4611.

Scalbert, D. (2017). Nuclear polaron beyond the mean-field approximation. *Phys. Rev. B*, **95**, 245209.

Scarlino, P., Kawakami, E., Stano, P., Shafiei, M., Reichl, C., Wegscheider, W., and Vandersypen, L. M. K. (2014). Spin-relaxation anisotropy in a GaAs quantum dot. *Phys. Rev. Lett.*, **113**, 256802.

Schofield, S. R., Curson, N. J., Simmons, M. Y., Rueß, F. J., Hallam, T., Oberbeck, L., and Clark, R. G. (2003). Atomically precise placement of single dopants in Si. *Phys. Rev. Lett.*, **91**, 136104.

Schrieffer, J. R. and Wolff, P. A. (1966). Relation between the Anderson and Kondo Hamiltonians. *Phys. Rev.*, **149**, 491.

Schulten, K. and Wolynes, P. G. (1978). Semiclassical description of electron spin motion in radicals including the effect of electron hopping. *J. Chem. Phys.*, **68**, 3292.

Segarra, C., Planelles, J., Climente, J. I., and Rajadell, F. (2015). Anisotropy of spin–orbit induced electron spin relaxation in [001] and [111] grown GaAs quantum dots. *New J. Phys.*, **17**, 033014.

Semenov, Y. G. and Kim, K. W. (2003). Effect of an external magnetic field on electron-spin dephasing induced by hyperfine interaction in quantum dots. *Phys. Rev. B*, **67**, 073301.

Shabaev, A., Efros, Al. L., Gammon, D., and Merkulov, I. A. (2003). Optical readout and initialization of an electron spin in a single quantum dot. *Phys. Rev. B*, **68**, 201305.

Shchepetilnikov, A. V., Frolov, D. D., Nefyodov, Yu. A., Kukushkin, I. V., Smirnov, D. S., Tiemann, L., Reichl, C., Dietsche, W., and Wegscheider, W. (2016). Nuclear magnetic resonance and nuclear spin relaxation in AlAs quantum well probed by ESR. *Phys. Rev. B*, **94**, 241302.

Sherman, E. Y. and Lockwood, D. J. (2005). Spin relaxation in quantum dots with random spin-orbit coupling. *Phys. Rev. B*, **72**, 125340.

Shklovskii, B. I. and Efros, A. L. (1984). *Electronic properties of doped semiconductors*. Springer Series in Solid-State Sciences. Springer-Verlag, Berlin/Heidelberg.

Shulman, R. G., Mays, J. M., and McCall, D. W. (1955). Nuclear magnetic resonance in semiconductors. I. Exchange broadening in InSb and GaSb. *Phys. Rev.*, **100**, 692.

Shulman, R. G., Wyluda, B. J., and Anderson, P. W. (1957). Nuclear magnetic resonance in semiconductors. II. Quadrupole broadening of nuclear magnetic resonance lines by elastic axial deformation. *Phys. Rev.*, **107**, 953.

Shumilin, A. V. and Kabanov, V. V. (2015). Kinetic equations for hopping transport and spin relaxation in a random magnetic field. *Phys. Rev. B*, **92**, 014206.

Shumilin, A. V., Sherman, E. Y., and Glazov, M. M. (2016). Spin dynamics of hopping electrons in quantum wires: Algebraic decay and noise. *Phys. Rev. B*, **94**, 125305.

Simin, D., Soltamov, V. A., Poshakinskiy, A. V., Anisimov, A. N., Babunts, R. A., Tolmachev, D. O., Mokhov, E. N., Trupke, M., Tarasenko, S. A., Sperlich, A., Baranov, P. G., Dyakonov, V., and Astakhov, G. V. (2016). All-optical DC nanotesla magnetometry using silicon vacancy fine structure in isotopically purified silicon carbide. *Phys. Rev. X*, **6**, 031014.

Sinitsyn, N. A., Li, Yan, Crooker, S. A., Saxena, A., and Smith, D. L. (2012). Role of nuclear quadrupole coupling on decoherence and relaxation of central spins in quantum dots. *Phys. Rev. Lett.*, **109**, 166605.

Sinitsyn, N. A. and Pershin, Y. V. (2016). The theory of spin noise spectroscopy: a review. *Rep. Progr. Phys.*, **79**, 106501.

Skiba-Szymanska, J., Chekhovich, E. A., Nikolaenko, A. E., Tartakovskii, A. I., Makhonin, M. N., Drouzas, I., Skolnick, M. S., and Krysa, A. B. (2008). Overhauser effect in individual InP/$Ga_xIn_{1-x}P$ dots. *Phys. Rev. B*, **77**, 165338.

Sklyanin, E. K. (1989). Separation of variables in the Gaudin model. *Journal of Soviet Mathematics*, **47**, 2473.

Sleator, T., Hahn, E. L., Hilbert, C., and Clarke, J. (1985, Oct). Nuclear-spin noise. *Phys. Rev. Lett.*, **55**, 1742.

Slichter, C. P. (1990). *Principles of magnetic resonance*. Springer, Berlin/Heidelberg.

Smirnov, D. S. (2015). Spin noise of localized electrons interacting with optically cooled nuclei. *Phys. Rev. B*, **91**, 205301.

Smirnov, D. S., Glasenapp, Ph., Bergen, M., Glazov, M. M., Reuter, D., Wieck, A. D., Bayer, M., and Greilich, A. (2017). Nonequilibrium spin noise in a quantum dot ensemble. *Phys. Rev. B*, **95**, 241408.

Smirnov, D. S. and Glazov, M. M. (2012). Spin coherence generation and detection in spherical nanocrystals. *J. Phys.: Condens. Mat.*, **24**, 345302.

Smirnov, D. S., Glazov, M. M., Ivchenko, E. L., and Lanco, L. (2015). Theory of optical spin control in quantum dot microcavities. *Phys. Rev. B*, **92**, 115305.

Smirnov, D. S. and Golub, L. E. (2017). Electrical spin orientation, spin-galvanic, and spin-Hall effects in disordered two-dimensional systems. *Phys. Rev. Lett.*, **118**, 116801.

Sokolov, P. S., Petrov, M. Yu., Mehrtens, T., Müller-Caspary, K., Rosenauer, A., Reuter, D., and Wieck, A. D. (2016). Reconstruction of nuclear quadrupole interaction in (In,Ga)As/GaAs quantum dots observed by transmission electron microscopy. *Phys. Rev. B*, **93**, 045301.

Sørensen, J. L., Hald, J., and Polzik, E. S. (1998). Quantum noise of an atomic spin polarization measurement. *Phys. Rev. Lett.*, **80**, 3487.

Spatzek, S., Varwig, S., Glazov, M. M., Yugova, I. A., Schwan, A., Yakovlev, D. R., Reuter, D., Wieck, A. D., and Bayer, M. (2011). Generation and detection of mode-locked spin coherence in (In,Ga)As/GaAs quantum dots by laser pulses of long duration. *Phys. Rev. B*, **84**, 115309.

Stanek, D., Raas, C., and Uhrig, G. S. (2013). Dynamics and decoherence in the central spin model in the low-field limit. *Phys. Rev. B*, **88**, 155305.

Stanek, D., Raas, C., and Uhrig, G. S. (2014). From quantum-mechanical to classical dynamics in the central-spin model. *Phys. Rev. B*, **90**, 064301.

Starosielec, S., Fainblat, R., Rudolph, J., and Hägele, D. (2010). Two-dimensional higher order noise spectroscopy up to radio frequencies. *Rev. Sci. Instr.*, **81**, 125101.

Stöckmann, H. J. (2006). *Quantum chaos: An introduction.* Cambridge University Press.

Strand, J., Lou, X., Adelmann, C., Schultz, B. D., Isakovic, A. F., Palmstrøm, C. J., and Crowell, P. A. (2005). Electron spin dynamics and hyperfine interactions in $Fe/Al_{0.1}Ga_{0.9}As/GaAs$ spin injection heterostructures. *Phys. Rev. B*, **72**, 155308.

Sundfors, R. K. (1969*a*). Determination of the gradient-elastic tensors for $A^{\text{iii}}B^{\text{v}}$ compounds using nuclear acoustic resonance. *Phys. Rev.*, **177**, 1221.

Sundfors, R. K. (1969*b*). Exchange and quadrupole broadening of nuclear acoustic resonance line shapes in the III-V semiconductors. *Phys. Rev.*, **185**, 458.

Sundfors, R. K. (1974). Experimental gradient-elastic tensors and chemical bonding in III-V semiconductors. *Phys. Rev. B*, **10**, 4244.

Syperek, M., Yakovlev, D. R., Yugova, I. A., Misiewicz, J., Sedova, I. V., Sorokin, S. V., Toropov, A. A., Ivanov, S. V., and Bayer, M. (2011). Erratum: Long-lived electron spin coherence in CdSe/Zn(S,Se) self-assembled quantum dots [*Phys. Rev. B* **84**, 085304 (2011)]. *Phys. Rev. B*, **84**, 159903.

Takagahara, T. (2010). Theory of unitary spin rotation and spin-state tomography for a single electron and two electrons. *J. Opt. Soc. Am. B*, **27**, A46.

Takei, N., Takeuchi, M., Eto, Y., Noguchi, A., Zhang, P., Ueda, M., and Kozuma, M. (2010). Faraday rotation with a single-nuclear-spin qubit in a high-finesse optical cavity. *Phys. Rev. A*, **81**, 042331.

Takeuchi, M., Takano, T., Ichihara, S., Takasu, Y., Kumakura, M., Yabuzaki, T., and Takahashi, Y. (2006). Paramagnetic Faraday rotation with spin-polarized ytterbium atoms. *Applied Physics B*, **83**, 107.

Tarasenko, S. A. (2005). Optical orientation of electron spins by linearly polarized light. *Phys. Rev. B*, **72**, 113302.

Tarasenko, S. A., Poshakinskiy, A. V., Simin, D., Soltamov, V. A., Mokhov, E. N., Baranov, P. G., Dyakonov, V., and Astakhov, G. V. (2018). Spin and optical properties of silicon vacancies in silicon carbide - A Review, *Phys. Status Solidi B* **255**, 1700258.

Tartakovskii, A. (2012). *Quantum dots: Optics, electron transport and future Applications.* Cambridge University Press, Cambridge.

Tartakovskii, A. I., Wright, T., Russell, A., Fal'ko, V. I., Van'kov, A. B., Skiba-Szymanska, J., Drouzas, I., Kolodka, R. S., Skolnick, M. S., Fry, P. W., Tahraoui, A., Liu, H.-Y., and Hopkinson, M. (2007). Nuclear spin switch in semiconductor quantum dots. *Phys. Rev. Lett.*, **98**, 026806.

Taylor, J. M., Petta, J. R., Johnson, A. C., Yacoby, A., Marcus, C. M., and Lukin, M. D. (2007). Relaxation, dephasing, and quantum control of electron spins in double quantum dots. *Phys. Rev. B*, **76**, 035315.

Testelin, C., Bernardot, F., Eble, B., and Chamarro, M. (2009). Hole–spin dephasing time associated with hyperfine interaction in quantum dots. *Phys. Rev. B*, **79**, 195440.

Testelin, C., Eble, B., Bernardot, F., Karczewski, G., and Chamarro, M. (2008). Signature of the Overhauser field on the coherent spin dynamics of donor-bound electrons in a single CdTe quantum well. *Phys. Rev. B*, **77**, 235306.

Tribollet, J., Bernardot, F., Menant, M., Karczewski, G., Testelin, C., and Chamarro, M. (2003). Interplay of spin dynamics of trions and two-dimensional electron gas in a *n*-doped CdTe single quantum well. *Phys. Rev. B*, **68**, 235316.

Twardowski, A., Pokita, E., and Gaj, J. A. (1980). Valence band spin-orbit splitting in CdTe and $Cd_{1-x}Mn_xTe$ and giant Zeeman effect in the Γ_7 band of $Cd_{1-x}Mn_xTe$. *Solid State Comm.*, **36**, 927.

Uhrig, G. S., Hackmann, J., Stanek, D., Stolze, J., and Anders, F. B. (2014). Conservation laws protect dynamic spin correlations from decay: Limited role of integrability in the central spin model. *Phys. Rev. B*, **90**, 060301.

Urbaszek, B., Braun, P.-F., Amand, T., Krebs, O., Belhadj, T., Lemaítre, A., Voisin, P., and Marie, X. (2007). Efficient dynamical nuclear polarization in quantum dots: Temperature dependence. *Phys. Rev. B*, **76**, 201301.

Urbaszek, B., Marie, X., Amand, T., Krebs, O., Voisin, P., Maletinsky, P., Högele, A., and Imamoglu, A. (2013). Nuclear spin physics in quantum dots: An optical investigation. *Rev. Mod. Phys.*, **85**, 79.

Van de Walle, C. G. and Blöchl, P. E. (1993). First-principles calculations of hyperfine parameters. *Phys. Rev. B*, **47**, 4244.

Van Vleck, J. H. (1940). Paramagnetic relaxation times for titanium and chrome alum. *Phys. Rev.*, **57**, 426.

Van Vleck, J. H. (1948). The dipolar broadening of magnetic resonance lines in crystals. *Phys. Rev.*, **74**, 1168.

Varshalovich, D. A., Moskalev, A. N., and Khersonskii, V. K. (1988). *Quantum theory of angular momentum.* World Scientific Publishing.

Varwig, S., Evers, E., Greilich, A., Yakovlev, D. R., Reuter, D., Wieck, A. D., Meier, T., Zrenner, A., and Bayer, M. (2016). Advanced optical manipulation of carrier spins in (In,Ga)As quantum dots. *Appl. Phys. B*, **122**, 17.

Varwig, S., Schwan, A., Barmscheid, D., Müller, C., Greilich, A., Yugova, I. A., Yakovlev, D. R., Reuter, D., Wieck, A. D., and Bayer, M. (2012). Hole spin precession in a (In,Ga)As quantum dot ensemble: From resonant spin amplification to spin mode locking. *Phys. Rev. B*, **86**, 075321.

Vas'ko, F. T. (1980). Spin splitting in the spectrum of two-dimensional electrons due to the surface potential. *JETP Lett.*, **30**, 541.

Vasko, F. T. and Raichev, O. E. (2005). *Quatum kinetic theory and applications. Electrons, photons, phonons*. Springer, Heidelberg.

Vidal, M., Durnev, M. V., Bouet, L., Amand, T., Glazov, M. M., Ivchenko, E. L., Zhou, P., Wang, G., Mano, T., Kuroda, T., Marie, X., Sakoda, K., and Urbaszek, B. (2016). Hyperfine coupling of hole and nuclear spins in symmetric (111)-grown GaAs quantum dots. *Phys. Rev. B*, **94**, 121302.

Vladimirova, M., Cronenberger, S., Scalbert, D., Ryzhov, I. I., Zapasskii, V. S., Kozlov, G. G., Lemaître, A., and Kavokin, K. V. (2018). Spin temperature concept verified by optical magnetometry of nuclear spins. *Phys. Rev. B* **97**, 041301(R).

Voon, L. C. Lew Y., and Willatzen, M. (2009). *The k p method.* Springer Science + Business Media.

Vurgaftman, I., Meyer, J. R., and Ram-Mohan, L. R. (2001). Band parameters for III-V compound semiconductors and their alloys. *J. Appl. Phys.*, **89**, 5815.

Walls, D. F. and Milburn, G. J. (2008). *Quantum optics, Second Edition*, Springer.

Wang, X. J., Buyanova, I. A., Zhao, F., Lagarde, D., Balocchi, A., Marie, X., Tu, C. W., Harmand, J. C., and Chen, W. M. (2009). Room-temperature defect-engineered spin filter based on a non-magnetic semiconductor. *Nat. Mater.*, **8**, 198.

Warburton, R. J., Schaflein, C., Haft, D., Bickel, F., Lorke, A., Karrai, K., Garcia, J. M., Schoenfeld, W., and Petroff, P. M. (2000). Optical emission from a charge-tunable quantum ring. *Nature*, **405**, 926.

Watson, T. F., Weber, B., Hsueh, Y.-L., Hollenberg, L. C. L., Rahman, R., and Simmons, M. Y. (2017). Atomically engineered electron spin lifetimes of 30 s in silicon. *Sci. Adv.*, **3**, e1602811.

Weisbuch, C. and Lampel, G. (1974). Spin-dependent recombination and optical spin orientation in semiconductors. *Solid State Communs.*, **14**, 141.

Wenckebach, W. T. (2008). The solid effect. *Applied Magnetic Resonance*, **34**, 227.

Werner, H.-J., Schulten, Z., and Schulten, K. (1977). Theory of the magnetic field modulated geminate recombination of radical ion pairs in polar solvents: Application to the pyrene–n, n-dimethylaniline system. *J. Chem. Phys.*, **67**, 646.

Wiegand, J., Smirnov, D. S., Hübner, J., Glazov, M. M., and Oestreich, M. (2018). Spin and reoccupation noise in a single quantum dot beyond the fluctuation-dissipation theorem. *Phys. Rev. B* **97**, 081403(R).

Wilamowski, Z. and Jantsch, W. (2004). Suppression of spin relaxation of conduction electrons by cyclotron motion. *Phys. Rev. B*, **69**, 035328.

Wilamowski, Z., Jantsch, W., Malissa, H., and Rössler, U. (2002). Evidence and evaluation of the Bychkov–Rashba effect in SiGe/Si/SiGe quantum wells. *Phys. Rev. B*, **66**, 195315.

Winkler, R. (2003). *Spin–orbit Coupling effects in two-dimensional electron and hole systems.* Springer.

Wolff, P. A. (1988). Theory of bound magnetic polarons in semimagnetic semiconductors. In *Semiconductors and semimetals* (ed. J. K. Furdyna and J. Kossut), Vol. 25, pp. 413. Elsevier.

Wolfowicz, G., Simmons, S., Tyryshkin, A. M., George, R. E., Riemann, H., Abrosimov, N. V., Becker, P., Pohl, H.-J., Lyon, S. A., Thewalt, M. L. W., and Morton, J. J. L. (2012). Decoherence mechanisms of ^{209}Bi donor electron spins in isotopically pure ^{28}Si. *Phys. Rev. B*, **86**, 245301.

Woods, L. M., Reinecke, T. L., and Lyanda-Geller, Y. (2002). Spin relaxation in quantum dots. *Phys. Rev. B*, **66**, 161318.

Wu, M. W., Jiang, J. H., and Weng, M.Q. (2010). Spin dynamics in semiconductors. *Phys. Rep.*, **493**, 61.

Xiao, M., House, M. G., and Jiang, H. W. (2010). Measurement of the spin relaxation time of single electrons in a silicon metal-oxide-semiconductor-based quantum dot. *Phys. Rev. Lett.*, **104**, 096801.

Xu, X., Yao, W., Sun, B., Steel, D. G., Bracker, A. S., Gammon, D., and Sham, L. J. (2009). Optically controlled locking of the nuclear field via coherent dark-state spectroscopy. *Nature*, **459**, 1106.

Yafet, Y. (1963). *g*-factors and spin-lattice relaxation of conduction electrons. In *Solid state physics* (ed. F. Seitz and D. Turnbull), p. 2. Academic Press, New York.

Yakovlev, D. R. and Bayer, M. (2017). Coherent spin dynamics of carriers. In *Spin physics in semiconductors, Second Edition.* (ed. M. I. Dyakonov), Springer Series in Solid-State Sciences, Vol. 157, p. 155. Springer International Publishing.

Yang, C. H., Rossi, A., Ruskov, R., Lai, N. S., Mohiyaddin, F. A., Lee, S., Tahan, C., Klimeck, G., Morello, A., and Dzurak, A. S. (2013). Spin-valley lifetimes in a silicon quantum dot with tunable valley splitting. *Nat. Comm.*, **4**, 2069.

Yang, L., Glasenapp, P., Greilich, A., Reuter, D., Wieck, A. D., Yakovlev, D. R., Bayer, M., and Crooker, S. A. (2014). Two-colour spin noise spectroscopy and fluctuation correlations reveal homogeneous linewidths within quantum-dot ensembles. *Nat. Commun.*, **5**, 4949.

Yang, W. and Sham, L. J. (2012). Collective nuclear stabilization in single quantum dots by noncollinear hyperfine interaction. *Phys. Rev. B*, **85**, 235319.

Yao, W., Liu, R.-B., and Sham, L. J. (2006). Theory of electron spin decoherence by interacting nuclear spins in a quantum dot. *Phys. Rev. B*, **74**, 195301.

Yu, P. Y. and Cardona, M. (2010). *Fundamentals of semiconductors.* Springer, Berlin/Heidelberg.

Yugova, I. A., Glazov, M. M., Ivchenko, E. L., and Efros, Al. L. (2009*a*). Pump-probe Faraday rotation and ellipticity in an ensemble of singly charged quantum dots. *Phys. Rev. B*, **80**, 104436.

Yugova, I. A., Glazov, M. M., Yakovlev, D. R., Sokolova, A. A., and Bayer, M. (2012). Coherent spin dynamics of electrons and holes in semiconductor quantum wells and quantum dots under periodical optical excitation: Resonant spin amplification versus spin mode locking. *Phys. Rev. B*, **85**, 125304.

Yugova, I. A., Greilich, A., Yakovlev, D. R., Kiselev, A. A., Bayer, M., Petrov, V. V., Dolgikh, Y. K., Reuter, D., and Wieck, A. D. (2007*a*). Universal behavior of the electron *g* factor in GaAs/Al_xGa_{1-x}As quantum wells. *Phys. Rev. B*, **75**, 245302.

Yugova, I. A., Greilich, A., Zhukov, E. A., Yakovlev, D. R., Bayer, M., Reuter, D., and Wieck, A. D. (2007*b*). Exciton fine structure in InGaAs/GaAs quantum dots revisited by pump-probe Faraday rotation. *Phys. Rev. B*, **75**, 195325.

Yugova, I. A., Sokolova, A. A., Yakovlev, D. R., Greilich, A., Reuter, D., Wieck, A. D., and Bayer, M. (2009*b*). Long-term hole spin memory in the resonantly amplified spin coherence of InGaAs/GaAs quantum well electrons. *Phys. Rev. Lett.*, **102**, 167402.

Zakharchenya, B. P., Novikov, V. A., and Fleisher, V. G. (1977). Quadrupole effects and singularities of NMR of optically oriented nuclei in semiconducting solid solutions. *JETP Lett.*, **26**, 203.

Zapasskii, V. S. (2013). Spin-noise spectroscopy: from proof of principle to applications. *Adv. Opt. Photon.*, **5**, 131.

Zapasskii, V. S., Greilich, A., Crooker, S. A., Li, Y., Kozlov, G. G., Yakovlev, D. R., Reuter, D., Wieck, A. D., and Bayer, M. (2013). Optical spectroscopy of spin noise. *Phys. Rev. Lett.*, **110**, 176601.

Zavoisky, E. K. (1945). Relaxation of liquid solutions for perpendicular fields. *J. Phys. (USSR)*, **9**, 211.

Zhang, W., Dobrovitski, V. V., Al-Hassanieh, K. A., Dagotto, E., and Harmon, B. N. (2006). Hyperfine interaction induced decoherence of electron spins in quantum dots. *Phys. Rev. B*, **74**, 205313.

Zheludev, N. I., Brummell, M. A., Harley, R. T., Malinowski, A., Popov, S. V., Ashenford, D. E., and Lunn, B. (1994). Giant specular inverse Faraday effect in $Cd_{0.6}Mn_{0.4}Te$. *Solid State Comm.*, **89**, 823.

Zhukov, E. A., Yakovlev, D. R., Bayer, M., Glazov, M. M., Ivchenko, E. L., Karczewski, G., Wojtowicz, T., and Kossut, J. (2007). Spin coherence of a two-dimensional electron gas induced by resonant excitation of trions and excitons in CdTe/(Cd,Mg)Te quantum wells. *Phys. Rev. B*, **76**, 205310.

Zhukov, E. A., Yakovlev, D. R., Glazov, M. M., Fokina, L., Karczewski, G., Wojtowicz, T., Kossut, J., and Bayer, M. (2010). Optical control of electron spin coherence in CdTe/(Cd,Mg)Te quantum wells. *Phys. Rev. B*, **81**, 235320.

Zieliński, M., Don, Y., and Gershoni, D. (2015). Atomistic theory of dark excitons in self-assembled quantum dots of reduced symmetry. *Phys. Rev. B*, **91**, 085403.

Zinovieva, A. F., Dvurechenskii, A. V., Stepina, N. P., Deryabin, A. S., Nikiforov, A. I., Rubinger, R. M., Sobolev, N. A., Leitão, J. P., and Carmo, M. C. (2008). Spin resonance of electrons localized on Ge/Si quantum dots. *Phys. Rev. B*, **77**, 115319.

Zook, J. D. (1964). Piezoelectric scattering in semiconductors. *Phys. Rev.*, **136**, A869.

Zutic, I., Fabian, J., and Sarma, S. D. (2004). Spintronics: Fundamentals and applications. *Rev. Mod. Phys.*, **76**, 323.

Zwanenburg, F. A., Dzurak, A. S., Morello, A., Simmons, M. Y., Hollenberg, L. C. L., Klimeck, G., Rogge, S., Coppersmith, S. N., and Eriksson, M. A. (2013). Silicon quantum electronics. *Rev. Mod. Phys.*, **85**, 961.

Index

The manufacturer's authorised representative in the EU for product safety is Oxford University Press España S.A. of el Parque Empresarial San Fernando de Henares, Avenida de Castilla, 2 – 28830 Madrid (www.oup.es/en or product.safety@oup.com). OUP España S.A. also acts as importer into Spain of products made by the manufacturer.

www.ingramcontent.com/pod-product-compliance
Lightning Source LLC
LaVergne TN
LVHW011820200926
848484LV00005B/1252
* 9 7 8 0 1 9 8 8 0 7 3 0 8 *